Plant Breeding and Genetics: A Molecular Approach

Plant Breeding and Genetics: A Molecular Approach

Editor: Kiara Woods

MURPHY & MOORE
www.murphy-moorepublishing.com

www.murphy-moorepublishing.com

MURPHY & MOORE

Cataloging-in-Publication Data

Plant breeding and genetics : a molecular approach / edited by Kiara Woods.
p. cm.
Includes bibliographical references and index.
ISBN 978-1-63987-752-2
1. Plant genetics. 2. Plant breeding. 3. Genetics. 4. Breeding. I. Woods, Kiara.
QK981 .P75 2023
581.35--dc23

Murphy & Moore Publishing
1 Rockefeller Plaza,
New York City,
NY 10020, USA

ISBN 978-1-63987-752-2

Contents

Preface

Plant breeding or plant genetics refers to the application of genetic principles for producing new breeds of plants that are more useful to humans. It deals with heredity in plants, mechanisms of hereditary transmission, and variation of inherited characteristics. Genomic selection is one the emerging methods of molecular breeding used for crop improvement. This method assists in the identification of superior genotypes with higher breeding value in segregating breeding populations. Marker-assisted selection (MAS) is an another molecular breeding method in which morphological, biochemical or DNA markers are used for indirect selection of a genetic determinant or determinants of a trait of interest. Some of the desirable traits include disease resistance, tolerance of abiotic stress, productivity, and quality. This book includes some of the vital pieces of works being conducted across the world, on various topics related to plant genetics and molecular breeding. Scientists and students actively engaged in plant breeding and genetics research will find it full of crucial and unexplored concepts.

This book is a comprehensive compilation of works of different researchers from varied parts of the world. It includes valuable experiences of the researchers with the sole objective of providing the readers (learners) with a proper knowledge of the concerned field. This book will be beneficial in evoking inspiration and enhancing the knowledge of the interested readers.

In the end, I would like to extend my heartiest thanks to the authors who worked with great determination on their chapters. I also appreciate the publisher's support in the course of the book. I would also like to deeply acknowledge my family who stood by me as a source of inspiration during the project.

Editor

1

Developments and Prospects in Imperative Underexploited Vegetable Legumes Breeding

Sandeep Kaur Dhaliwal [1], Akshay Talukdar [2], Ashish Gautam [3], Pankaj Sharma [1], Vinay Sharma [4] and Prashant Kaushik [5,6,*]

1 Department of Plant Breeding and Genetics, Punjab Agricultural University, Ludhiana 141004, India; sandeep2-pbg@pau.edu (S.K.D.); pankaj-pbg@pau.edu (P.S.)

2 Division of Genetics, Indian Agricultural Research Institute, New Delhi 110012, India; akshay.talukdar@icar.gov.in

3 Department of Genetics and Plant Breeding, G.B. Pant University of Agriculture and Technology, Pantnagar, Uttarakhand 263145, India; gautam.ashish801@gmail.com

4 International Crops Research Institute for the Semi-Arid Tropics (ICRISAT), Hyderabad 502324, India; S.Vinay@cgiar.org

5 Instituto de Conservación y Mejora de la Agrodiversidad Valenciana, Universitat Politècnica de València, 46022 Valencia, Spain

6 Nagano University, Ueda 386-0031, Japan

* Correspondence: prakau@doctor.upv.es

Abstract: Vegetable legumes are an essential source of carbohydrates, vitamins, and minerals, along with health-promoting bioactive chemicals. The demand for the use of either fresh or processed vegetable legumes is continually expanding on account of the growing consumer awareness about their well-balanced diet. Therefore, sustaining optimum yields of vegetable legumes is extremely important. Here we seek to present d etails of prospects of underexploited vegetable legumes for food availability, accessibility, and improved livelihood utilization. So far research attention was mainly focused on pulse legumes' performance as compared to vegetable legumes. Wild and cultivated vegetable legumes vary morphologically across diverse habitats. This could make them less known, underutilized, and underexploited, and make them a promising potential nutritional source in developing nations where malnutrition still exists. Research efforts are required to promote underexploited vegetable legumes, for improving their use to feed the ever-increasing population in the future. In view of all the above points, here we have discussed underexploited vegetable legumes with tremendous potential; namely, vegetable pigeon pea (*Cajanus cajan*), cluster bean (*Cyamopsis tetragonoloba*), winged bean (*Psophocarpus tetragonolobus*), dolichos bean (*Lablab purpureus*), and cowpea (*Vigna unguiculata*), thereby covering the progress related to various aspects such as pre-breeding, molecular markers, quantitative trait locus (QTLs), genomics, and genetic engineering. Overall, this review has summarized the information related to advancements in the breeding of vegetable legumes which will ultimately help in ensuring food and nutritional security in developing nations.

Keywords: underexploited legumes; vegetable breeding; pre-breeding; molecular markers; QTLs

---

1. Introduction

Agriculture has been under growing pressure to produce sufficient food, feed, and biofuel on scarce land for the planet's predicted nine billion people by 2050 [1]. Growing adequate food for the ever-increasing population in the climate change scenario is a major challenge for food security. In this case, underexploited vegetable legumes make an erudite argument for contribution to rural people's dietary needs. These days, consideration of underexploited vegetable legumes is growing new protein sources to meet the ever-increasing demand for vegetable proteins [2]. The term underexploited/underutilized crop refers to the group of cultivated and wild species that have limited global market potential and are sometimes deemed as under-used [3]. Globally, underexploited legumes are known as nutritious resources and can be intended to improve health [4] and minimize disease risks [5]. They have a specific profile with high nutrient and protein content alternatives to maintain farmers' livelihoods and soil protection [6]. In comparison, underexploited legumes are a less utilized potential source to improve protein and micronutrient content in comparison to cereals which provided ample calories but insufficient micronutrients [7]. Legumes are mainly cultivated as pulse (seeds) or animal fodder. However, some members of the legume family are cultivated for their pods and immature seeds to cook as a vegetable [8].

Vegetable legumes have unique organoleptic qualities and are usually regarded as important sources of carbohydrates, minerals, vitamins, and health-promoting bioactive compounds. In contrast, legumes are also known to carry some antinutritional factors like lectins, phytic acid, saponins, and vicine [9]. Moreover, legumes are characterized by their ability to develop in a symbiotic relationship with nitrogen-fixing bacteria and therefore are also used as soil-enriching green manure [10].

Pigeon pea (*Cajanus cajan*) cultivation is mostly confined to South Asia and to East Africa, which is one of the most malnourished regions of the globe. Whereas, the cluster bean or guar is an incredibly drought-tolerant annual legume crop grown because of its use as a vegetable, green manure, and forage [11]. On the other hand, winged bean is important in that nearly all components of the plant can be consumed, from the seeds, pods, and flowers, to the foliage as well as the tuberous roots, with the stems as well as leaves used as fodder [12]. Therefore, in countries where protein deficiency is high, or access to meat protein is low, winged bean is a candidate for helping to diversify diets and significantly improve nutrition.

Furthermore, legumes, when consumed as vegetables, contain more water compared to pulses. Therefore, the soluble carbohydrates are higher and starch content is lower in the vegetable legumes, making them much more palatable compared to dry pulses [13]. Additionally, vegetable legumes are more abundant sources of health-promoting compounds like carotenoids, vitamin A, chlorophyll, phenolics, and vitamin C. For that reason, their consumption is generally supposed to make for far healthier nutrition [14]. Besides, vegetable legumes are short-season crops with a short shelf life [15]. On the other hand, the usage of processed vegetable legume products is continuously growing due to the increasing awareness of their well-balanced nutrition and high health-promoting compounds content [16].

Not many efforts have incorporated legumes for their beneficial use as vegetable legumes. So, there is an utmost need to improve vegetable legumes with the help of modern breeding technologies. There are many underexploited species of vegetable legumes which are available as local accessions and landraces and are being consumed as vegetables [17]. So, these accessions are the valuable genetic variability basket from which we can extract the traits of interest for the improvement of vegetable legumes. Traits like high yield, early podding, year-round availability, long and large green pods, better shelf life, and biotic and abiotic stress resistance are among the most desired characteristics in the vegetable legume breeding programs [18]. Econometric research awareness and policymakers' attentions on underexploited legumes will be needed for further diversification of nutritional profiles and enhancements of human nutrition. There are currently about 150 cultivated crops, and only 30 edible species are often used for global diets, the majority of which are cereal-based, and which in developing countries rely especially on rainfed agriculture [19]. Almost all of these crops cannot withstand abiotic stresses due to global climate change [20]. However, underexploited legumes have tremendous potential to withstand harsh conditions that cannot be ignored, which will help in mitigating nutritional insecurity. Over the past decade, it has been seen that cultivated legumes having a narrow genetic base and continuous use of a few elite breeding lines are the key causes affecting genetic improvement in breeding programs. To fulfill the needs of plant-based micronutrients and rejuvenation of soil health, breeding programs need to adopt a new approach. Crop wild relatives (CWRs) have become an ideal source of novel alleles for a range of important traits needed for improvement in breeding programs. In this regard, recently, in a pigeon pea breeding program utilizing wild *Cajanus platycarpus* sp., a stable promising trait-specific introgression line (IL) CPL 87119 has been identified. It showed higher potential for yield and nutrient-rich traits with a broad genetic base [21]. Efforts are required in this direction to increase the quality of other underexploited legumes.

In recent years, advances in next-generation sequencing (NGS) methods and a steep decrease in sequencing cost offers an incredible opportunity for the improvement of vegetable legumes [22]. Moreover, the number of publications that deal with the breeding aspects of underexploited vegetable legumes are diminutive in comparison to papers dealing with grain legumes [23]. To best of our knowledge, a review paper centered on highlighting the breeding development and thereby, outlining the crucial breakthroughs in underexploited vegetable legumes is lacking in the international scientific literature. Further, combining the available knowledge on this subject will help in developing an in-depth understanding of the breeding of vegetable legumes. Recognizing this gap, this review paper provides a detailed overview of all elements connected to the underexploited vegetable legumes.

2. Crop Wild Relatives

Comprehending the relationship of crop plants and their wild relatives is a tremendous focus of plant breeders. This expertise is of excellent worth in dissecting the process of crop domestication by determining and employing wild relatives for crop development. Breeding programs, as well as germplasm characterization research, over the years have discovered that the cultivated plants, generally, have a relatively reduced tolerance to stresses compared to their wild relatives [24]. The one-dimensional prospect for enhanced yield has been hypothesized to guide metabolic supply allocation in the direction of accelerated progress, thereby overlooking other traits. On the other hand, breeding bottlenecks have relatively reduced inherited deviation of contemporary vegetation as well as led to the loss of genes created by crop wild relatives (CWRs). Although genes are identified for disease and insect pest resistance, they have been seen as negatively correlated with yield [25]. However, it has been discovered that breeding for disease and insect pest resistance traits might be attained without having the demand for crop yield compromised. CWRs present several arrays of attributes with the chance to minimize the amount of yield loss as a direct result of biotic and abiotic stresses. The traits present in the CWRs can be introduced into cultivated varieties using conventional breeding approaches (if there is sexual compatibility), transgenesis, and more. In this direction, introgression of characteristics of interest originating out of a CWR to a cultivated type via consistent breeding would encounter linkage drag [26]. Below we have discussed the breeding objectives and use of CWRs for the six underexploited vegetable legumes addressed in this review in detail.

2.1. Vegetable Pigeon Pea (Cajanus cajan)

Vegetable pigeon pea possesses favorable agronomic qualities compared to other main grain legumes, and its wild relatives show promise in providing vital adaptive traits [27]. Higher investment in phenotypic and genotypic characterization and evaluation for the adaptive traits present in the CWR, symbolize equally immediate steps for the improvement of vegetable pigeon pea. Additional unrepresented species like *Cajanus crassus* and *Cajanus scarabaeoides*, are essential for developing germplasm collections for the improvement of important traits of the cultivated pigeon pea [28]. As strategies for the effective utilization of extensive diversity of plant genetic resources, conservation, collection, and accessibility of even more distant relatives of vegetable pigeon pea will be rewarding. In this direction, *C. scarabaeoides*, as well as *C. platycarpus*, are recognized as demonstrating potential related to the adaptation to climatic change [29].

2.2. Cluster Bean (Cyamopsis Tetragonoloba)

The genus *Cyamopsis* has four important members, *C. tetragonoloba*, *Cyamopsis serrata*, *Cyamopsis senegalensis*, and *Cyamopsis dentata* [30]. It has been accepted that the cultivated *C. tetragonoloba*, was developed from *C. senegalensis*, which is a drought tolerant African species. Breeding programs are mainly focused on breeding for high nutrition and dietary fiber, but also for improving the gum content (galactomannans) in the endosperm (90%). Wild relatives *C. serrata*, *C. senegalensis*, and *C. tetragonoloba* are diploid with chromosome number 2n = 2x = 14 [31]. Recently genome size was determined for the three *Cyamopsis* species with the help of flow cytometry. It was observed that the genome size of wild species, *C. serrata*, was approximately double (979.6 Mbp) that of cultivated cluster bean *C. tetragonoloba* (580.9 Mbp) whereas *C. senegalensis* (943.4 Mbp) had genome size intermediate between these two species. This information is critical to further implement specific tools for crossing wild relatives, which are the storehouse of many useful genes [32].

2.3. Winged Bean (Psophocarpus tetragonolobus)

The untamed progenitor of winged bean has remained somewhat enigmatic due to the absence of wild *Psophocarpus* in Asia, leading to one suggestion that the true wild progenitor is now extinct [33]. Morphological phylogenetic analyses of the nine species in the genus have come to mixed conclusions. Probably the closest wild species to winged bean scanned by the most recent morphological analysis places winged bean alongside *Psophocarpus scandens* and *Psophocarpus palustris*. Few efforts to cross winged bean with various members of the genus have been reported, however, one profitable cross between winged bean as well as *P. scandens* has been manufactured following many attempts [34]. Yet, molecular phylogenetic analysis is lacking. Identifying the true progenitor(s) may assist in the breeding of winged bean and could be necessary to understand the genetic changes associated with domestication. Relatively few studies have investigated the domestication genetics of legumes, except for scientific studies of the winged bean [35]. Therefore, little is known about the genes as well as alleles that were under selection by early farmers. This may contribute to the observation that genetic enhancement of legumes remains slow between distant relative and other crops.

2.4. Dolichos Bean (Lablab purpureus)

The dolichos bean is unquestionably of African origin; the only taxon known is subspecies *uncinatus*, which is widespread in tropical Africa. In Africa, *Lablab's* wild ancestor grows in hilly areas and coastal lowlands in southern, eastern, and western Africa. Its beans are too small and are not eaten. The cultivated form is known in Egypt, Sudan, and both East and West Africa [36]. Hence, it is essential to evaluate the breeding potential of parents and to select good combiners in the *Lablab* bean. Good results of any breeding program rely on an assortment of parents [37]. Nature is depicted by this study and the magnitude of gene action regarding evolving connected characteristics of parents and their offspring. Gene actions may show heterosis in F_1 or linkage in other generations [38]. Certainly, *Lablab's* undomesticated ancestor is still scattered across, and endemic to, much of tropical Africa [39]. Additionally, the wild forms collected from India and analyzed through molecular marker research were discovered genetically positioned intermediately between wild and cultivated forms. Still, there is a lot of consideration, which thinks the origin centers are Africa and Asia [40].

2.5. Cowpea (Vigna Unguiculata)

Cowpea is indigenous to West Africa and weedy and wild types exist in many parts of the region. It typically suffers from considerable losses caused by diseases and pests [41]. Though its tolerance to drought and heat is superior to other crops, predictions for climate change in the region suggest that there is a need for germplasm with even higher levels of adaptation to these abiotic and biotic stresses [42]. To be able to enhance tolerance to these stresses, this project seeks to tap the genetic diversity present in the wild relatives of cowpea. Cowpea wild relatives are the storehouse of important genes for disease and insect resistance, and the genes of abiotic stress resistance [43]. Moreover, CWRs of cowpea are important for maintaining genetic diversity and protecting against loss of germplasms because of genetic vulnerability. Several reports have shown that the weedy subspecies of cowpea (*V. unguiculata* subsp. *dekindtiana*, *stenophylla*, etc.) are easy to hybridize with the popular cultivated varieties. F_1 hybrids are also known to use a degree of vigor over the parent genotypes. The effective crossing between cultivated cowpea (*V. unguiculata*) varieties and their wild distant relatives varies based on genotypes and species [44]. Members of the var. *pubescens* have been known to confer some degree of insect resistance on cowpea owing to the presence of hairs (hence the title *pubescens*) on the plants. Transferring the hairiness trait from the wild lines to the cultivated varieties is essential for the development of insect resistance and consequently, avoidance of pathogens transmitted by such insects [45]. In the past, many studies were focused on determining the cross-compatibility between cultivated cowpea and wild relatives, finding out the reproductive potential, and also heterosis of the F_1 hybrids from these crosses [46].

3. Pre-Breeding

Pre-breeding activities in mobilizing novel alleles for cultivar development from wild relatives have been routinely involved in breeding programs [47]. Pre-breeding includes all activities directed at the identification of attractive crop traits including genes, as well as the consequent transfer of theirs into an excellent set of parents for extra choice [48]. The procedure identifies helpful characteristic(s) or maybe genes that could be exploited in cultivar growth. For vegetable legumes improvement, enough genetic diversity is present in wild relatives and landraces, which carry several helpful genes for cultivar improvement [49]. Pre-breeding activities must be initiated to produce new genetic variability using wild relatives and promising landraces for usage by the breeders in crop advancement programs. Pre-breeding must concentrate on the constant source of significant variability into the breeding pipeline to build new high yielding cultivars with a broad genetic base [50]. Though pre-breeding is helpful to enrich the primary gene pool for cultivar improvement, it is a time-consuming and challenging affair also. In this direction, linkage drag associated with wild relatives can make the pre-breeding tasks far more troublesome. Genomic-assisted pre-breeding is going to help to conquer the linkage drag and can facilitate the focused transfer of valuable genes/segments from wild relatives [51]. Therefore, it is important to understand all the above-mentioned factors for pre-breeding to exploit the available genetic variability for developing improved cultivars with a broad genetic base in vegetable legumes.

4. Molecular Markers from Diversity to QTLs

The directed evolution towards the improvement of existing germplasm requires tracking the desired traits to bring them together. Earlier, the phenotype served as a tracker of traits which is now replaced by more reliable DNA markers. The amalgam of breeding and molecular biology has made deeper insight into traits possible, as the genome is dug with more markers, one gets closer to a gene controlling trait. The complete bouquet of genes, quantitative trait loci (QTLs), and molecular markers linked to traits are put together for reliable marker-assisted breeding. The advancement in marker technology is slower in legumes, particularly for vegetable type than cereals, earning them the title of orphan crops. In the evolutionary line of molecular markers, the first generation markers, namely, restriction fragment length polymorphism (RFLP), random amplification of polymorphic DNA (RAPD), and amplified fragment length polymorphism (AFLP), generating information of many loci in one go, have been employed mainly in diversity analysis of pigeon pea [52], cluster bean [53], winged bean [54], dolichos bean [55], and cowpea [56,57]. On the other hand, sequence-based markers, namely, simple sequence repeats (SSRs), single nucleotide polymorphism (SNPs), and their modifications, being more reliable and reproducible, are employed in linkage mapping, trait mapping, and fine-mapping studies. SSR marker systems are available in pigeon pea [58], cluster bean [59], winged bean [33], and cowpea [60], whereas inter-species SSRs have been used in dolichos bean [61,62]. SNPs are markers of choice owing to their ubiquitous nature and abundance in the genome. SNPs have been identified in pigeon pea [63], cluster bean [64], winged bean [65], dolichos bean [66], and cowpea [67] for potent use in genomics assisted breeding. Once the choice of marker system is established, it draws a path for mapping traits of interest leading towards fine mapping and cloning. A selected set of mapped QTLs in pigeon pea and cowpea being comparatively rich in molecular resources along with important marker systems in cluster bean, winged bean, and dolichos bean, which are growing towards mapping studies are compiled (Table 1).

The pigeon pea crop has a rich set of molecular markers and QTLs for its improvement. Dense molecular linkage maps have been developed using SSR markers [68] and SNP markers [69]. A consensus genetic map was developed using previously published maps and four maps generated from four mapping populations. The consensus map comprised 339 SSRs spanning over a genetic distance of 1059 cM [68]. The consensus map is available, but traits of vegetable pigeon pea such as pod color, pod size, pod weight, and tenderness remained undiscovered on a genomic scale. The SSR and SNP markers can be used for reducing generations for obtaining desired recombinants through marker-assisted backcross breeding and de novo mapping studies. The QTLs mapped in pulse pigeon pea can be transferred to vegetable types such as sterility mosaic resistance, determinacy, earliness, Fusarium wilt resistance, and fertility restoration are compiled in Table 1. Fusarium wilt is a deadly disease limiting crop yield in Eastern and Southern Africa, whereas sterility mosaic virus is another devastating disease transmitted by a mite (*Aceria cajani*). The use of genomic resistance is a cost-effective alternative against chemical control for fungal diseases and viral vectors. Three QTLs, namely, *qFW11.1*, *qFW11.2*, and *qFW11.3* for Fusarium wilt resistance [70] and a candidate gene *CcLG11* for sterility mosaic virus were mapped which can serve as potent resistance donors [71].

The development of an efficient marker system is slower in cluster bean, ultimately reducing growth in revealing genetic control of loci across the genome. The orphan crop gained the focus of breeders mainly due to gum produced by it which is useful commercially in textile and other allied industries. Thus, its use as a vegetable crop remained shadowed. However, the advancements in marker technology over the years could be to utilized in the improvement of the crop for vegetable purpose. In the HES 1401 cultivar a total of 16,476 expressed *sequence* tags (EST) were reported, and it is the first step in the omics era for the crop [72]. Subsequently, these EST sequences were explored for the existence of SSRs with MISA (microsatellite identification tool) software (Thomas Thiel @ the Plant Genome Resources Center) which resulted in obtaining 187 SSRs [73]. Later on, the same set of EST was utilized to develop SSRs and was used to validate 32 diverse genotypes [74]. Recently, a total of 73,934 sequences were developed in the GG-4 variety by Mi seq NGS technology [59]. Later in the process, these sequences were mined for SSRs and resulting in finding 15,399 SSRs [59]. In future, marker development studies in vegetable legumes would need to be explored to generate genomic resources and genic markers as has been done in other crops.

Winged bean is a neglected crop in terms of genomic resources and molecular breeding. Molecular characterization was attempted for 24 accessions of winged bean using 13 RAPD and 7 ISSR markers [75]. It was found that ISSR markers were more promising in comparison with RAPD markers. It was an era of development of omic resources in other crops when this study was conducted for cluster bean, this clearly indicates its neglect in molecular breeding. The further studies also used ISSR markers for diversity analysis [55], until the breakthrough study in which transcriptome sequencing was performed on the "Ibadan Local-1" cultivar and 1900 SSRs markers were discovered [76]. More transcriptomes were sequenced after this study by Vatanparast et al. [65] and Wong et al. [77] giving rise to 12,956 and 9682 SSRs, respectively. Besides, a total of 5190 SNPs was also generated in the study [73]. Among these, a total of 20 microsatellite primer pairs were validated on 53 accessions of cluster bean for their use in molecular studies [33]. There is still plenty of scope for development of saturated linkage maps of SSR and SNP markers for the improvement of cluster bean.

Dolichos bean has a variety of uses such as vegetable, medicinal plant, and fodder, but its molecular resources remain unrevealed. The first genetic linkage was made harboring 127 RFLP and 91 RAPD using the second filial generation of cross "Rongai (cultivated) and CPI 24973 (wild species)". A total of 17 linkage groups spanning 1610 cM were generated in the study [78]. Due to scarcity of microsatellite markers in a crop, SSRs from related crops were tested for transferability for use in molecular breeding. Transferability was tested for 50 SSRs from soybean [79], genic SSRs from cowpea [80], 42 SSRs from soybean, *Medicago truncatula*, and chickpea [61], and 134 SSRs from French bean, mung bean, cowpea, faba bean, and moth bean [62]. The transferable SSRs from related crops were used for mapping photoperiod insensitivity and determinate growth habit in dolichos bean using bulk segregant analysis [81]. Bulks and population derived from the cross of two phenotypic extremes, namely, GNIB21 and GP189, result in identification of a *PvTFLy1*, a locus controlling the determinate habit of growth linked with photoperiod sensitivity. These traits are found linked in common bean and soybean as well, and thought to be controlled by mutation of *Dt1* and *E3* homologs, in dolichos bean. Although transferable markers were reliable, there was the urgency of generating markers from its genome to cover the entire genome. Therefore, a total of 459 ESTs of dolichos bean have been obtained from the National Center for Biotechnology Information (NCBI) and searched for the presence of microsatellites. Thus, 22 SSRs were discovered in a total of 420 unigenes and validated on a set of 24 accessions of dolichos bean [82]. Association mapping is a strategy of using historic linkages to associate markers with traits. It is particularly useful in dolichos bean owing to its self-pollinated nature, as a result of which fewer recombination and widely spaced markers could be used. A set of 234 SSRs (mostly in-house designed) were used in association with mapping for identifying genetic control of days to 50% flowering, fresh pod number per plant, and fresh pod weight per plant. Three markers, namely, KTD 200, KTD 130, and KTD 273 were found to be associated with traits, respectively explaining more than 10 percent of phenotypic variation in each case [83].

Cowpea is a much-explored crop at the molecular level in comparison to the crops discussed above. Dense molecular maps and QTLs are available for its genetic improvement. The first genetic linkage map of vegetable cowpea (asparagus bean) comprising 191 SNP and 184 SSR loci distributed on 11 linkage groups was developed covering 745 cM of total length [84]. The development of such genetic maps is a foundation stone for future breeding work. Pod tenderness and sweet taste are essential traits of vegetable cowpea, QTL mapping of these traits using backcross and F_2 generation of the cross of JP81610 and JP89083 was performed. Pod tenderness was found to be under genetic control of three QTLs explaining up to 50 percent phenotypic variation whereas two QTLs were obtained for total soluble solids (TSS) [85]. The QTLs for tender pods on LG 7 co-localized with pod length QTLs, which played an important role in the domestication of cowpea. In another study, major horticulturally important traits were mapped using recombinant inbred line RILs and it was found that QTLs for days to flowering, nodes to the first flower, leaf senescence were clustered together on LG11 whereas QTLs for pod number per plants were scattered on various linkage groups [86]. Pod length and biotic and abiotic stresses faced by cowpea were mapped using genome-wide association studies (GWAS) which take into account historical linkages to map the traits that are compiled in Table 1. These mapping efforts not only detected the genomic regions and candidate genes but also enabled the development and validation of trait linked markers and facilitate their use in future breeding programs.

Table 1. Molecular markers and mapped quantitative trait loci (QTLs) in vegetable legumes.

Crop	Molecular Marker/QTL	Source	Trait/Objective	Reference
Pigeon pea	*qSMD4* major QTL and minor QTLs	F_2 (ICP 8863 × ICPL 20097 TTB 7 × ICP 7035)	Sterility mosaic resistance	[87]
	13 QTLs for six traits	(Pusa Dwarf × HDM04-1)	Earliness, plant type, high-density linkage map	[88]
	339 SSR, 4 QTLs	F_2 (ICPB 2049 × ICPL 99050, ICPA 2043 × ICPR 3467, ICPA 2039 × ICPR 2447, ICPA 2043 × ICPR 2671)	Linkage map, fertility restoration	[68]
	CcTFL1 gene	F_2 (ICPL 85010 × ICP 15774)	Determinacy	[89]
	*C.cajan*_01839 for sterility mosaic, *C. cajan*_03203 for Fusarium wilt	RILs (ICPL 20096 × ICPL 332)	Fusarium wilt, sterility mosaic disease	[90]
	421 hypervariable SSRs from a genome sequence	94 genotypes	Diversity Analysis, Hybrid Purity Testing, Trait Mapping	[91]
	3 major QTLs (CcLG11)	F_2, RIL (ICPL 20096 × ICPL 332, ICPL 20097 × ICP 8863, ICP 8863 × ICPL 87119)	Sterility mosaic resistance	[71]
	Dt1 locus, Indel marker from*CcTFL1* gene	F_2 (ICP 5529 × ICP 11605)	Determinacy	[92]
	547 SNP (bead-array), 319 SNP (RAD), 65 SSR	F_2 (Asha × UPAS, Pusa Dwarf × H2001-4, Pusa Dwarf × HDM04-1)	Molecular linkage map	[69]
	CcLG08 carry major QTL	F_2 (ICPA 2039 × ICPL 87119	Fertility restoration	[92]
	CcLG07 (8 QTLs), SNP S7_14185076 (linked to 4 traits)	BC (ICPL 87119 × ICPW 15613, ICPL 87119 × ICPW 29)	Yield related traits	[93]
Cluster bean	16,476 EST	HES 1401	cDNA library from seeds	[72]
	L19, *D1*, *AB7* and *QLTY 3* (Bacterial blight) and *OPQ 20*,*OPD10*, *OPD14*,*OPQ 12*,*OPAC 8* and *OPF 9* (drought tolerance) QTLs	HG 75 × PNB (Bacterial blight) and HG 563 × PNB (drought tolerance)	Mapping Bacterial blight resistance and drought tolerance	[94]
	5 RAPD	35 genotypes	RAPD and ISSR cloning and sequencing	[95]
	100 SSRs	32 genotypes	Validation of SSRs	[74]
	15,399 SSRs	GG-4 variety	Sequencing by Miseq NGS	[59]
Winged bean	13 RAPD and 7 ISSR	24 accessions	Molecular characterization	[75]
	100 ISSR	45 accessions	Diversity analysis	[54]
	1900 SSRs	Ibadan Local-1	Transcriptome sequencing -Illumina HiSeq 2500	[76]
	12,956 SSRs, 5190 SNPs	2 accessions	Transcriptome sequencing- Roche 454 Genome Sequencer FLX	[65]
	9682 SSR	6 accessions	Transcriptome sequencing-Illumina MiSeq	[77]
	20 SSRs	53 accessions	Primer design from in house assembled transcriptome using primer3	[33]
Dolichos bean	127 RFLP, 91 RAPD	Rongai (cultivar) × CPI 24973 (wild) -17 linkage groups, 1610 cM	F_2 population for genetic linkage map	[78]
	41 main effect QTLs (22 for growth phenological traits and 19 for fruit traits)	Meidou2012 × 'Nanhui23	Growth phenological and fruit traits	[96]
	40 QTLs (8.1 to 55.0% variation)	(Meidou2012 × Nanhui 23).	Inflorescence length traits	[97]
	21 SSR	13 genotypes	Transferability of SSRs from French bean/diversity analysis	[98]
	60 SNPs, 16 InDels.	Sequencing polymorphic genic segments of 9 parents	Allele-specific PCR primers	[99]
	22 SSRs	420 unigenes	479 EST from NCBI for SSR mining	[82]
	42 SSRs		Transferability of markers from soybean, *Medicago truncatula*, green gram chickpea	[61]
	134 SSRs	143 genotypes	Transferability of SSRs from French bean, mung bean, cowpea, faba bean, and moth bean/diversity analysis	[62]
	9 QTLs using 234 SSRs	64 accessions for GWAS	Fresh pod yield mapping	[83]
	PvTFLy1 locus	GNIB21 × GP189	Photoperiod responsive flowering	[81]

Table 1. *Cont.*

Crop	Molecular Marker/QTL	Source	Trait/Objective	Reference
Cowpea/Asparagus bean	191 SNP and 184 SSR loci	RILs (ZN016 × Zhijiang282)	Molecular linkage map	[84]
	3 QTLs for pod tenderness, 2 QTLs for total soluble solid	F_2 and BC (P81610 × JP89083)	Pod tenderness and total soluble solid	[85]
	Major QTLs on LG 11	RILs (ZN016 × ZJ282)	Days to first flowering (FLD), leaf senescence (LS), nodes to first flower (NFF), and pod number per plant (PN)	[86]
	39 SNPs using GWAS	95 accessions of asparagus bean	Drought tolerance	[100]
	18 SNPs using GWAS	95 asparagus bean, 4 African cowpea accessions	Fusarium wilt	[101]
	QTLS on LG 1,4,7	F_2 and BC (JP81610 × TVnu-457)	Pod fiber content and pod shattering	[102]
	72 SNPs using GWAS	RILs (ZN016 × Zhijiang282)	Pod length	[103]
	17,996 SNPs using RAD sequencing, QTLS on LG4, 5, 6, 7, 9, 10, 11	$F_{2:3}$ (Green pod cowpea × Xiabao II)	High-density SNP map and yield traits	[104]
	5225 SNP markers by SLAR-seq	F_2 (Dubai bean × Ningjiang 3)	High-density map by sequencing	[105]
	Ruv2 locus	F_2 and RILs (ZN016 × Zhijiang282)	Rust resistance	[106]
	qCel7.1, *qHem7.1*, and *qLig7.1*	F_2 (JP81610 × TVnu-457)	Pod fiber content	[107]

5. Genomic and Transcriptomic Resources

Breeding objectives for crops, including vegetable legumes, are constant over the years but approaches towards achieving the goals are ever-changing. The availability of genomic and transcriptomic resources has changed the ways of shaping genomes and creating innovative possibilities to alter the genome for the desired phenotype (Table 2). The era of genomics was revolutionary for legumes by sequencing of model legume species such as *Glycine max* [108] which served as the legume genome reference. The dissection of sequenced genomes of model plants aids the understanding of evolution, important gene families, and re-arrangements in the structure of chromosomes in related crops [109]. Pigeon pea genome [110], dolichos bean [111], and cowpea [112] have been sequenced providing insights into agriculturally essential genes.

The sequenced genomes of legumes and the model species would accelerate genomic advancements through comparative genomics in cluster bean, dolichos bean, and winged bean, which are still at a rudimentary stage. Furthermore, genome sequencing was one of the important milestones in comparative genetics making it easier for scientists to compare the genomes, transfer of traits and markers, identifying orthologous and paralogous genes, and more in-depth insight into evolution and domestication [105]. Legumes stand as shuffled, deleted, and doubled genomes from one common ancestor with a common monophyletic family making transferability of genomic information possible [113]. Comparative genome analysis of asparagus bean with soybean, adzuki bean, and mung bean using SLAF (specific length amplified fragment sequencing) markers revealed conserved genomic regions, and offer support in assembling genome sequence [105]. Salinity tolerance mechanisms of *Strophostyles helvola*, a wild inhabitant of beaches in North America was revealed by transcriptome sequencing [114].

More and more genomes get sequenced owing to the reduction in the cost of sequencing and new reference genomes become available, stimulating resequencing projects [115]. Resequencing of wild and cultivated germplasm in various crops has been initiated to get closer to genes underlying essential traits. One of the cost-effective resequencing or de novo sequencing strategy is the reduced representation library approach [116]. One of the widely used dimensions of this technology is genotyping by sequencing (GBS); GBS has been used for mapping the traits of interest by deep sequencing of parents and multiplexed sequencing of large mapping populations in one go. Fusarium wilt resistance and fertility restoration was mapped in pigeon pea [70,117], and potyvirus resistance mapping [118] and mapping aphid resistance in cowpea, using GBS [119]. The technology has enormous potential in mapping and genomic studies of cluster bean, winged bean, and dolichos bean owing to lack of genomic tools, as the platform is flexible. The sequencing of the population under investigation is carried out, which exempts it from ascertainment bias [120].

The primary outcome of genomic and transcriptomic studies is a large set of SNP markers that can be used for high throughput genotyping assays. SNPs fit best for most high throughput genotyping because they are omnipresent in eukaryotic genomes, cost-effective, automated platforms, and allele calling and data analysis are simple owing to their bi-allelic nature [121]. In pigeon pea, the 56k Axiom SNP chip was used for mapping seed quality and high-selfing flower traits [122]. The genotyping assay has also been used for mapping pod length in cowpea [103] and understanding molecular mechanisms governing incompatible and compatible reaction against *Striga* in cowpea [123]. In parallel, the latest pigeon pea SNP chip "CcSNPnks" has enormous potential for use in mapping studies as the SNPs originate from unique genes, conserved genes of pigeon pea with related crops and other agriculturally important genes [124]. These developed chips can also be employed in the winged bean, cluster bean, dolichos bean, and other orphan legumes for testing their suitability for molecular studies. Adopting such genomics and transcriptomic methods could overcome several limitations of traditional breeding and improve the precision and efficiency of crop breeding procedures.

Table 2. Transcriptomic and valuable genomic resources in legumes (pigeon pea, cluster bean, winged bean, dolichos bean, and cowpea).

Crop	Objective	Description	Genetic Improvement of Vegetable Type	Platform	Reference
Pigeon pea	Transcriptome seq	50,566 SSRs, 12,000 SNPs, 0.12 million unique sequences and 150.8 million sequence reads	Enhancing genomic resources	Roche FLX/454	[125]
	RNA-seq	1.696 million reads, 3771 SSRs	To target protein-coding and regulatory genes	Roche 454 GS-FLX	[126]
	Gene expression atlas (CcGEA)	590.84 million paired-end data from RNA-Seq, 28 793 genes, regulatory genes, i.e., pollen-specific (*SF3*), sucrose–proton symporter	To target protein-coding and regulatory genes	Illumina HiSeq 2000	[127]
	Comparative transcriptome	*Cajanus cajan* (L.) and *Cajanus platycarpus* (Benth.) sequence revealed 0.11 million transcripts, 82% annotated	Valuable data from wild sources	Illumina Hi-Seq 2500	[128]
	WRKY characterization	94 WRKY genes characterized and validated phylogenetically three groups (I, II, III)	Elucidating stress-responsive machinery	qRT-PCR	[129]
	Axiom SNP array	56K SNPs from 104 genotypes	SNP genotyping	Axiom Affymetrix	[130]
	CcSNPnkssnp chip for Affymetrix GeneTitan	62k SNPs from conserved, unique, and stress resistance genes	SNP typing	Illumina Hiseq	[124]
Cluster bean	seedling (Ibadan Local-1)	1900 SSRs and 1800 conserved orthologous loci	Stimulating genomics accelerated breeding in winged bean	Illumina HiSeq 2500	[76]
	RNA-Seq	5773 SSR, 3594 SNPs, 62,146 unigenes with mean 679 bp length, and 11,000 genes annotated for biochemical pathways	To target protein-coding and regulatory genes	Illumina HiSeq 2500	[131]

Table 2. *Cont.*

Crop	Objective	Description	Genetic Improvement of Vegetable Type	Platform	Reference
	RNA-Seq	127,706 transcripts, 48,007 non-redundant unigenes, 79% annotations,8687 SSRs	To target protein coding and regulatory genes	Illumina paired end sequencing	[132]
	CbLncRNAdb database	lncRNAs, miRNAs identification, and characterization	Understanding the stress mechanism	http://cabgrid.res.in/cblncrnadb.	[133]
	Whole-genome sequencing	1859 SSRs from 1091 scaffolds constituting 60% genome of the cluster bean	Towards complete genome assembly	Illumina and Oxford nanopore	[134]
	Whole-genome assembly	1.2 Gb genomic reads comprising 50% genome of cluster bean (Illumina and Oxford nanopore)	Towards complete genome assembly	Illumina HiSeq 2500	[135]
	Genome sequencing of GG-4	15,399 SSRs generated	Towards complete genome assembly	Illumina MiSeq	[59]
Winged bean	CPP34 (PI 491423) and CPP37 (PI 639033) accessions	16,115 total contigs, 12,956 SSRs and 5190 SNPs developed	To target protein-coding and regulatory genes	Roche 454 Genome Sequencer FLX	[65]
	Tissue specific (leaf, pod root, and reproductive tissues)	198,554 contigs, 24,598 SSR motifs detected	Library of various tissues available for digging important traits	Illumina MiSeq	[77]
	Tannin controlling genes	1235 contigs expressed differentially	Identification of candidates	Illumina Nextseq 500	[136]
Dolichos bean	ORCAE-AOCC	Genomic portal for orphan crops such as dolichos bean	Information for molecular studies		[137]
Cowpea	Chilling tolerance	ICE1-CBF3-COR id cold-responsive cascade present in asparagus bean	Engineering cold-tolerant genotypes	Illumina Hiseq2500	[138]
	Molecular mechanism of chilling injury	Redox reactions enzymes, energy metabolism enzymes, and transcription factors, i.e., *WRKY*, *MYB*, *bHLH*, *NAC*, and *ERF* are involved in chilling injury	To plan genetic improvement by an understanding mechanism of chilling injury	Illumina HiSeq2500	[139]
	Transformable cowpea genotypes	Tissue-specific data special emphasis on reproductive organs	Genetic improvement and mapping studies	Illumina Hiseq 2500	[140]
	SNP chip-Cowpea iSelect Consortium Array	51 128 SNPs obtained by WGS sequencing of 37 different cowpea accessions	High throughput genotyping	Illumina HiSeq 2500	[141]

6. Transgenics and Genome Editing

Plant breeding offers extensive opportunities for the creation of desirable variation through hybridization and mutation. The scale of hybridization is limited, and the transfer of genomic information is impossible across reproductively incompatible genotypes by conventional techniques. Genetic engineering serves the purpose of transferring alien genes across species which otherwise are not feasible through conventional breeding. The routinely employed transformation technique in legumes is *Agrobacterium*-mediated gene transfer [142] owing to their dicot nature [143]. The major bottleneck in legume transgenic is the regeneration of explants due to their recalcitrant nature [144]. Among various explants, the use of young embryonic axes [145], cotyledonary nodes [146], and immature tissues and preconditioning of seedling with thidiazuron [147] have proved successful recovery. A few selected examples of economically important traits integrated into legume genome include, rice chitinase pigeon pea [146], *dhdps-r1* (increased lysine) pigeon pea [148], *P5CSF129A* (salt-tolerant) pigeon pea [149], *cry1Ac* pigeon pea [150], *cry1AcF* pigeon pea [151], *cry1AcF* dolichos bean [152], *aAI-1* (insect resistant) cowpea [153], *cry1Ab* cowpea [154], and soybean isoflavone synthase gene in cowpea [155]. On the other hand, there are no reports of the transformation of cluster bean and winged bean for economically important traits in the context of vegetable type characteristics.

Although, the regeneration and transformation protocols in both crops have been standardized [35]. In winged bean, successful organogenesis has been obtained from callus derived from cotyledons [156], epicotyls [157], excised segments of leaf [158], and protoplasts [159]. Further, in cluster bean, cotyledonary nodes, cotyledons [160], hypocotyls, and epicotyls have reported successful regeneration [161]. In view of available protocols for transformation and improvement, cluster bean and winged bean can be exploited for genetic engineering for useful, economical traits such as biotic and abiotic resistance genes, quality traits, yield-enhancing genes, and growth habit controlling genes. However, some anti-nutrient traits from these legumes have been extracted, such as cowpea trypsin inhibitor against rice stem borers in transformed rice plants [162]. The winged bean lysine-rich protein (WBLRP) isolated from winged bean was patented [163] and hexaploid wheat was transformed by WBLRP comprising expression vector which increased lysine content by 2-3-fold in transgenic wheat [164].

One of the promising techniques of post-transcriptional gene silencing is RNA interference (RNAi) in which ds-RNA molecules prevent gene expression, conferring resistance to pathogenic nucleic acids and regulating the expression of protein translating mRNAs [165]. This technology has emerged as a promising technique in plants to fight against invading pathogenic viruses. The host plant is engineered to express ds-RNA, which inhibits expression of the complementary gene in pathogens [166]. RNAi is a potent technology for insect resistance in legumes through silencing genes essential for insect survival. In tobacco, RNAi mediated gene silencing was achieved against *Helicoverpa armigera* through vector construct carrying *HaAce1* gene (*H. armigera* acetylcholinesterase) in the backbone of *HaAce1*-preamiRNA1 from *Arabidopsis* controlled by CaMV 35S promoter against *H.a armigera* [167]. With the availability of cloned sequences of insect and pathogen genes, a similar approach can be employed in vegetable legumes for insect and pathogen resistance [168]. The transgenic crops face biosafety issues that vary from country to country and struggle through the journey from lab to land [169]. Genome editing is emerging as a widely adopted targeted approach and does not fall under the category of genetically engineered crops in the USA [170]. However, some countries still lack clarity between the two technologies (Figure 1).

Genome editing can be accomplished by site-specific double-strand breaks in DNA caused by homing endonucleases (HEs) [171], zinc finger nucleases (ZFNs) [172], transcription activator-like effector nucleases (TALENs) [173], and clustered regularly interspaced short palindromic repeat and CRISPR-associated protein (CRISPR-Cas type II) [174]. The *homology-directed repair* (HDR) is a template mediated repair technique, in which exogenous template sequences can be supplied to introduce desired sequence change. Mostly plant viruses serve the purpose of delivery of template sequence as in the case of potato, a geminivirus replicon (GVR) is employed to deliver sequence-specific nucleases (SSNs) targeting acetolactate synthase 1 (ALS1) gene and customized repair templates constructed to induce point mutation for herbicide resistance within the locus [175]. Among the various technologies for inducing double-strand breaks, CRISPR-Cas has proven to be more site-specific with the least off-targets, easier to use, and thus widely adopted for genome editing [176]. The CRISPR-Cas system is ready to be exploited in legumes for desirable mutations in the gene of interest with optimization of the protocol in model species *Arabidopsis* and *Nicotiana benthamiana*. The pea early browning virus was used as a delivery system of Cas 9 and guide RNA in model species, and this virus is known to cause disease in 30 species along with the members of family Leguminosae [177,178]. Therefore, the same virus can be engineered for legume vegetables to express desirable guide RNA sequence homologous to the site to be mutated. In future, using genome editing methods will lead to the development of non-genetically modified crops, with desired traits.

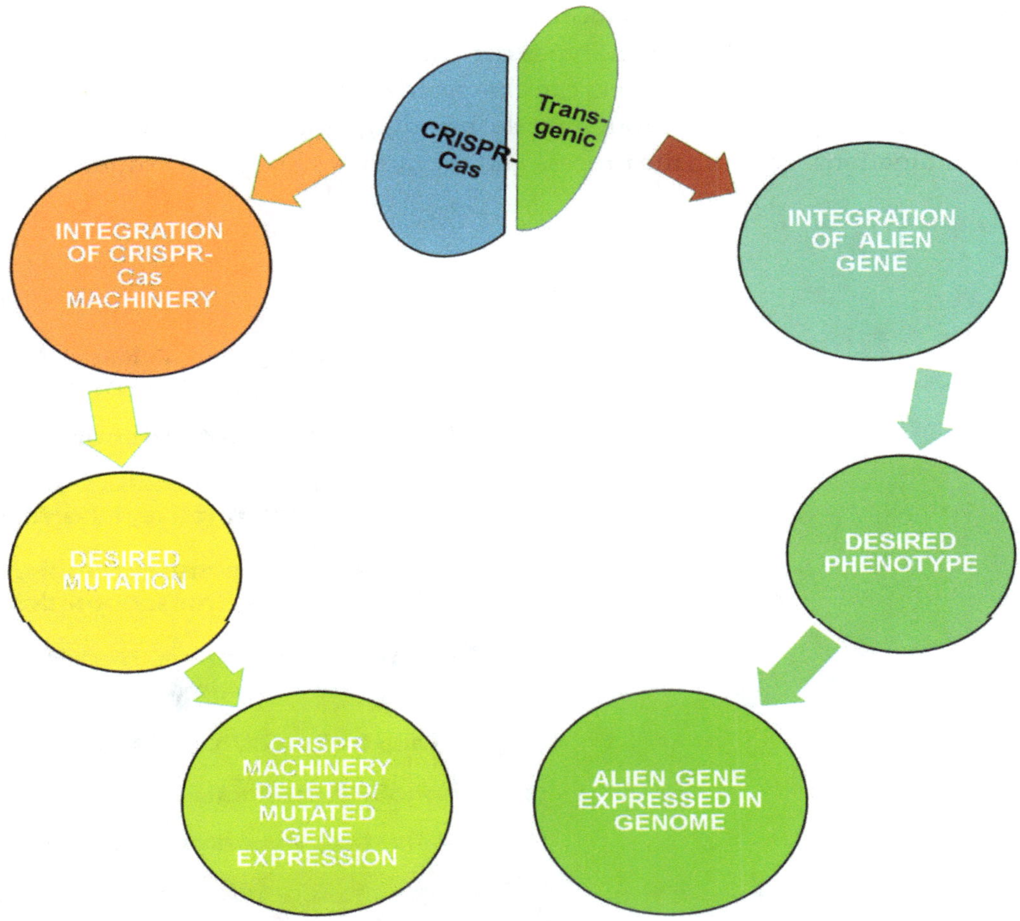

Figure 1. Illustration of CRISPR-Cas versus transgenic methodology and outcomes.

7. Future Prospects

Most developed legumes used for food are consumed as grain seeds known as pulses. Nevertheless, the growth of some legume species is aimed at the consumption of theirs as vegetables [179]. From a health perspective, legumes are considered valuable sources of plant protein, carbohydrates, essential minerals, vitamins, and phytochemicals. Importantly, legume vegetables comprise a low-fat diet with high proportions of digestible proteins. Vegetable legumes provide several raw materials for products ranging from coatings for cloth and paper to eco-friendly plastics as well as biofuel. Legumes draw the interest of scientists seeking to exploit their nutritional resources [180]. Legumes have extreme diversity and various stress tolerance capabilities; it might enhance food and food security in low-income areas of Africa [20]. Exploring underexploited legumes' nutritional value can overcome malnutrition, especially in Sub-Saharan Africa, in developing nations [181]. Similarly, nutrient strength can be analyzed and used in formulations for the food-to-food approach [182]. Study especially on wild relatives of underexploited legumes for qualitative and quantitative traits and their domestication will contribute significantly to nutrition security. In this respect, underexploited legumes of reasonable nutritional value can be harnessed with other local food products for recipe development based on nutritional requirements [181].

Average yields of legumes have changed significantly in recent decades. This achievement was due in part, to the breeding of better performing hybrids developed by combining CWRs inbred plant genomes. Moreover, in recent decades breakthroughs in genomic technologies and availability of vegetable legumes' draft genome sequence knowledge accelerates the breeding vision. However, to support the world's increasing population, it will be essential to sustain the rate of increase in vegetable legumes production. This challenge will probably be addressed through better farming, much more reliable seed supplies, plus more stable markets as well by the application of genomics. Most commercially important plant phenotypes depend on the interactions of large numbers of genes. With the advent of genomics tools, breeders can characterize the allelic characteristics of their particular germplasm in detail that is exquisite throughout the breeding program and therefore retaining the many useful allele combinations. Overall, the process and its various steps can be summarized as in Figure 2.

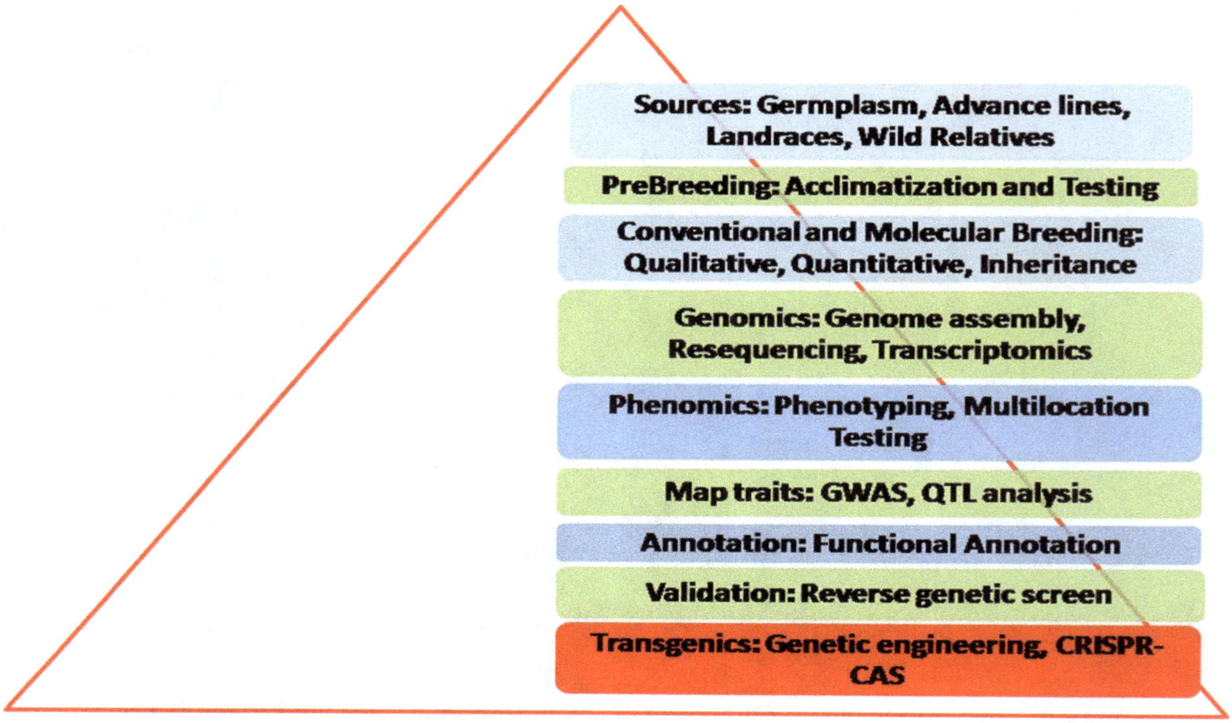

Figure 2. Stages and best approaches for consideration in vegetable legumes improvement.

Integration of genomics tools with conventional breeding methods would assist breeders in their attempts to design and properly choose the best combinations of chromosome segments, alleles, and genes offered in the related species to fulfill the requirements for crop production enhancement.

Author Contributions: P.K. conceived of and designed the project; P.K. supervised the study; S.K.D., V.S., P.K., A.G., A.T. and P.S. wrote the paper. P.K. checked and corrected the final draft. All authors have read and agreed to the published version of the manuscript.

Acknowledgments: The authors are thankful to the anonymous reviewers for their careful reading of the manuscript and for providing insightful suggestions.

Abbreviations

QTLs	Quantitative trait loci
NGS	Next-generation sequencing
CWRs	Crop wild relatives
RFLP	Restriction fragment length polymorphism
RAPD	Random amplification of polymorphic DNA

AFLP	Amplified fragment length polymorphism
SSRs	Simple sequence repeats
SNPs	Single nucleotide polymorphism
EST	Expressed *sequence* tag
ISSR	Inter simple sequence repeat
NCBI	National Center for Biotechnology Information
GWAS	Genome-wide association studies
SLAF	Specific length amplified fragment sequencing
GBS	Genotyping-by-sequencing
WBLRP	Winged bean lysine-rich protein
RNAi	RNA interference
HEs	Homing endonucleases
ZFNs	Zinc finger nucleases
CRISPR-Cas	Clustered regularly interspaced short palindromic repeat and CRISPR-associated protein
HDR	*Homology directed repair*
GVR	Geminivirus replicon

References

1. Godfray, H.C.J.; Beddington, J.R.; Crute, I.R.; Haddad, L.; Lawrence, D.; Muir, J.F.; Pretty, J.; Robinson, S.; Thomas, S.M.; Toulmin, C. Food security: The challenge of feeding 9 billion people. *Science* **2010**, *327*, 812–818. [CrossRef]
2. Palai, J.B.; Jena, J.; Maitra, S. Prospects of underutilized food legumes in sustaining pulse needs in India–A review. *Crop. Res.* **2019**, *54*, 82–88.
3. Harouna, D.V.; Venkataramana, P.B.; Ndakidemi, P.A.; Matemu, A.O. Under-exploited wild *Vigna* species potentials in human and animal nutrition: A review. *Glob. Food Sec.* **2018**, *18*, 1–11. [CrossRef]
4. Agrawal, A.; Pandey, A.; Varaprasad, K.S.; Tyagi, R.K.; Khetarpal, R.K. Regional expert consultation on underutilized crops for food and nutrition security in asia and the pacific. *IJPGR* **2018**, *31*, 194–195. [CrossRef]
5. Clydesdale, F. Functional foods: Opportunities & challenges. *Food Technol.* **2004**, *58*, 35–40.
6. Mall, T.P. Diversity of potential orphan plants in health management and climate change mitigation from Bahraich (Uttar Pradesh). *Int. J. Curr. Res. Biosci. Plant Biol.* **2017**, *4*, 106–145. [CrossRef]
7. Mabhaudhi, T.; Chimonyo, V.G.; Chibarabada, T.P.; Modi, A.T. Developing a roadmap for improving neglected and underutilized crops: A case study of South Africa. *Front. Plant Sci.* **2017**, *8*, 2143. [CrossRef]
8. Blair, M.W.; Wu, X.; Bhandari, D.; Zhang, X.; Hao, J. Role of legumes for and as horticultural crops in sustainable agriculture. In *Organic Farming for Sustainable Agriculture*; Springer: Cham, Switzerland, 2016; pp. 185–211.
9. Krupa, U. Main nutritional and antinutritional compounds of bean seeds-a review. *Polish J. Food Nutr. Sci.* **2008**, *58*, 149–155.
10. Sharma, V.; Bhattacharyya, S.; Kumar, R.; Kumar, A.; Ibañez, F.; Wang, J.; Guo, B.; Sudini, H.K.; Gopalakrishnan, S.; DasGupta, M.; et al. Molecular Basis of Root Nodule Symbiosis between Bradyrhizobium and 'Crack-Entry'Legume Groundnut (*Arachis hypogaea* L.). *Plants* **2020**, *9*, 276. [CrossRef]
11. Akhtar, L.H.; Bukhari, S.; Salah-ud-Din, S.; Minhas, R. Response of new guar strains to various row spacings. *Pak. J. Agric. Sci.* **2012**, *49*, 469–471.
12. OchiaiYanagi, S. Properties of winged bean (*Psophocarpus tetragonolobus*) protein in comparison with soybean (*Glycine max*) and common bean (*Phaseolus vulgaris*) protein. *Agric. Biol. Chem.* **1983**, *47*, 2273–2280. [CrossRef]
13. Kelly, J.D.; Cichy, K.A.; Siddiq, M.; Uebersax, M.A. Dry bean breeding and production technologies. In *Dry Beans and Pulses: Production, Processing, and Nutrition*; Wiley-Blackwell: Chichester, UK, 2013; pp. 23–54.
14. Taïbi, K.; Taïbi, F.; Abderrahim, L.A.; Ennajah, A.; Belkhodja, M.; Mulet, J.M. Effect of salt stress on growth, chlorophyll content, lipid peroxidation and antioxidant defence systems in *Phaseolus vulgaris* L. *S. Afr. J. Bot.* **2016**, *105*, 306–312. [CrossRef]

15. Coleman, E. *Four-Season Harvest: Organic Vegetables from Your Home Garden all Year Long*; Chelsea Green Publishing: Hartford, VM, USA, 2012.
16. Sá, A.G.; Moreno, Y.M.; Carciofi, B.A. Plant proteins as high-quality nutritional source for human diet. *Trends Food Sci. Technol.* **2020**, *97*, 170–184.
17. Kumar, P.; Kaushik, P. Evaluation of genetic diversity in cultivated and exotic germplasm sources of Faba Bean using important morphological traits. *BioRxiv* **2020**, *24*, 918284.
18. Kumar, J.; Choudhary, A.K.; Gupta, D.S.; Kumar, S. Towards exploitation of adaptive traits for climate-resilient smart pulses. *IJMS* **2019**, *20*, 2971. [CrossRef]
19. Popoola, J.; Ojuederie, O.; Omonhinmin, C.; Adegbite, A. Neglected and underutilized legume crops: Improvement and future prospects. In *Recent Advances in Grain Crops Research*; IntechOpen: London, UK, 2019; pp. 3–9.
20. Ertiro, B.T.; Azmach, G.; Keno, T.; Chibsa, T.; Abebe, B.; Demissie, G.; Wegary, D.; Wolde, L.; Teklewold, A.; Worku, M. Fast-tracking the development and dissemination of a drought-tolerant maize variety in Ethiopia in response to the risks of climate change. In *The Climate-Smart Agriculture Papers*; Springer: Cham, Switzerland, 2019; pp. 79–86.
21. Sharma, S.; Paul, P.J.; Sameer Kumar, C.V.; Nimje, C. Utilizing Wild *Cajanus platycarpus*, a Tertiary Genepool Species for Enriching Variability in the Primary Genepool for Pigeonpea Improvement. *Front. Plant Sci.* **2020**, *11*, 1055. [CrossRef]
22. Kulski, J.K. Next-Generation Sequencing—An Overview of the History, Tools, and "Omic" Applications. In *Next Generation Sequencing—Advances, Applications and Challenges*; Kulski, J.K., Ed.; IntechOpen: London, UK, 2016; pp. 1–3.
23. Ebert, A.W. Potential of underutilized traditional vegetables and legume crops to contribute to food and nutritional security, income and more sustainable production systems. *Sustainability* **2014**, *6*, 319–335. [CrossRef]
24. Relatives, W.C. *Genomic and Breeding Resources by Ch. Kole*; Springer: Berlin/Heidelberg, Germany, 2011.
25. Hoisington, D.; Khairallah, M.; Reeves, T.; Ribaut, J.M.; Skovmand, B.; Taba, S.; Warburton, M. Plant genetic resources: What can they contribute toward increased crop productivity? *Proc. Natl. Acad. Sci. USA* **1999**, *96*, 5937–5943. [CrossRef]
26. Stewart, C.N., Jr. *Genetically Modified Planet: Environmental Impacts of Genetically Engineered Plants*; Oxford University Press: Oxford, UK, 2004.
27. Saxena, K.; Chauhan, Y.; Sameer Kumar, C.V.; Hingane, A.; Kumar, R.; Saxena, R.; Rao, G.V.R. *Developing Improved Varieties of Pigeonpea, In Achieving Sustainable Cultivation of Grain Legumes Volume 2: Improving Cultivation of Particular Grain Legumes*; Sivasankar, S., Ed.; Burleigh Dodds Science Publishing: Cambridge, UK, 2018; pp. 1–30.
28. Sharma, P.; Singh, I.; Sirari, A.; Khosla, G.; Singh, G.; Ludhar, N.K.; Singh, S. Inheritance and molecular mapping of restorer-of-fertility (Rf) gene in A2 hybrid system in pigeonpea (*Cajanus cajan*). *Plant Breed.* **2019**, *138*, 741–743. [CrossRef]
29. Ayenan, M.A.T.; Danquah, A.; Ahoton, L.E.; Ofori, K. Utilization and farmers' knowledge on pigeonpea diversity in Benin, West Africa. *J. Ethnobiol. Ethnomed.* **2017**, *13*, 37. [CrossRef]
30. Hymowitz, T. The trans-domestication concept as applied to guar. *Econo. Bot.* **1972**, *26*, 49–60. [CrossRef]
31. Krishna, G.; Reddy, P.S.; Ramteke, P.W.; Rambabu, P.; Tawar, K.B.; Bhattacharya, P. Agrobacterium-mediated genetic transformation of pigeon pea [*Cajanus cajan* (L.) Millsp.] for resistance to legume pod borer *Helicoverpaarmigera*. *J. Crop Sci. Biotechnol.* **2011**, *14*, 197–204. [CrossRef]
32. Tyagi, A.; Sharma, P.; Saxena, S.; Sharma, R.; Mithra, S.A.; Solanke, A.U.; Singh, N.K.; Sharma, T.R.; Gaikwad, K. The genome size of clusterbean (*Cyamopsis tetragonoloba*) is significantly smaller compared to its wild relatives as estimated by flow cytometry. *Gene* **2019**, *707*, 205–211. [CrossRef]
33. Yang, S.; Grall, A.; Chapman, M.A. Origin and diversification of winged bean (*Psophocarpus tetragonolobus* (L.) DC.), a multipurpose underutilized legume. *Am. J. Bot.* **2018**, *105*, 888–897. [CrossRef]
34. Tanzi, A.S.; Eagleton, G.E.; Ho, W.K.; Wong, Q.N.; Mayes, S.; Massawe, F. Winged bean (*Psophocarpus tetragonolobus* (L.) DC.) for food and nutritional security: Synthesis of past research and future direction. *Planta* **2019**, *250*, 911–931. [CrossRef]
35. Mohanty, C.S.; Singh, V.; Chapman, M.A. Winged bean: An underutilized tropical legume on the path of improvement, to help mitigate food and nutrition security. *Sci. Hortic.* **2020**, *260*, 108789. [CrossRef]

36. Maass, B.L.; Knox, M.R.; Venkatesha, S.C.; Angessa, T.T.; Ramme, S.; Pengelly, B.C. Lablab purpureus—A Crop Lost for Africa? *Trop. Plant Biol.* **2010**, *3*, 123–135. [CrossRef] [PubMed]
37. Rony, M.B.U.; Islam, A.A.; Rasul, M.G.; Zakaria, M. Genetic analysis of yield and related characters of Lablab Bean. *J. Nep. Agric. Res.* **2019**, *5*, 1–21. [CrossRef]
38. Evans, D.M.; Gillespie, N.A.; Martin, N.G. Biometrical genetics. *Biol. Psychol.* **2002**, *61*, 33–51. [CrossRef]
39. Maass, B. Domestication, origin and global dispersal of *Lablab purpureus* (L.) Sweet (Fabaceae): Current understanding. *Legume Perspec.* **2016**, *13*, 5–8.
40. Berthaud, J.; Clément, J.C.; Emperaire, L.; Louette, D.; Pinton, F.; Sanou, J.; Second, G. The Role of Local Level Gene Flow in Enhancing and Maintaining Genetic Diversity. In *Broadening the Genetic Base of Crop Production*; CABI Publishing in Association with FAO and IPGRI: Wallingford, UK, 2001; pp. 81–103.
41. Boukar, O.; Fatokun, C. Strategies in cowpea breeding. In *New Approaches to Plant Breeding of Orphan Crops in Africa*; FAO: Rome, Italy, 2009; pp. 69–92.
42. Kouam, E.B.; Pasquet, R.S.; Campagne, P.; Tignegre, J.B.; Thoen, K.; Gaudin, R.; Gepts, P. Genetic structure and mating system of wild cowpea populations in West Africa. *BMC Plant Biol.* **2012**, *12*, 113. [CrossRef] [PubMed]
43. Amorim, L.L.; Ferreira-Neto, J.R.; Bezerra-Neto, J.P.; Pandolfi, V.; de Araújo, F.T.; da Silva Matos, M.K.; Santos, M.G.; Kido, E.A.; Benko-Iseppon, A.M. Cowpea and abiotic stresses: Identification of reference genes for transcriptional profiling by qPCR. *Plant Methods* **2018**, *14*, 88. [CrossRef] [PubMed]
44. Nwosu, D.J.; Awa, E.N. Cross-compatibility between some cultivated cowpea varieties and a wild relative (subsp. *Dekindtiana* Var *Pubescens*). *J. Scienti. Res.* **2013**, *5*, 195–200. [CrossRef]
45. Sariah, J.E. Enhancing Cowpea (*Vigna Unguiculata* L.) Production through Insect pest Resistant Line in East Africa. Ph.D. Thesis, University of Copenhagen, Faculty of Life Sciences, København, Denmark, 2010.
46. Andersson, M.S.; De Vicente, M.C. *Gene Flow between Crops and Their Wild Relatives*; JHU Press: Baltimore, MA, USA, 2010.
47. Shimelis, H.; Laing, M. Timelines in conventional crop improvement: Pre-breeding and breeding procedures. *Aust. J. Crop Sci.* **2012**, *6*, 1542.
48. Stannard, C.; Moeller, N.I. *Identifying Benefit Flows: Studies on the Potential Monetary and Non-Monetary Benefits Arising from the International Treaty on Plant Genetic for food and Agriculture*; Food and Agriculture Organization of the United Nations: Rome, Italy, 2013.
49. Jarvis, D.I.; Hodgkin, T. Wild relatives and crop cultivars: Detecting natural introgression and farmer selection of new genetic combinations in agroecosystems. *Mol. Ecol.* **1999**, *8*, S159–S173. [CrossRef]
50. Sharma, S.; Upadhyaya, H.D.; Varshney, R.K.; Gowda, C.L.L. Pre-breeding for diversification of primary gene pool and genetic enhancement of grain legumes. *Front. Plant Sci.* **2013**, *4*, 309. [CrossRef]
51. Nass, L.L.; Paterniani, E. Pre-breeding: A link between genetic resources and maize breeding. *Sci. Agric.* **2000**, *57*, 581–587. [CrossRef]
52. Malviya, N.; Yadav, D. RAPD analysis among pigeon pea [*Cajanus cajan* (L.) Mill sp.] cultivars for their genetic diversity. *J. Genet. Eng. Biotechnol.* **2010**, *1*, 1–9.
53. Gresta, F.; Mercati, F.; Santonoceto, C.; Abenavoli, M.R.; Ceravolo, G.; Araniti, F.; Anastasi, U.; Sunseri, F. Morpho-agronomic and AFLP characterization to explore guar (*Cyamopsis tetragonoloba* L.) genotypes for the Mediterranean environment. *Ind. Crop Prod.* **2016**, *86*, 23–30. [CrossRef]
54. Chen, D.; Yi, X.; Yang, H.; Zhou, H.; Yu, Y.; Tian, Y.; Lu, X. Genetic diversity evaluation of winged bean (*Psophocarpus tetragonolobus* (L.) DC.) using inter-simple sequence repeat (ISSR). *Genet. Resour. Crop Evol.* **2015**, *62*, 823–828. [CrossRef]
55. Kinmani, E.N.; Wachira, F.N.; Kinyua, M.G. Molecular diversity of Kenyan lablab bean (*Lablab purpureus* L. Sweet) accessions using amplified fragment length polymorphism markers. *Am. J. Plant Sci.* **2012**, *3*, 3–32.
56. Fang, J.; Chao, C.C.; Roberts, P.A.; Ehlers, J.D. Genetic diversity of cowpea [*Vigna unguiculata* (L.) Walp.] in four West African and USA breeding programs as determined by AFLP analysis. *Genet. Resour. Crop Evol.* **2007**, *54*, 1197–1209. [CrossRef]
57. Zannouou, A.; Kossou, D.K.; Ahanchede, A.; Zoundjihékpon, J.; Agbicodo, E.; Struik, P.C.; Sanni, A. Genetic variability of cultivated cowpea in Benin assessed by random amplified polymorphic DNA. *Afr. J. Biotechnol.* **2008**, *7*, 24.

58. Varshney, R.K.; Chen, W.; Li, Y.; Bharti, A.K.; Saxena, R.K.; Schlueter, J.A.; Donoghue, M.T.; Azam, S.; Fan, G.; Whaley, A.M.; et al. Draft genome sequence of pigeonpea (*Cajanus cajan*), an orphan legume crop of resource-poor farmers. *Nat. Biotechnol.* **2012**, *30*, 83–89. [CrossRef] [PubMed]
59. Kumar, S.; Palve, A.S.; Patel, S.K.; Selvanayagam, S.; Sharma, R.; Rathore, A. Development of genomic microsatellite markers in cluster bean using next-generation DNA sequencing and their utility in diversity analysis. *Curr. Plant Biol.* **2020**, *21*, 2214–6628. [CrossRef]
60. Ogunkanmi, L.A.; Ogundipe, O.T.; Ng, N.Q.; Fatokun, C.A. Genetic diversity in wild relatives of cowpea (*Vigna unguiculata*) as revealed by simple sequence repeats (SSR) markers. *J. Food Agric. Environ.* **2008**, *6*, 253–268.
61. Shivakumar, M.S.; Ramesh, S. Transferability of cross legume species/genera SSR markers to Dolichos Bean (*Lablab purpureus* L. Sweet) var. Lignosus. *Ecialisesp* **2015**, *49*, 263–265.
62. Rai, N.; Kumar, S.; Singh, R.K.; Rai, K.K.; Tiwari, G.; Kashyap, S.P.; Singh, M.; Rai, A.B. Genetic diversity in Indian bean (*Lablab purpureus*) accessions as revealed by quantitative traits and cross-species transferable SSR markers. *Indian J. Agric. Sci.* **2016**, *86*, 654–660.
63. Saxena, R.K.; Von Wettberg, E.; Upadhyaya, H.D.; Sanchez, V.; Songok, S.; Saxena, K.; Kimurto, P.; Varshney, R.K. Genetic diversity and demographic history of Cajanus spp. illustrated from genome-wide SNPs. *PLoS ONE* **2014**, *9*, e88568. [CrossRef]
64. Thakur, O.; Randhawa, G.S. Identification and characterization of SSR, SNP and InDel molecular markers from RNA-Seq data of guar (*Cyamopsis tetragonoloba*, L. Taub.) roots. *BMC Genom.* **2018**, *19*, 951. [CrossRef]
65. Vatanparast, M.; Shetty, P.; Chopra, R.; Doyle, J.J.; Sathyanarayana, N.; Egan, A.N. Transcriptome sequencing and marker development in winged bean (*Psophocarpus tetragonolobus*; *Leguminosae*). *Sci. Rep.* **2016**, *6*, 29070. [CrossRef] [PubMed]
66. Venkatesha, S.C.; Ganapathy, K.N.; Gowda, M.B.; Gowda, P.R.; Mahadevu, P.; Girish, G.; Ajay, B.C. Variability and Genetic Structure among Lablab Bean Collections of India and their Relationship with Exotic Accessions. *Vegetos* **2013**, *26*, 121–130. [CrossRef]
67. Muchero, W.; Diop, N.N.; Bhat, P.R.; Fenton, R.D.; Wanamaker, S.; Pottorff, M.; Hearne, S.; Cisse, N.; Fatokun, C.; Ehlers, J.D.; et al. A consensus genetic map of cowpea [*Vigna unguiculata* (L) Walp.] and synteny based on EST-derived SNPs. *Proc. Natl. Acad. Sci. USA* **2009**, *106*, 18159–18164. [CrossRef] [PubMed]
68. Bohra, A.; Saxena, R.K.; Gnanesh, B.N.; Saxena, K.; Byregowda, M.; Rathore, A.; KaviKishor, P.B.; Cook, D.R.; Varshney, R.K. An intra-specific consensus genetic map of pigeonpea [*Cajanus cajan* (L.) Millspaugh] derived from six mapping populations. *Theor. Appl. Genet.* **2012**, *125*, 1325–1338. [CrossRef]
69. Arora, S.; Mahato, A.K.; Singh, S.; Mandal, P.; Bhutani, S.; Dutta, S.; Kumawat, G.; Singh, B.P.; Chaudhary, A.K.; Yadav, R.; et al. A high-density intraspecific SNP linkage map of pigeonpea (*Cajanascajan* L. Millsp.). *PLoS ONE* **2017**, *12*, e0179747. [CrossRef]
70. Saxena, R.K.; Singh, V.K.; Kale, S.M.; Tathineni, R.; Parupalli, S.; Kumar, V.; Garg, V.; Das, R.R.; Sharma, M.; Yamini, K.N.; et al. Construction of genotyping-by-sequencing based high-density genetic maps and QTL mapping for fusarium wilt resistance in pigeonpea. *Sci. Rep.* **2017**, *7*, 1–11. [CrossRef]
71. Saxena, R.K.; Kale, S.M.; Kumar, V.; Parupali, S.; Joshi, S.; Singh, V.; Garg, V.; Das, R.R.; Sharma, M.; Yamini, K.N.; et al. Genotyping-by-sequencing of three mapping populations for identification of candidate genomic regions for resistance to sterility mosaic disease in pigeonpea. *Sci. Rep.* **2017**, *7*, 1–10. [CrossRef]
72. Naoumkina, M.; Torres-Jerez, I.; Allen, S.; He, J.; Zhao, P.X.; Dixon, R.A.; May, G.D. Analysis of cDNA libraries from developing seeds of guar (*Cyamopsis tetragonoloba* (L.) Taub). *BMC Plant Biol.* **2007**, *7*, 1–2. [CrossRef]
73. Kuravadi, A.N.; Tiwari, P.B.; Tanwar, U.K.; Tripathi, S.K.; Dhugga, K.S.; Gill, K.S.; Randhawa, G.S. Identification and characterization of EST-SSR markers in clusterbean (*Cyamopsis spp.*) *Crop Sci.* **2014**, *54*, 1097–1102. [CrossRef]
74. Kumar, S.; Parekh, M.J.; Patel, C.B.; Zala, H.N.; Sharma, R.; Kulkarni, K.S.; Fougat, R.S.; Bhatt, R.K.; Sakure, A.A. Development and validation of EST-derived SSR markers and diversity analysis in cluster bean (*Cyamopsis tetragonoloba*). *J. Plant Biochem. Biot.* **2015**, *25*, 263–269. [CrossRef]
75. Mohanty, C.S.; Verma, S.; Singh, V.; Khan, S.; Gaur, P.; Gupta, P.; Nizar, M.A.; Dikshit, N.; Pattanayak, R.; Shukla, A.; et al. Characterization of winged bean (*Psophocarpus tetragonolobus* (L.) DC.) based on molecular, chemical and physiological parameter. *Am. J. Mol. Biol.* **2013**, *3*, 187–197. [CrossRef]
76. Chapman, M.A. Transcriptome sequencing and marker development for four underutilized legumes. *Appl. Plant Sci.* **2015**, *3*, 1400111. [CrossRef] [PubMed]

77. Wong, Q.N.; Tanzi, A.S.; Ho, W.K.; Malla, S.; Blythe, M.; Karunaratne, A.; Massawe, F.; Mayes, S. Development of gene-based SSR markers in winged bean (*Psophocarpus tetragonolobus* (L.) DC.) for diversity assessment. *Genes* **2017**, *8*, 100. [CrossRef] [PubMed]
78. Konduri, V.; Godwin, I.D.; Liu, C.J. Genetic mapping of the *Lablab purpureus* genome suggests the presence of "cuckoo" gene(s) in this species. *Theor. Appl. Genet.* **2000**, *100*, 866–871. [CrossRef]
79. Yao, L.M.; Zhang, L.D.; Hu, Y.L.; Wang, B.; Wu, T.L. Characterization of novel soybean derived simple sequence repeat markers and their transferability in hyacinth bean [*Lablab purpureus* (L.) Sweet]. *Indian J. Genet. Plant Breed.* **2012**, *72*, 46.
80. Venkatesha, S.C.; Byregowda, M.; Mahadevu, P.; Mohan Rao, A.; Kim, D.J.; Ellis, T.H.N.; Knox, M.R. Genetic diversity within *Lablab purpureus* and the transferability of gene-specific markers from a range of legume species. *Plant Genet. Resou.* **2007**, *5*, 154–171. [CrossRef]
81. Ramtekey, V.; Bhuriya, A.; Ayer, D.; Parekh, V.; Modha, K.; Kale, B.; Vadodariya, G.; Patel, R. Molecular tagging of photoperiod responsive flowering in Indian bean [*Lablab purpureus* (L.) Sweet]. *Indian J. Genet.* **2019**, *79*, 269.
82. Zhang, G.; Xu, S.; Mao, W.; Gong, Y.; Hu, Q. Development of EST-SSR markers to study genetic diversity in hyacinth bean (*Lablab purpureus* L.). *Plant Omics* **2013**, *6*, 295–301.
83. Vaijayanthi, P.V.; Ramesh, S.; Gowda, M.B.; Rao, A.M.; Keerthi, C.M. Genome-wide marker-trait association analysis in a core set of Dolichos bean germplasm. *Plant Genet. Res.* **2018**, *17*, 1–11. [CrossRef]
84. Xu, P.; Wu, X.; Wang, B.; Liu, Y.; Ehlers, J.D.; Close, T.J.; Roberts, P.A.; Diop, N.N.; Qin, D.; Hu, T.; et al. A SNP and SSR based genetic map of asparagus bean (*Vigna. unguiculata* ssp. sesquipedialis) and comparison with the broader species. *PLoS ONE* **2011**, *6*, e15952. [CrossRef]
85. Kongjaimun, A.; Somta, P.; Tomooka, N.; Kaga, A.; Vaughan, D.A.; Srinives, P. QTL mapping of pod tenderness and total soluble solid in yardlong bean [*Vigna unguiculata* (L.) Walp. subsp.*unguiculata*cv.-gr. *sesquipedalis*]. *Euphytica* **2013**, *189*, 217–223. [CrossRef]
86. Xu, P.; Wu, X.; Wang, B.; Hu, T.; Lu, Z.; Liu, Y.; Qin, D.; Wang, S.; Li, G. QTL mapping and epistatic interaction analysis in asparagus bean for several characterized and novel horticulturally important traits. *BMC Genet.* **2013**, *14*, 4. [CrossRef]
87. Gnanesh, B.N.; Bohra, A.; Sharma, M.; Byregowda, M.; Pande, S.; Wesley, V.; Saxena, R.K.; Saxena, K.B.; Kishor, P.K.; Varshney, R.K. Genetic mapping and quantitative trait locus analysis of resistance to sterility mosaic disease in pigeonpea [*Cajanus cajan* (L.) Millsp.]. *Field Crops. Res.* **2011**, *123*, 53–61. [CrossRef]
88. Kumawat, G.; Raje, R.S.; Bhutani, S.; Pal, J.K.; Mithra, A.S.; Gaikwad, K.; Sharma, T.R.; Singh, N.K. Molecular mapping of QTLs for plant type and earliness traits in pigeonpea (*Cajanus cajan* L. Millsp.). *BMC Genet.* **2012**, *13*, 1–11. [CrossRef] [PubMed]
89. Mir, R.R.; Kudapa, H.; Srikanth, S.; Saxena, R.K.; Sharma, A.; Azam, S.; Saxena, K.; Penmetsa, R.V.; Varshney, R.K. Candidate gene analysis for determinacy in pigeonpea (*Cajanus* spp.). *Theor. Appl. Genet.* **2014**, *127*, 2663–2678. [CrossRef] [PubMed]
90. Singh, V.K.; Khan, A.W.; Saxena, R.K.; Kumar, V.; Kale, S.M.; Sinha, P.; Chitikineni, A.; Pazhamala, L.T.; Garg, V.; Sharma, M.; et al. Next-generation sequencing for identification of candidate genes for Fusarium wilt and sterility mosaic disease in pigeonpea (*Cajanus cajan*). *Plant Biotechnol. J.* **2016**, *14*, 1183–1194. [CrossRef]
91. Bohra, A.; Jha, R.; Pandey, G.; Patilm, P.G.; Saxena, R.K.; Singh, I.P.; Singh, D.; Mishra, R.K.; Mishra, A.; Singh, F.; et al. New hypervariable ssr markers for diversity analysis, hybrid purity testing and trait mapping in pigeonpea [*Cajanus cajan* (L.) Millspaugh]. *Front. Plant Sci.* **2017**, *8*, 377. [CrossRef]
92. Saxena, R.K.; Obala, J.; Sinjushin, A.; Kumar, C.S.; Saxena, K.B.; Varshney, R.K. Characterization and mapping of Dt1 locus which co-segregates with CcTFL1 for growth habit in pigeonpea. *Theor. Appl. Genet.* **2017**, *130*, 1773–1784. [CrossRef]
93. Saxena, R.K.; Kale, S.; Mir, R.R.; Mallikarjuna, N.; Yadav, P.; Das, R.R.; Molla, J.; Sonnappa, M.; Ghanta, A.; Narasimhan, Y.; et al. Genotyping-by-sequencing and multilocation evaluation of two interspecific backcross populations identify QTLs for yield-related traits in pigeonpea. *Theor. Appl. Genet.* **2020**, *133*, 737–749. [CrossRef]
94. Bajaj, A. Molecular Mapping of Bacterial Blight Resistance Gene, Drought Tolerant QTL (s) and Genetic Diversity Analysis in Clusterbean {*Cyamopsis Tetragonoloba* (L) Taub}. Ph.D. Thesis, Chaudhary Charan Singh Haryana Agricultural University, Hisar, India, 2011.
95. Sharma, P.; Kumar, V.; Raman, K.; Tiwari, K. A set of SCAR markers in cluster bean (*Cyamopsis tetragonoloba* L. Taub) genotypes. *Adv. Bio. Biotechnol.* **2014**, *5*, 131–141.

96. Yuan, J.; Yang, R.; Wu, T. Bayesian mapping QTL for fruit and growth phenological traits in *Lablab purpureus* (L.) Sweet. *Afr. J. Biotechnol.* **2009**, *8*, 167–175.
97. Yuan, J.; Wang, B.; Wu, T.L. Quantitative trait loci (QTL) mapping for inflorescence length traits in *Lablab purpureus* (L.) sweet. *Afr. J. Biotechnol.* **2011**, *10*, 3558–3566.
98. Shivachi, A.; Kiplagat, K.O.; Kinyua, G.M. Microsatellite analysis of selected Lablab purpureus genotypes in Kenya. *Rwanda J.* **2012**, *28*, 39–52. [CrossRef]
99. Vanktesha, C. Molecular Characterization and Development of Mapping Populations for Construction of Genetic Map in Dolichos Bean [*Lablab Purpureus* L. (sweet)]. Ph.D. Thesis, University of Agricultural Sciences, Banglore, India, 2013.
100. Xu, P.; Moshelion, M.; Wu, X.; Halperin, O.; Wang, B.; Luo, J.; Wallach, R.; Wu, X.; Lu, Z.; Li, G. Natural variation and gene regulatory basis for the responses of asparagus beans to soil drought. *Front. Plant Sci.* **2015**, *6*, 891. [CrossRef]
101. Wu, X.; Wu, X.; Xu, P.; Wang, B.; Lu, Z.; Li, G. Association mapping for fusarium wilt resistance in chinese asparagus bean germplasm. *Plant Genome* **2015**, *8*, 1–6. [CrossRef]
102. Suanum, W.; Somta, P.; Kongjaimun, A.; Yimram, T.; Kaga, A.; Tomooka, N.; Takahashi, Y.; Srinives, P. Co-localization of QTLs for pod fiber content and pod shattering in F2 and backcross populations between yard long bean and wild cowpea. *Mol. Breed.* **2016**, *36*, 80. [CrossRef]
103. Xu, P.; Wu, X.; Muñoz-Amatriaín, M.; Wang, B.; Wu, X.; Hu, Y.; Huynh, B.L.; Close, T.J.; Roberts, P.A.; Zhou, W.; et al. Genomic regions, cellular components and gene regulatory basis underlying pod length variations in cowpea (*V. unguiculata* L. Walp). *Plant Biotechnol. J.* **2017**, *15*, 547–557. [CrossRef]
104. Pan, L.; Wang, N.; Wu, Z.; Guo, R.; Yu, X.; Zheng, Y.; Xia, Q.; Gui, S.; Chen, C. A high density genetic map derived from rad sequencing and its application in qtl analysis of yield-related traits in *Vigna unguiculata*. *Front. Plant Sci.* **2017**, *8*, 1544. [CrossRef]
105. Huang, H.; Tan, H.; Xu, D.; Tang, Y.; Niu, Y.; Lai, Y.; Tie, M.; Li, H. High-density genetic map construction and comparative genome analysis in asparagus bean. *Sci. Rep.* **2018**, *8*, 1–9. [CrossRef]
106. Wu, X.; Li, G.; Wang, B.; Hu, Y.; Wu, X.; Wang, Y.; Lu, Z.; Xu, P. Fine mapping Ruv2, a new rust resistance gene in cowpea (*Vigna unguiculata*), to a 193-kb region enriched with NBS-type genes. *Theor. Appl. Genet.* **2018**, *131*, 2709–2718. [CrossRef]
107. Watcharatpong, P.; Kaga, A.; Chen, X.; Somta, P. Narrowing down a major QTL region conferring pod fiber contents in yardlong bean (*Vigna unguiculata*), a vegetable cowpea. *Genes* **2020**, *11*, 363. [CrossRef] [PubMed]
108. Schmutz, J.; Cannon, S.B.; Schlueter, J.; Ma, J.; Mitros, T.; Nelson, W.; Hyten, D.L.; Song, Q.; Thelen, J.J.; Cheng, J.; et al. Genome sequence of the palaeopolyploid soybean. *Nature* **2010**, *463*, 178–183. [CrossRef] [PubMed]
109. Cannon, S.B.; May, G.D.; Jackson, S.A. Three sequenced legume genomes and many crop species: Rich opportunities for translational genomics. *Plant Physiol.* **2009**, *151*, 970–977. [CrossRef] [PubMed]
110. Singh, N.K.; Gupta, D.K.; Jayaswal, P.K.; Mahato, A.K.; Dutta, S.; Singh, S.; Bhutani, S.; Dogra, V.; Singh, B.P.; Kumawat, G.; et al. The first draft of the pigeonpea genome sequence. *J. Plant Biochem. Biotechnol.* **2012**, *21*, 98–112. [CrossRef]
111. Chang, Y.; Liu, H.; Liu, M.; Liao, X.; Sahu, S.K.; Fu, Y.; Song, B.; Cheng, S.; Kariba, R.; Muthemba, S.; et al. The draft genomes of five agriculturally important African orphan crops. *GigaScience* **2019**, *8*, giy152. [CrossRef]
112. Xia, Q.; Pan, L.; Zhang, R.; Ni, X.; Wang, Y.; Dong, X.; Gao, Y.; Zhang, Z.; Kui, L.; Li, Y.; et al. The genome assembly of asparagus bean, *Vigna unguiculata* ssp. sesquipedialis. *Sci. Data* **2019**, *6*, 1–10. [CrossRef]
113. Cronk, Q.; Ojeda, I.; Pennigton, R.T. Legume comparative genomics: Progress in phylogenetics and phylogenomics. *Curr. Opin. Plant Biol.* **2009**, *9*, 99–103. [CrossRef]
114. Zhang, H.; Zuelsdorf, C.; Penneys, D.; Fan, S.; Kofsky, J.; Song, B.H. Transcriptome profiling of a beach-adapted wild legume for dissecting novel mechanisms of salinity tolerance. *Sci. Data* **2018**, *5*, 180290. [CrossRef]
115. Huang, X.; Feng, Q.; Qian, Q.; Zhao, Q.; Wang, L.; Wang, A.; Guan, J.; Fan, D.; Weng, Q.; Huang, T. High-throughput genotyping by whole-genome resequencing. *Genome Res.* **2009**, *19*, 1068–1076. [CrossRef]
116. Baird, N.A.; Etter, P.D.; Atwood, T.S.; Currey, M.C.; Shiver, A.L.; Lewis, Z.A.; Selker, E.U.; Cresko, W.A.; Johnson, E.A. Rapid SNP discovery and genetic mapping using sequenced RAD markers. *PLoS ONE* **2008**, *3*, e3376. [CrossRef]
117. Saxena, R.K.; Patel, K.; Kumar, C.S.; Tyagi, K.; Saxena, K.B.; Varshney, R.K. Molecular mapping and inheritance of restoration of fertility (Rf) in A4 hybrid system in pigeonpea (*Cajanus cajan* (L.) Millsp.). *Theor. Appl. Genet.* **2018**, *131*, 1605–1614. [CrossRef] [PubMed]

118. Hart, J.P.; Griffiths, P.D. Genotyping-by-sequencing enabled mapping and marker development for the *by-2* potyvirus resistance allele in common bean. *Plant Genome* **2015**, *8*, 1–4. [CrossRef] [PubMed]
119. Qin, J.; Shi, A.; Mou, B.; Bhattarai, G.; Yang, W.; Weng, Y.; Motes, D. Association mapping of aphid resistance in USDA cowpea (*Vigna unguiculata* L. Walp.) core collection using SNPs. *Euphytica* **2017**, *213*, 36. [CrossRef]
120. Thomson, M.J. High-throughput SNP genotyping to accelerate crop improvement. *Plant Breed. Biotechnol.* **2014**, *2*, 195–212. [CrossRef]
121. Chao, S.; Zhang, W.; Akhunov, E.; Sherman, J.; Ma, Y.; Luo, M.C.; Dubcovsky, J. Analysis of gene-derived SNP marker polymorphism in US wheat (*Triticum aestivum* L.) cultivars. *Mol. Breed.* **2009**, *23*, 23–33. [CrossRef]
122. Yadav, P.; Saxena, K.B.; Hingane, A.; Kumar, C.S.; Kandalkar, V.S.; Varshney, R.K.; Saxena, R.K. An "Axiom Cajanus SNP Array" based high density genetic map and QTL mapping for high-selfing flower and seed quality traits in pigeonpea. *BMC Genom.* **2019**, *20*, 1–10. [CrossRef]
123. Huang, K.; Mellor, K.E.; Paul, S.N.; Lawson, M.J.; Mackey, A.J.; Timko, M.P. Global changes in gene expression during compatible and incompatible interactions of cowpea (*Vigna unguiculata* L.) with the root parasitic angiosperm *Striga gesnerioides*. *BMC Genom.* **2012**, *13*, 402. [CrossRef]
124. Singh, S.; Mahato, A.K.; Jayaswal, P.K.; Singh, N.; Dheer, M.; Goel, P.; Raje, R.S.; Yasin, J.K.; Sreevathsa, R.; Rai, V.; et al. A 62K genic-SNP chip array for genetic studies and breeding applications in pigeonpea (*Cajanus cajan* L. Millsp.). *Sci. Rep.* **2020**, *10*, 1–4. [CrossRef]
125. Dubey, A.N.; Farmer, A.N.; Schlueter, J.E.; Cannon, S.B.; Abernathy, B.R.; Tuteja, R.E.; Woodward, J.I.; Shah, T.R.; Mulasmanovic, B.E.; Kudapa, H.I.; et al. Defining the transcriptome assembly and its use for genome dynamics and transcriptome profiling studies in pigeonpea (*Cajanus cajan* L.). *DNA Res.* **2011**, *18*, 153–164. [CrossRef]
126. Dutta, S.; Kumawat, G.; Singh, B.P.; Gupta, D.K.; Singh, S.; Dogra, V.; Gaikwad, K.; Sharma, T.R.; Raje, R.S.; Bandhopadhya, T.K.; et al. Development of genic-SSR markers by deep transcriptome sequencing in pigeonpea [*Cajanus cajan* (L.) Millspaugh]. *BMC Plant Biol.* **2011**, *11*, 17. [CrossRef]
127. Pazhamala, L.T.; Purohit, S.; Saxena, R.K.; Garg, V.; Krishnamurthy, L.; Verdier, R.; Varshney, R.K. Gene expression atlas of pigeonpea and its application to gain insights into genes associated with pollen fertility implicated in seed formation. *J. Exp. Bot.* **2017**, *68*, 2037–2054. [CrossRef] [PubMed]
128. Rathinam, M.; Mishra, P.; Vasudevan, M.; Budhwar, R.; Mahato, A.; Prabha, A.L.; Singh, N.K.; Rao, U.; Sreevathsa, R. Comparative transcriptome analysis of pigeonpea, *Cajanus cajan* (L.) and one of its wild relatives *Cajanus platycarpus*(Benth.) Maesen. *PLoS ONE* **2019**, *14*, e0218731. [CrossRef] [PubMed]
129. Singh, A.; Singh, P.K.; Sharma, A.K.; Singh, N.K.; Sonah, H.; Deshmukh, R.; Sharma, T.R. Understanding the role of the WRKY gene family under stress conditions in pigeonpea (*Cajanus Cajan* L.). *Plants* **2019**, *8*, 214. [CrossRef] [PubMed]
130. Saxena, R.K.; Rathore, A.; Bohra, A.; Yadav, P.; Das, R.R.; Khan, A.W.; Singh, V.K.; Chitikineni, A.; Singh, I.P.; Kumar, C.S.; et al. Development and application of high-density Axiom Cajanus SNP array with 56K SNPs to understand the genome architecture of released cultivars and founder genotypes. *Plant Genome* **2018**, *11*, 1–10. [CrossRef] [PubMed]
131. Tanwar, U.K.; Pruthi, V.; Randhawa, G.S. RNA-seq of guar (*Cyamopsis tetragonoloba*, l. taub.) leaves: De novo transcriptome assembly, functional annotation and development of genomic resources. *Front. Plant Sci.* **2017**, *8*, 91. [CrossRef] [PubMed]
132. Rawal, H.C.; Kumar, S.; Mithra, S.V.; Solanke, A.U.; Nigam, D.; Saxena, S.; Tyagi, A.; Yadav, N.R.; Kalia, P.; Singh, N.P.; et al. High quality unigenes and microsatellite markers from tissue specific transcriptome and development of a database in clusterbean (*Cyamopsis tetragonoloba*, L. Taub). *Genes* **2017**, *8*, 313. [CrossRef] [PubMed]
133. Sahu, S.; Rao, A.R.; Pandey, J.; Gaikwad, K.; Ghoshal, S.; Mohapatra, T. Genome-wide identification and characterization of lncRNAs and miRNAs in cluster bean (*Cyamopsis tetragonoloba*). *Gene* **2018**, *667*, 112–121. [CrossRef]
134. Tribhuvan, K.U.; Mithra, S.V.A.; Sharma, P.; Das, A.; Kumar, K.; Tyagi, A.; Solanke, A.U.; Sharma, R.; Jadhav, P.V.; Raveendran, M.; et al. Identification of genomic SSRs in cluster bean (*Cyamopsis tetragonoloba*) and demonstration of their utility in genetic diversity analysis. *Ind. Crops Prod.* **2019**, *133*, 221–231. [CrossRef]
135. Grigoreva, E.; Ulianich, P.; Ben, C.; Gentzbittel, L.; Potokina, E. First Insights into the Guar (*Cyamopsis tetragonoloba* (L.) Taub.) Genome of the 'Vavilovskij 130'Accession, Using Second and Third-Generation Sequencing Technologies. *Rus. J. Genet.* **2019**, *55*, 1406–1416. [CrossRef]

136. Singh, V.; Goel, R.; Pande, V.; Asif, M.H.; Mohanty, C.S. De novo sequencing and comparative analysis of leaf transcriptomes of diverse condensed tannin-containing lines of underutilized *Psophocarpus tetragonolobus* (L.) DC. *Sci. Rep.* **2017**, *7*, 44733. [CrossRef]
137. Yssel, E.J.; Kao, S.; Peer, Y.V.; Sterck, L. ORCAE-AOCC: A centralized portal for the annotation of african orphan crop genomes. *Genes* **2019**, *10*, 950. [CrossRef] [PubMed]
138. Tan, H.; Huang, H.; Tie, M.; Tang, Y.; Lai, Y.; Li, H. Transcriptome profiling of two asparagus bean (*Vigna Unguiculata* subsp. sesquipedalis) cultivars differing in chilling tolerance under cold stress. *PLoS ONE* **2016**, *11*, e0151105. [CrossRef] [PubMed]
139. Zuo, J.; Wang, Y.; Zhu, B.; Luo, Y.; Wang, Q.; Gao, L. sRNAome and transcriptome analysis provide insight into chilling response of cowpea pods. *Gene* **2018**, *671*, 142–151. [CrossRef] [PubMed]
140. Spriggs, A.; Henderson, S.T.; Hand, M.L.; Johnson, S.D.; Taylor, J.M.; Koltunow, A. Assembled genomic and tissue-specific transcriptomic data resources for two genetically distinct lines of Cowpea (*Vigna unguiculata* (L.) Walp). *Gates Open Res.* **2018**, *2*, 7. [CrossRef]
141. Muñoz-Amatriaín, M.; Mirebrahim, H.; Xu, P.; Wanamaker, S.I.; Luo, M.; Alhakami, H.; Alpert, M.; Atokple, I.; Batieno, B.J.; Boukar, O.; et al. Genome resources for climate-resilient cowpea, an essential crop for food security. *Plant J.* **2017**, *89*, 1042–1054. [CrossRef]
142. Sharma, K.K.; Lavanya, M.; Anjaiah, V. *Agrobacterium* mediated production of transgenic (*Cajanus cajan* L. Millsp.) expressing the synthetic *BT cry1Ab* gene. *Vitr. Cell. Dev. Bio. Plant* **2006**, *42*, 165–173. [CrossRef]
143. Eapen, S. Advances in development of transgenic pulse crops. *Biotechnol. Adv.* **2008**, *26*, 162–168. [CrossRef]
144. Dita, M.A.; Rispail, N.; Prats, E.; Rubiales, D.; Singh, K.B. Biotechnology approaches to overcome biotic and abiotic stress constraints in legumes. *Euphytica* **2006**, *147*, 1–24. [CrossRef]
145. Krishnamurthy, K.V.; Suhasini, K.; Sagare, A.P.; Meixner, M.; De Kathen, A.; Pickardt, T.; Schieder, O. *Agrobacterium* mediated transformation of chickpea (*Cicer arietinum* L.) embryo axes. *Plant Cell Rep.* **2000**, *19*, 235–240. [CrossRef]
146. Kumar, S.M.; Kumar, B.K.; Sharma, K.K.; Devi, P. Genetic transformation of pigeonpea with rice chitinase gene. *Plant Breed.* **2004**, *123*, 485–489. [CrossRef]
147. Bakshi, S.; Roy, N.K.; Sahoo, L. Seedling preconditioning in thidiazuron enhances axillary shoot proliferation and recovery of transgenic cowpea plants. *PCTOC* **2012**, *110*, 77–91. [CrossRef]
148. Thu, T.T.; Dewaele, E.; Trung, L.Q.; Claeys, M.; Jacobs, M.; Angenon, G. Increasing lysine levels in pigeonpea (*Cajanus cajan* (L.) Millsp.) seeds through genetic engineering. *PCTOC* **2007**, *91*, 135–143. [CrossRef]
149. Surekhs, C.; Kumari, N.K.; Aruna, V.; Suneetha, G.; Arundhati, A.; Kishor, P.B. Expression of the *Vigna aconitifolia* P5CSF129A gene in transgenic pigeonpea enhances proline accumulation and salt tolerance. *Plant Cell Rep.* **2014**, *116*, 27–36.
150. Krishnan, S.G.; Dwivedi, N.K.; Singh, J.P. Primitive weedy forms of guar, adak guar: Possible missing link in the domestication of guar [*Cyamopsis tetragonoloba* (L.) Taub.]. *Genet. Resour. Crop Evol.* **2011**, *58*, 961. [CrossRef]
151. Ramu, S.V.; Rohini, S.; Keshavareddy, G.; Gowri Neelima, M.; Shanmugam, N.B.; Kumar, A.R.V.; Sarangi, S.K.; Ananda Kumar, P.; Udayakumar, M. Expression of a synthetic cry1AcF gene in transgenic Pigeon pea confers resistance to *Helicoverpa armigera*. *J. Appl. Entomol.* **2012**, *136*, 675–687. [CrossRef]
152. Keshamma, E.; Sreevathsa, R.; Kumar, A.M.; Reddy, K.N.; Manjulatha, M.; Shanmugam, N.B.; Kumar, A.R.; Udayakumar, M. *Agrobacterium*-mediated in planta transformation of field bean (*Lablab purpureus* L.) and recovery of stable transgenic plants expressing the *cry1AcF* gene. *Plant Mol. Biol. Rep.* **2012**, *30*, 67–78. [CrossRef]
153. Solleti, S.K.; Bakshi, S.; Purkayastha, J.; Panda, S.K.; Sahoo, L. Transgenic cowpea (*Vigna unguiculata*) seeds expressing a bean α-amylase inhibitor 1 confers resistance to storage pests, bruchid beetles. *Plant Cell Rep.* **2008**, *27*, 1841–1850. [CrossRef]
154. Adesoye, A.; Machuka, J.; Togun, A. *CRY 1AB* transgenic cowpea obtained by nodal electroporation. *Afr. J. Biotechnol.* **2008**, *7*, 3200–3210.
155. Kaur, N.; Murphy, J.B. Enhanced isoflavone biosynthesis in transgenic cowpea (*Vigna unguiculata* L.) Callus. *Plant Mol. Biol. Biotechnol.* **2012**, *3*, 1–8.
156. Klu, G.Y.P.; Raemakers, C.J.J.M.; Jacobsen, E.; Van-Harten, A.M. Direct organogenesis and somatic embryogenesis in mature cotyledon explants of winged bean (*Psophocarpus tetragonolobus* (L.) DC) using cytokinin-based media. *Plant Genet. Res. Newsl.* **2002**, *2002*, 51–69.
157. Venketeswaran, S.; Dias, M.A.D.L.; Weyers, U.V. Organogenesis and somatic embryogenesis from callus of winged bean (*Psophocarpus tetragonolobus* (L.) DC.). *Acta Hortic.* **1990**, *280*, 202–206. [CrossRef]

158. Singh, V.; Chauhan, N.S.; Singh, M.; Idris, A.; Madanala, R.; Pande, V.; Mohanty, C.S. Establishment of an efficient and rapid method of multiple shoot regeneration and a comparative phenolics profile in in vitro and greenhouse-grown plants of *Psophocarpus tetragonolobus* (L.) DC. *Plant Signal. Behav.* **2014**, *9*, e970443. [CrossRef] [PubMed]
159. Wilson, V.M.; Haq, N.; Evans, P.K. Protoplast isolation, culture and plant regeneration in the winged bean, *Psophocarpus tetragonolobus* (L) DC. *Plant Sci.* **1985**, *41*, 61–68. [CrossRef]
160. Sheikh, W.A.; Dedhrotiya, A.T.; Khan, N.; Gargi, T.; Patel, J.B.; Acharya, S. An efficient in vitro regeneration protocol from cotyledon and cotyledonary node of cluster bean (*Cyamopsis tetragonoloba* L Taub). *Curr. Tren. Biotechnol. Pharm.* **2015**, *9*, 175–181.
161. Meghwal, M.K.; Kalaskar, S.R.; Rathod, A.H.; Tikka, S.B.S.; Acharya, S. Effect of plant growth regulators on in vitro regeneration from different explants in cluster bean [*Cyamopsis tetragonoloba* (L.) Taub]. *J. Cell Tissu. Res.* **2014**, *14*, 4647–4652.
162. Xu, D.; Xue, Q.; McElroy, D.; Mawal, Y.; Hilder, V.A.; Wu, R. Constitutive expression of a cowpea trypsin inhibitor gene, *CpTi*, in transgenic rice plants confers resistance to two major rice insect pests. *Mol. Breed.* **1996**, *2*, 167–173. [CrossRef]
163. Sun, S.S.; Xiong, L.; Jing, Y.; Liu, B. Lysine Rich Protein from Winged Bean. U.S. Patent No. 6184437, 6 February 2001.
164. Meng, C.; Chen, X.; Liang, R.; Yang, F.; Zhang, L.; Zhang, X.; Chen, T.; Sun, S.S. Expression of lysine-rich protein gene and analysis of lysine content in transgenic wheat. *Chin. Sci. Bull.* **2004**, *49*, 2053–2057. [CrossRef]
165. Wilson, R.C.; Doudna, J.A. Molecular mechanisms of RNA interference. *Ann. Rev. Biophys.* **2013**, *42*, 217–239. [CrossRef]
166. Prins, M.; Laimer, M.; Noris, E.; Schubert, J.; Wassenegger, M.; Tepfer, M. Strategies for antiviral resistance in transgenic plants. *Mol. Plant Pathol.* **2008**, *9*, 73. [CrossRef]
167. Saini, R.P.; Raman, V.; Dhandapani, G.; Malhotra, E.V.; Sreevathsa, R.; Kumar, P.A.; Sharma, T.R.; Pattanayak, D. Silencing of *HaAce1* gene by host-delivered artificial microRNA disrupts growth and development of *Helicoverpaarmigera*. *PLoS ONE* **2018**, *13*, e0194150. [CrossRef]
168. Gordon, K.H.; Waterhouse, P.M. RNAi for insect-proof plants. *Nature Biotechnol.* **2007**, *25*, 1231–1232. [CrossRef] [PubMed]
169. Devos, Y.; Maeseele, P.; Reheul, D.; Van Speybroeck, L.; De Waele, D. Ethics in the Societal Debate on Genetically Modified Organisms: A (Re)Quest for Sense and Sensibility. *J. Agric. Environ. Ethics* **2007**, *21*, 29–61. [CrossRef]
170. Walt, E. Gene edited CRISPER mushroom escapes US regulation. *Nature* **2016**, *532*, 293. [CrossRef] [PubMed]
171. Cohen-Tannoudji, M.; Robine, S.; Choulika, A.; Pinto, D.; El Marjou, F.; Babinet, C.; Louvard, D.; Jaisser, F. I-SceI-induced gene replacement at a natural locus in embryonic stem cells. *Mol. Cellu. Biol.* **1998**, *18*, 1444–1448. [CrossRef]
172. Bibikova, M.; Golic, M.; Golic, K.G.; Carroll, D. Targeted chromosomal cleavage and mutagenesis in Drosophila using zinc finger nucleases. *Genetics* **2002**, *161*, 1169–1175.
173. Christian, M.; Cermak, T.; Doyle, E.L.; Schmidt, C.; Zhang, F.; Hummel, A.; Bogdanove, A.J.; Voytas, D.F. Targeting DNA double-strand breaks with TAL effector nucleases. *Genetics* **2010**, *186*, 757–761. [CrossRef]
174. Mali, P.; Yang, L.; Esvelt, K.M.; Aach, J.; Guell, M.; DiCarlo, J.E.; Norville, J.E.; Church, G.M. RNA-guided human genome engineering via Cas9. *Science* **2013**, *339*, 823–826. [CrossRef]
175. Butler, N.M.; Baltes, N.J.; Voytas, D.F.; Douches, D.S. Geminivirus-mediated genome editing in potato (*Solanum tuberosum* L.) using sequence-specific nucleases. *Front. Plant Sci.* **2016**, *7*, 1045. [CrossRef]
176. Song, G.; Jia, M.; Chen, K.; Kong, X.; Khattak, B.; Xie, C.; Li, A.; Mao, L. CRISPR/Cas9: A powerful tool for crop genome editing. *Crop J.* **2016**, *4*, 75–82. [CrossRef]
177. Ali, Z.; Eid, A.; Ali, S.; Mahfouz, M.M. Pea early-browning virus -mediated genome editing via the CRISPR/Cas9 system in *Nicotiana benthamiana* and *Arabidopsis*. *Virus Res.* **2018**, *244*, 333–337. [CrossRef]
178. Zhang, Y.; Massel, K.; Godwin, I.D.; Gao, C. Applications and potential of genome editing in crop improvement. *Genome Biol.* **2018**, *19*, 210. [CrossRef] [PubMed]
179. Maphosa, Y.; Jideani, V.A. The role of legumes in human nutrition. *Funct. Food Improv. Health Adequate Food.* **2017**, *2*, 1–13.
180. Ruelle, M.L.; Asfaw, Z.; Dejen, A.; Tewolde-Berhan, S.; Nebiyu, A.; Tana, T.; Power, A.G. Inter-and intraspecific diversity of food legumes among households and communities in Ethiopia. *PLoS ONE* **2019**, *14*, e0227074. [CrossRef] [PubMed]

181. FAO; IFAD; UNICEF; WFP; WHO. State of Food Security and Nutrition in the World 2019. Safeguarding Against Economic Slowdowns and Downturns. 2019. Available online: http://www.fao.org/3/ca5162en/ca5162en (accessed on 10 June 2020).
182. Botelho, R.; Araújo, W.; Pineli, L. Food formulation and not processing level: Conceptual divergences between public health and food science and technology sectors. *Crit. Rev. Food Sci. Nutr.* **2018**, *58*, 639–650. [CrossRef] [PubMed]

2

Effects of *OsCDPK1* on the Structure and Physicochemical Properties of Starch in Developing Rice Seeds

Jian-Zhi Jiang [1], Chun-Hsiang Kuo [1], Bo-Hong Chen [1], Mao-Kei Chen [1], Choun-Sea Lin [2] and Shin-Lon Ho [1,*]

[1] Department of Agronomy, National Chiayi University, Chiayi 60004, Taiwan; ssdog27@yahoo.com.tw (J.-Z.J.); efab888@yahoo.com.tw (C.-H.K.); chen.po.hung9117@gmail.com (B.-H.C.); fritz29522879@yahoo.com.tw (M.-K.C.)

[2] Agricultural Biotechnology Research Center, Academia Sinica, Taipei 11529, Taiwan; cslin99@gate.sinica.edu.tw

* Correspondence: slho@mail.ncyu.edu.tw

Abstract: Overexpression of a constitutively active truncated form of *OsCDPK1* (*OEtr*) in rice produced smaller seeds, but a double-stranded RNA gene-silenced form of *OsCDPK1* (*Ri*) yielded larger seeds, suggesting that *OsCDPK1* plays a functional role in rice seed development. In the study presented here, we propose a model in which *OsCDPK1* plays key roles in negatively controlling the grain size, amylose content, and endosperm appearance, and also affects the physicochemical properties of the starch. The dehulled transgenic *OEtr* grains were smaller than the dehulled wild-type grains, and the *OEtr* endosperm was opaque and had a low amylose content and numerous small loosely packed polyhedral starch granules. However, the *OEtr* grain sizes and endosperm appearances were not affected by temperature, which ranged from low (22 °C) to high (31 °C) during the grain-filling phase. In contrast, the transgenic *Ri* grains were larger, had higher amylose content, and had more transparent endosperms filled with tightly packed polyhedral starch granules. This demonstrates that *OsCDPK1* plays a novel functional role in starch biosynthesis during seed development and affects the transparent appearance of the endosperm. These results improve our understanding of the molecular mechanisms through which the grain-filling process occurs in rice.

Keywords: rice; *OsCDPK1*; seed development, starch biosynthesis; endosperm appearance

---

1. Introduction

The quality of rice (*Oryza sativa* L.) grain is defined in terms of several main factors, including (i) eating and cooking qualities, and (ii) milling qualities and appearance [1]. Eating and cooking qualities are determined by the amylose content, amylopectin structure, gelatinization temperature, and pasting viscosity [2], and milling qualities and appearance correlate strongly with the transparency, flouriness, and chalkiness of the endosperm [3]. The filling and accumulation of starch granules in developing rice endosperms can accelerate at high temperatures, causing the starch in the endosperm cells to be packed loosely and the kernel to be chalky. Such grains crack easily during milling, yielding poor eating and cooking qualities [4–6].

It has been shown in many studies that chalky and less-transparent kernels contain more amylopectin and less amylose in the endosperm than do less-chalky and more-transparent kernels [7–9]. Eliminating chalkiness by regulating the amylopectin and amylose content ratios (by affecting biosynthesis) in the endosperm during the grain-filling phase is therefore a key way of improving grain quality. Two enzymes involved in amylose biosynthesis are ADP-glucose pyrophosphorylase and the Waxy gene-encoded granule-bound starch synthase I (GBSSI) [10,11]. The enzymes involved in the biosynthesis and modification of amylopectin are ADP-glucose pyrophosphorylase, soluble starch synthase, the starch-branching enzyme (BE), and the starch-debranching enzyme [10–12]. In higher plants, BE plays an essential role in amylopectin biosynthesis because it is the only enzyme that can add α-1,6-glucosidic linkages to polyglucans [13]. Three BE isoforms—BEI, BEIIa, and BEIIb—have been found in rice [10,14]. The rice mutant *amylose-extender*, which has a null mutation in *BEIIb*, has been found to alter the degree of polymerization (DP) of amylopectin, giving fewer short chains (DP $\leq$ 17) and more long chains (DP $\geq$ 18), the changes being related to the dose on the *amylose-extender* locus in the triploid endosperm cells. These results suggest that *BEIIb* might have critical effects on the amylopectin structure and the rheological properties of the starch [12]. Several floury endosperm rice mutants (*flo1–flo7*) have been isolated. Treating fertilized rice egg cells with the chemical mutagen *N*-methyl-*N*-nitrosourea yielded mutants *flo1* and *flo2*, which had floury endosperms [15,16]. The *flo2* mutant has been found, through map-based cloning, to be a member of the tetratricopeptide repeat-motif protein family. The gene mutation decreases the grain size and decreases the starch quality (by decreasing the amylose content) and also changes the fine structure of the amylopectin [17]. The *flo3* mutant was produced through applying gamma-irradiation and ethyl methansulfonate treatment, and had a low 16 kDa globulin content in the endosperm [18]. The white-core floury endosperm mutants *flo4* and *flo5* were produced through T-DNA insertional mutagenesis and were found to have *pyruvate orthophosphate dikinase B* and *starch synthase IIIa* (SSIIIa) gene mutations, respectively [19,20]. The *flo4* mutant endosperm had a low amylose content, suggesting that *pyruvate orthophosphate dikinase B* might play a role in regulating carbon metabolism during the grain filling process [19]. DP analysis of amylopectin in *flo5* was performed, and the amounts of DP 6–8 and DP 16–20 in the mutant endosperm were found to be decreased but the amounts of DP 9–15 and DP 22–29 had increased, indicating that *SSIIIa* strongly affects the chain-length distribution of amylopectin biosynthesized in developing rice grains [20]. The floury endosperm mutant *flo6* had a completely floury white endosperm, but the *flo7* endosperm was floury and white only at the peripheries, and not in the interior [21,22]. Map-based cloning demonstrated that *flo6* was an insertion mutation in the unknown function gene *Os03g0686900* [21] and that *flo7* was a deletion mutation in the unknown function gene *Os10g0463800* [22], suggesting that these genes may play vital roles in starch biosynthesis and granule formation in the endosperm during the grain-filling process.

Calcium ions (Ca^{2+}) are secondary messengers in plant cells, and are used in response to various environmental and developmental stimuli through temporal and spatial fluctuations of the cytosolic Ca^{2+} concentration [23,24]. Calcium-dependent protein kinases (CDPKs) are a major family of calcium sensors that have been characterized in various plant species. CDPKs are Ser/Thr protein kinases that are encoded by multigene families [25,26]. CDPKs have four functional domains: an N-terminal variable domain, a catalytic kinase domain, an autoinhibitory domain, and a calcium-binding EF-hands regulatory domain [27]. Under normal growth conditions (i.e., absence of Ca^{2+} signals), the autoinhibitory domain can interact with the kinase domain thereby inhibits kinase activity [24]. Deletion of the autoinhibitory domain and the Ca^{+2} binding domains from the coding region could bypass Ca^{2+} signal stimulation and resulted in produced a constitutively active enzyme of CDPKs [26,27]. It has been shown in many studies that CDPKs play important physiological roles in response to various environmental stresses and developmental processes [23,24,28–30]. Only in a few studies have CDPKs been shown to play a role in starch biosynthesis during the grain-filling process in rice. The rice *SPK* (which shares 79% of its amino acid sequence with *OsCDPK1*) has been found to encode a sucrose synthase kinase. Expression of antisense *SPK* in transgenic rice produced watery grains because large amounts of sucrose had accumulated in the endosperm due to low sucrose synthase activity, resulting in inefficient sucrose degradation [31]. This indicates that SPK may be a regulator in the starch biosynthesis pathway.

In the study presented here, *OsCDPK1* was found to play pivotal roles in rice-seed development, in the physicochemical properties of the starch produced, and in the appearance of the endosperm.

The pleiotropic effects on various agronomic traits in loss- and gain-of-function of *OsCDPK1* transgenic plants are also characterized.

2. Results

2.1. Phenotypic Changes in Transgenic Rice Plants with Overexpressing or Silenced OsCDPK1

In previous studies, to understand the physiological roles of *OsCDPK1*, the gene-overexpression and gene-silencing approaches were implemented. For the truncated form of *OsCDPK1* (*OEtr*), the coding sequences in which the autoinhibitory region and the calcium-binding domains had been removed were expressed under the control of maize ubiquitin gene promoter to generate the constitutively active form of the *OEtr* transgenic plants [32]. The transgenic plants in *OEtrs* yielded smaller seeds, whereas RNA-interference gene knockdown mutants (*Ris*) yielded larger seeds [32]. In this study, the various agronomic traits in the T4 transgenic lines of *OEtrs* (*OEtr-1*, *-3* and *-4*) and *Ris* (*Ri-1*, *-2* and *-3*) were analyzed. The results (Table 1) were consistent with our previous studies [32] and showed that the *Ris* (*Ri-1*, *-2* and *-3*) lines had, on average, 7.1% and 10.8% increases in plant height and 1000-grain weight, respectively, compared to those of WT (wild type). However, both examined traits decreased in the *OEtrs* (*OEtr-1*, *-3* and *-4*) lines compared with WT, with 5.8% and 21.1% reductions in plant height and 1000-grain weight, respectively. Furthermore, compared to WT, the heading date and growth duration were shorter in *Ris*, while *OEtrs* showed no significant difference. The average of the heading date of *Ris* (83.2 day) was 6 days shorter than that of WT (89.3 day) and the growth duration of *Ris* (104.8 day) was 10 days shorter than that of WT (115.2 day). The dehulled grain weight, starch content, and amylose content were higher in *Ris* and lower in *OEtrs* than those of WT; for example, the average weights of dehulled grain of *Ris* and *OEtrs* were 111.2% and 84.1%, respectively, of that of WT. The average percentage of starch content versus dehulled grain weight in WT, *OEtrs*, and *Ris* was 73.4%, 64.2%, and 78.1%, respectively. Moreover, the average amylose content in WT, *OEtrs*, and *Ris* was 23.2%, 15.1% and 26.2%, respectively.

Table 1. Comparison of agronomic traits between WT and transgenic lines.

Genotypes	Plant Height (cm)	Heading Day (day)	Growth Duration (day)	1000-Grain (g)	Dehulled Grain (mg/grain)	Starch Content (mg/grain)	Amylose Content (%)
WT	108.2 ± 2.5	89.3 ± 2.4	115.2 ± 3.3	23.2 ± 0.6	21.4 ± 0.3	15.7 ± 0.4 (73.4%) [a]	23.2 ± 0.3
OEtr-1	99.5 ± 3.5 *	92.1 ± 2.3	117.6 ± 2.7	18.2 ± 0.3 **	16.8 ± 0.2 **	10.5 ± 0.1 ** (62.5%) [a]	14.3 ± 0.3 *
OEtr-3	102.6 ± 2.6 *	92.4 ± 3.4 *	119.1 ± 3.8	18.6 ± 0.1 **	17.3 ± 0.4 **	11.8 ± 0.3 ** (68.2%) [a]	15.3 ± 0.2 *
OEtr-4	103. 6 ± 3.1 *	91.1 ± 2.7 *	119.4 ± 3.3	18.1 ± 0.3 **	16.6 ± 0.2 **	10.3 ± 0.1 ** (62.0%)[a]	15.6 ± 0.4 *
Ri-1	115.4 ± 4.6 *	83.6 ± 2.5 **	106.6 ± 3.1 *	25.6 ± 0.3 *	23.9 ± 0.3 *	18.7 ± 0.2 * (78.2%)[a]	25.8 ± 0.1 *
Ri-2	118.7 ± 3.8 *	82.7 ± 2.8 **	103.6 ± 2.1 *	25.3 ± 0.2 *	23.5 ± 0.2 *	18.5 ± 0.2 * (78.7%) [a]	26.6 ± 0.5 *
Ri-3	113.6 ± 4.3 *	83.2 ± 3.4 **	104.3 ± 3.6 *	26.2 ± 0.5 *	24.4 ± 0.2 *	18.9 ± 0.6 * (77.5%) [a]	26.2 ± 0.2 *

Mean values calculated from three independent transgenic lines. All data are presented as mean ± SE. Statistical significance is determined by *t*-test. Values in the same column indicate significant differences between WT and mutant lines at * $p < 0.05$ and ** $p < 0.01$. [a] Percentage of dehulled grain weight.

2.2. Ectopic Overexpression and Silencing of OsCDPK1 in Transgenic Rice Plants Yielded Opaque and Transparent Endosperms, Respectively

We examined the 55-day-old plant phenotypes in the T4 transgenic lines further, and the results were consistent with Table 1. The *Ri-1*, *-2* and *-3* plants were higher (mean value is 76.6 ± 6.18 cm) and the *OEtr-1*, *-3 and -4* plants were shorter (52.4 ± 3.58 cm) than the WT plants (63.3 ± 4.63 cm) (Figure 1A,B). At the grain-filling stage, the WT and transgenic plants were transferred to the growth room at an optimal temperature of a cycle of 25 °C for 16 h light and 20 °C for 8 h dark. Fifteen seeds were randomly selected from each individual line and dehulled. We found that the dehulled *Ri-1* grains were longer and that the *OEtr-1* grains were shorter than the WT grains (Figure 1E,F). We also found that the dehulled *OEtr-1* grains were all of the floury-kernel phenotype (Figure 1C,E).

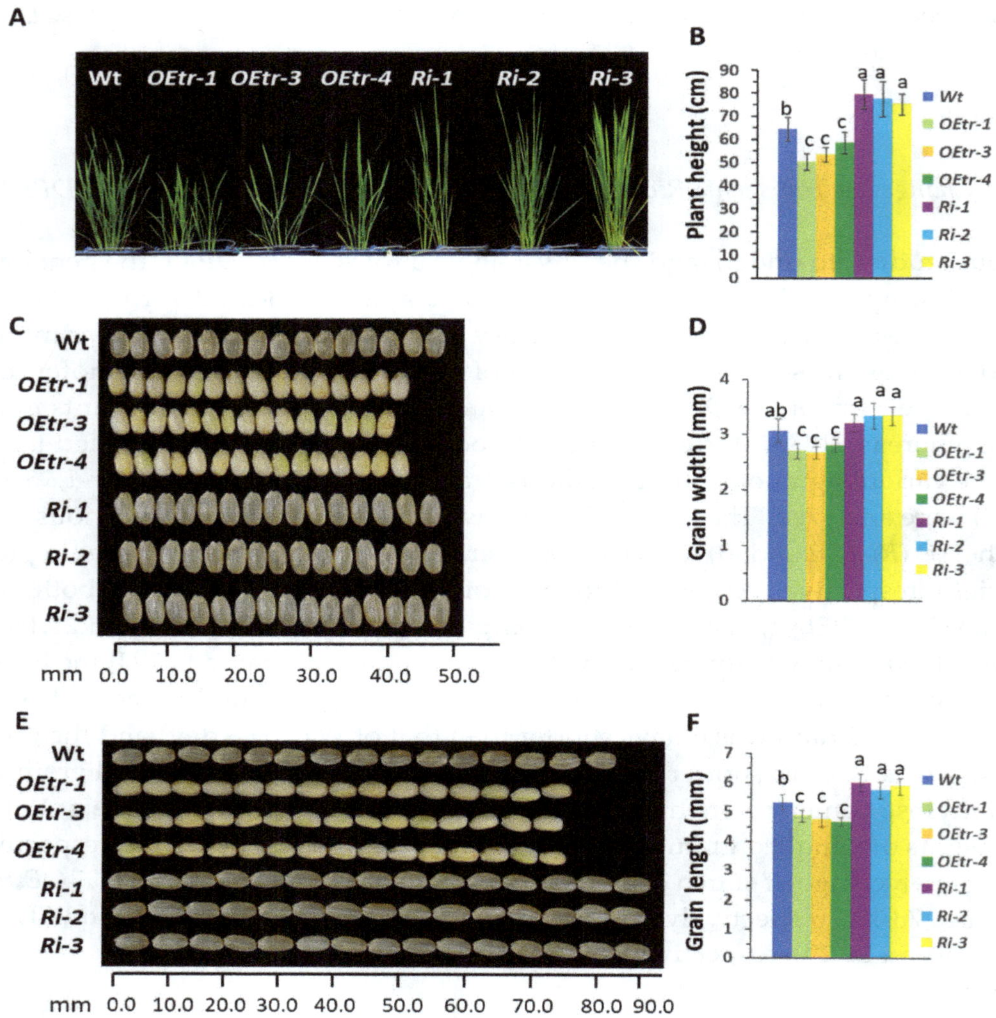

Figure 1. Plant heights and grain morphologies in the WT, the *OEtr-1*, *-3* and *-4* (*OEtrs*), and the *Ri-1*, *-2* and *-3* (*Ris*) plants. After 5 d of flowering, rice plants were transferred to a growth chamber and grown at an optimal temperature (25 °C for 16 h light, 20 °C for 8 h dark). (**A**) Heights of the 55-day-old WT, *OEtrs*, and *Ris* plants. (**B**) Quantification of the plant heights. Each error bar is the standard error for 15 individual plants. (**C**) Endosperm appearances and grain widths for the WT, *OEtrs*, and *Ris* plants. Fifteen grains per line were positioned in a row and measured. (**D**) Quantification of the grain widths. Each error bar is the standard error for 50 individual grains. (**E**) Endosperm appearances and grain lengths for the WT, *OEtrs*, and *Ris* plants. Fifteen grains per line were positioned in a row and measured. (**F**) Quantification of the grain lengths. Each error bar is the standard deviation ($n = 50$). Different letters above the bars indicate significant differences, identified by performing ANOVAs ($p < 0.01$).

We determined whether the temperature affected the endosperm appearance in the transgenic lines by growing rice plants at relatively low temperatures (LT; 22 °C for 16 h light, 20 °C for 8 h dark) and at relatively high temperatures (HT; 31 °C for 16 h light, 28 °C for 8 h dark) during the grain-filling process. The mature seeds were collected from the WT, the *Ri-1*, *Ri-2* and *Ri-3* (*Ris*), and the *OEtr-1*, *OEtr-3* and *OEtr-4* (*OEtrs*) plants. As shown in Figure 2, under LT and HT conditions, all the dehulled *OEtrs* grains had smaller and floury endosperms (100% in the *OEtr-1*, *OEtr-3* and *OEtr-4* lines). In contrast, under LT condition, most of the grains in the WT and *Ris* lines displayed transparency phenotype; the ratios of chalky grains were 16.2% in the WT grains, and 6.0%, 6.7%, and 7.1% in the *Ri-1*, *Ri-2* and *Ri-3* grains, respectively (Figure 2A and Table 2). However, the higher temperature caused a significant increase in the WT and *Ris* lines' endosperms of the chalky phenotype; the ratios of chalky grains were 63.6% in the WT grains, and 50.1%, 51.2% and 45.1% in the *Ri-1*, *Ri-2* and

Ri-3 grains, respectively (Figure 2A and Table 2). Illuminating the kernels with a backlight showed that the *OEtrs* grains all had opaque endosperms regardless of the temperature at which the plants were grown at LT or HT (Figure 2B). However, most of the WT and *Ris* grains displayed transparent endosperms under the LT condition (the *Ris* endosperms being more transparent than those of WT), the proportion of opaque endosperms increased to more than 50% in both the WT and *Ris* lines under HT growth condition (Figure 2B and Figure S1). Cross-sections of the endosperms (Figure 2C) indicated that the kernels in the WT and *Ris* displayed transparent phenotype under LT treatment and showed partial opaque phenotype under HT condition. However, in the *OEtrs*, all endosperms displayed completely floury appearance under both LT and HT conditions, regardless of the different temperature treatments used. These results suggest that *OsCDPK1* affects rice endosperm appearance in a temperature-independent manner.

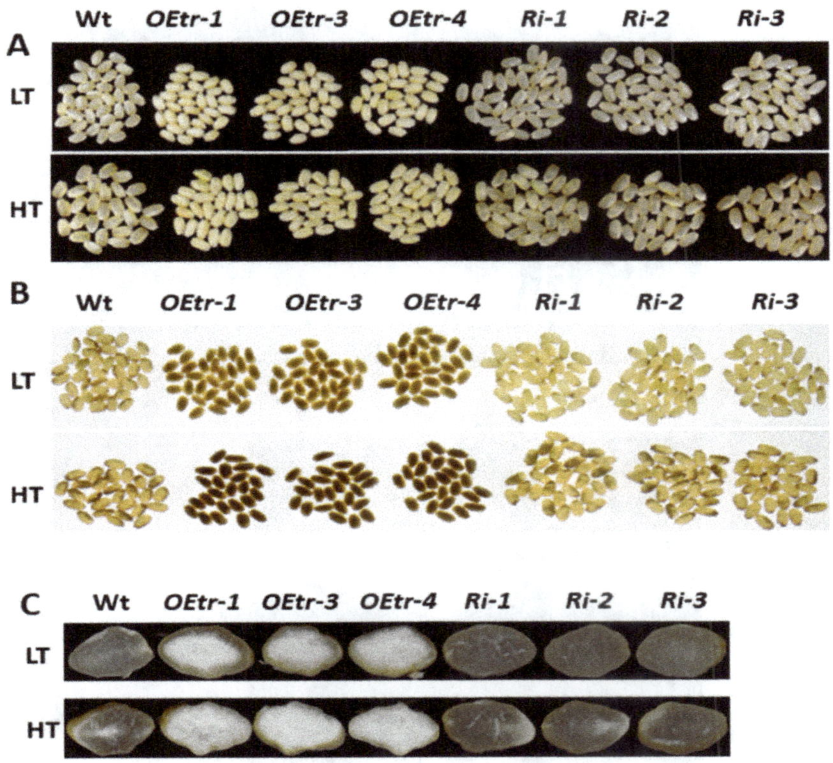

Figure 2. Effects of temperature on the appearances of the WT, *OEtrs*, and *Ris* endosperms. After 5 d of flowering, rice plants were transferred to a growth chamber and grown at a lower temperature (22 °C for 16 h light, 20 °C for 8 h dark) or a higher temperature (31 °C for 16 h light, 28 °C for 8 h dark). (**A**,**B**) Seeds harvested from the plants grown at the lower and higher temperatures, respectively, illuminated using (**A**) normal lighting and (**B**) backlighting. (**C**) Cross-sections of the endosperms of the seeds from plants grown at either lower or higher temperature. LT: lower temperature; HT: higher temperature.

Table 2. The ratios of chalky grains in wild type (TNG67), *OEtr-1*, and *Ri-1* lines growth under lower temperature (LT; 22 °C for 16 h light, 20 °C for 8 h dark) or higher temperature (HT; 31 °C for 16 h light, 28 °C for 8 h dark) during the rice grain-filling process.

Plant Species	Wild Type	*OEtr-1*	*OEtr-3*	*OEtr-4*	*Ri-1*	*Ri-2*	*Ri-3*
The ratios of chalky grains in LT (%)	16.2 ± 2.19	100	100	100	6.0 ± 0.61	6.7 ± 0.56	7.1 ± 0. 61
The ratios of chalky grains in HT (%)	63.6 ± 5.19	100	100	100	50.1 ± 3.61	51.2 ± 4.56	45.1 ± 3.21

Mean values calculated from 100 independent seeds. All data are presented as mean ± SE. Statistical significance is determined by *t*-test.

2.3. *Effect of OsCDPK1 on Starch Granule Morphology in Rice Endosperms*

Due to the similar transgenic plant phenotypes and grain morphology (Figures 1 and 2), the *OsCDPK1*-overexpressing line, *OEtr-1*, and the *OsCDPK1*-silencing line, *Ri-1*, were therefore selected for further study. We examined the starch granule morphology in the WT, *Ri-1*, and *OEtr-1* transgenic seeds by analyzing the endosperm cross-sections acquired from scanning electron microscopy. As shown in Figure 3, the three-dimensional structures of the starch granules in the endosperms were irregularly polygonal and polyhedral in all three grains types. The starch granules were large and tightly packed in the WT and *Ri-1* endosperms but small and loosely packed in the *OEtr-1* endosperm (Figure 3B), suggesting that *OsCDPK1* affects the starch granule size and packing density in developing rice seed.

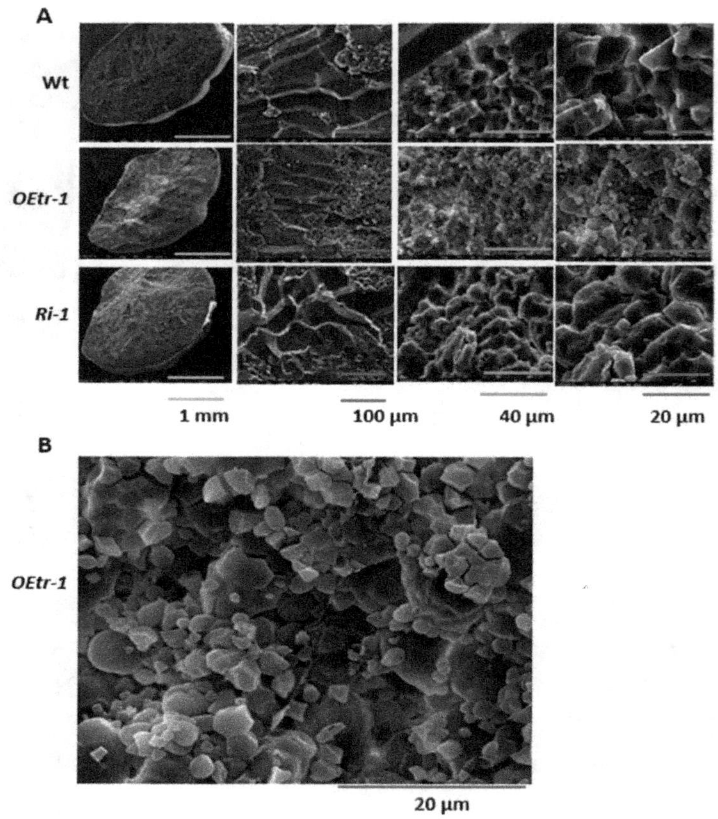

Figure 3. Scanning electron microscopy images of the structures of the starch granules in the rice endosperms. (**A**) The central areas of the cross-sections of mature endosperms from the WT plants (top panel), *OEtr-1* plants (middle panel), and *Ri-1* plants (bottom panel) were acquired using a scanning electron microscope. (**B**) A zoomed-in image of the 20 μm photo of *OEtr-1*. Scale bar colors: blue as 1 mm, red as 100 μm, green as 40 μm, purple as 20 μm.

2.4. *Effects of OsCDPK1 on Starch Properties and Gelatinization in the Endosperm*

We examined the apparent amylose content in the endosperm by reacting 20 mg of each rice endosperm powder sample with 1 N NaOH to gelatinize the starch. The amylose content was measured using a colorimetric method using an I_2/KI solution [12,33]. The *OEtr-1* samples had less affinity than the other samples for iodine and were light purple, whereas the *Ri-1* and WT samples were dark blue and light blue, respectively (Figure 4A and Figure S2). As described in Table 1, the average amylose content in WT, *OEtrs*, and *Ris* was 23.2%, 15.1% and 26.2%, respectively. Here, the apparent amylose content in WT, *OEtr-1* and *Ri-1* were examined in detail to dissect the roles of *OsCDPK1* in starch biosynthesis. The absorption spectra of the I_2/KI stained solutions and found strong absorbance between 480 and 720 nm for both the *Ri-1* and WT samples, with maxima at 620 nm, but stronger absorbance was found for *Ri-1* than for WT at the same wavelength (Figure 4B). *OEtr-1* absorbance was weaker and decreased as the wavelength increased. We compared the results to the absorbances of potato amylose standards to allow the apparent amylose contents to be determined. The apparent amylose contents of the WT, *OEtr-1*, and *Ri-1* samples were 23.35%, 14.74%, and 26.15%, respectively (Table S2). These results demonstrate that the *Ri-1* seed endosperms had higher amylose content than the WT seed endosperms and that the *OEtr-1* seed endosperms had lower amylose content than the WT seed endosperms. Because the WT, *OEtr-1*, and *Ri-1* amylose contents were different, we investigated starch gelatinization at different urea concentrations (0–9 M). A 20-mg aliquot of a rice endosperm powder was mixed with 1 mL of urea solution, and the mixture was allowed to react for 24 h. The mixture was then centrifuged and the degree of gelatinization determined by measuring the sediment volume. Starch gelatinization started at urea concentrations of 3.0–4.0 M (Figure 4C and Figure S3). In 4.0 M urea, the *Ri-1* sediment volume was 5.3% higher than the WT sediment volume, whereas the *OEtr-1* sediment volume was 36.4% lower than the WT sediment volume (Figure 4D). These results indicate that *OsCDPK1* affects the physicochemical properties of the starch in rice endosperms.

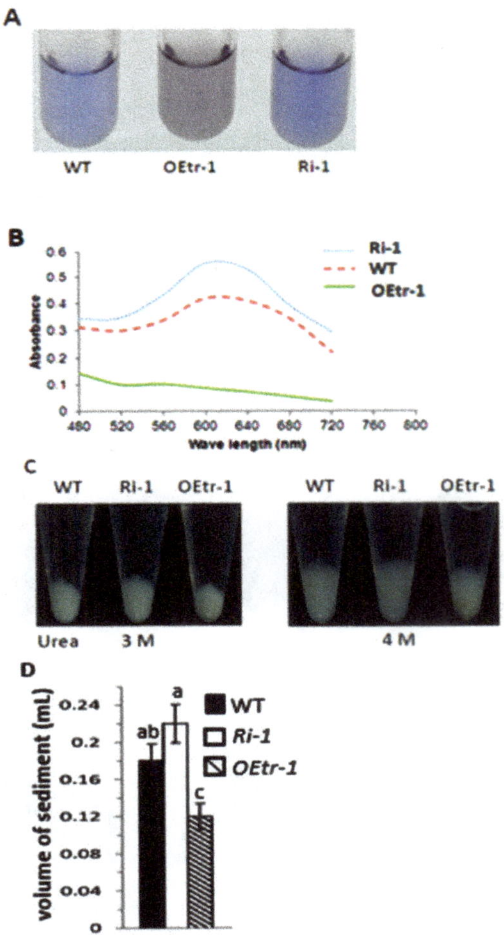

Figure 4. Iodine-staining and gelatinization properties of the starch in the rice endosperms. A 20-mg aliquot of endosperm powder was treated with 1 N NaOH as described in Section 4. (**A**) Supernatants of the iodine-stained WT, *OEtr-1*, and *Ri-1* samples. (**B**) Starch–iodine absorbance spectra of the supernatants. (**C**) Effects of using 3.0 and 4.0 M urea solutions on the gelatinization characteristics of the WT, *OEtr-1*, and *Ri-1* endosperm starch. A 20-mg aliquot of rice powder was mixed in an Eppendorf tube with 1 mL of urea solution and the mixture was shaken for 24 h at 25 °C. The mixture was centrifuged, and the volume of the gelatinized starch sediment was measured. (**D**) Quantification of the gelatinization volume. Different letters above the bars indicate significant differences, identified by performing ANOVAs ($p < 0.05$). Each value is the mean ± SD of three independent measurements.

2.5. OsCDPK1 Expression Profiles in Developing Rice Seeds

Our results have demonstrated that *OsCDPK1* affects rice seed development. It is necessary to track changes in *OsCDPK1* gene expression during rice seed development. The *OsCDPK1::GUS* transgenic line was generated using a *GUS* (*β-glucuronidase*) reporter gene controlled by the *OsCDPK1* promoter (−1706 to +301 bp, i.e., a total of 2007 bp upstream of the translational start site) containing the first intron (607 bp, in the 5′-untranslated region) (Figure 5A). As shown in Figure 5B, strong GUS staining was observed in the ovaries and anthers before flowering, but weaker staining was observed in the styles and lemma. At 1 DAF (days after flowering), strong GUS activity was found only in the ovaries and styles, weaker activity was found in the lemma, and no GUS activity was found in the anthers and stigma. Between 2 and 5 DAF, concentrated GUS staining was found in the rachilla and both ends of the developing seeds and weak staining was found in the lemma. The blue color gradually expanded from both ends toward the central parts of the developing seeds between 6 and 7 DAF, and at 8 DAF the entire seeds were thoroughly stained blue. Staining gradually decreased afterwards, but remained strong between 10 and 14 DAF, then decreased quickly after 14 DAF and had completely gone by 18 DAF. These results suggest that *OsCDPK1* was expressed in a particular temporal and spatial way, predominantly in the middle stage of rice seed development.

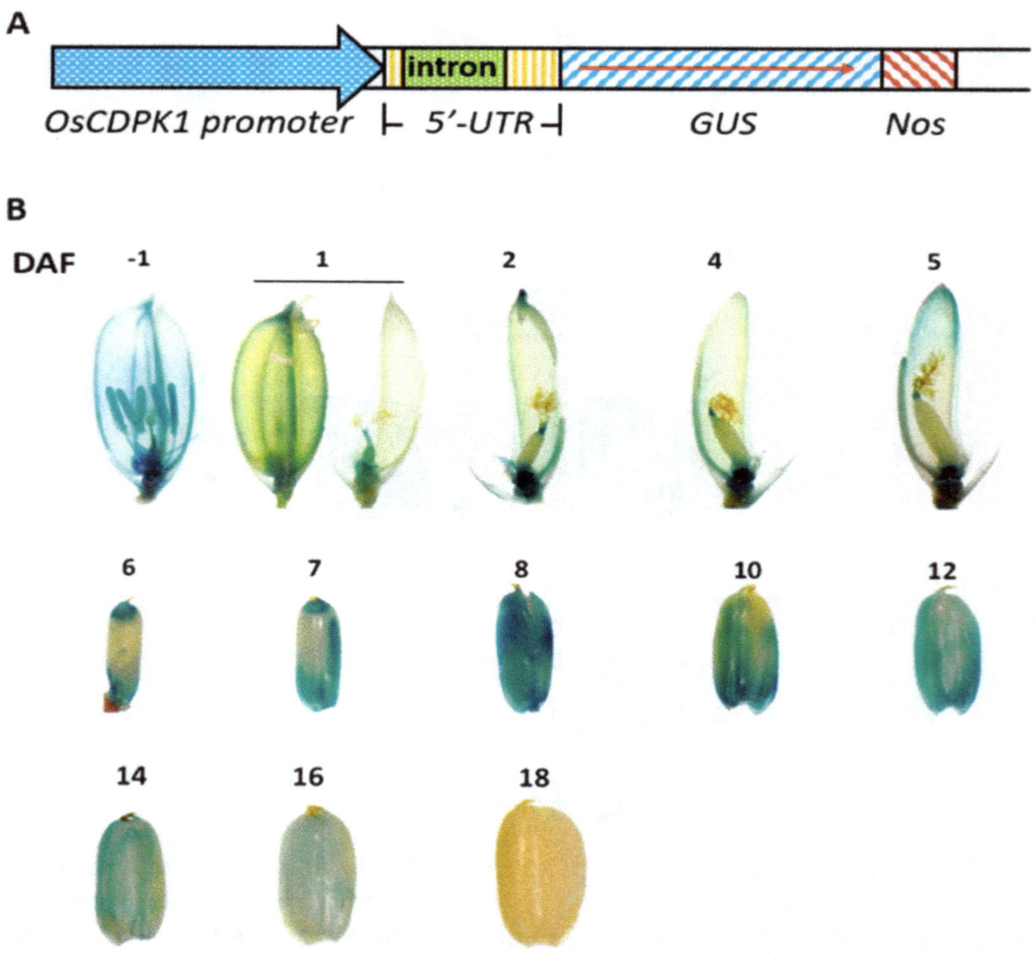

Figure 5. Histochemical GUS activity staining in flowers and developing rice seeds. (**A**) Map of the *OsCDPK1::GUS* expression construct. (**B**) Rice spikelets were collected before flowering and 1–18 days after flowering (DAF). The lemma and palea were partially or completed removed from each spikelet or developing seed before staining. The stained spikelets or immature grains were preserved in 70% ethanol and photographed. −1 DAF means before flowering.

2.6. Effects of OsCDPK1 on the Levels of Starch-Biosynthesis-Related Genes in Developing Rice Seeds

We further investigated the roles of *OsCDPK1* in rice-seed development by analyzing the expression patterns of 12 genes involved in starch biosynthesis. These genes were granule-bound starch synthase (*OsGBSSI*), starch synthase (*OsSSI, OsSSIIa, OsSSIIb, OsSSIIc, OsSSIIIa,* and *OsSSIIIb*), branching enzyme (*OsBEI*), ADP-glucose pyrophosphorylase large subunit (*OsAGPLI, OsAGPLII,* and *OsAGPLIII*), and a small subunit of ADP-glucose pyrophosphorylase (*OsAGPSIIb*). The developing WT, *Ri-1*, and *OEtr-1* rice seeds at 5 and 12 DAF were analyzed. Total RNA was isolated from the dehulled embryo-less half seeds and subjected to quantitative RT-PCR. The relative expression levels of *OsAGPLI, OsAGPSIIb, OsGBSSI, OsSSIIc,* and *OsSSIIIa* were significantly up-regulated in *Ri-1* and down-regulated in *OEtr-1* at 12 DAF, but there was no significant difference at 5 DAF compared with the genes in WT plants (Figure 6). Expression of the seven other genes in the transgenic lines was not significantly different at 5 or 12 DAF from expression in the WT plants. These results suggest that *OsCDPK1* might be involved in regulating starch-biosynthesis-related genes in the mid-development stage of the rice seed.

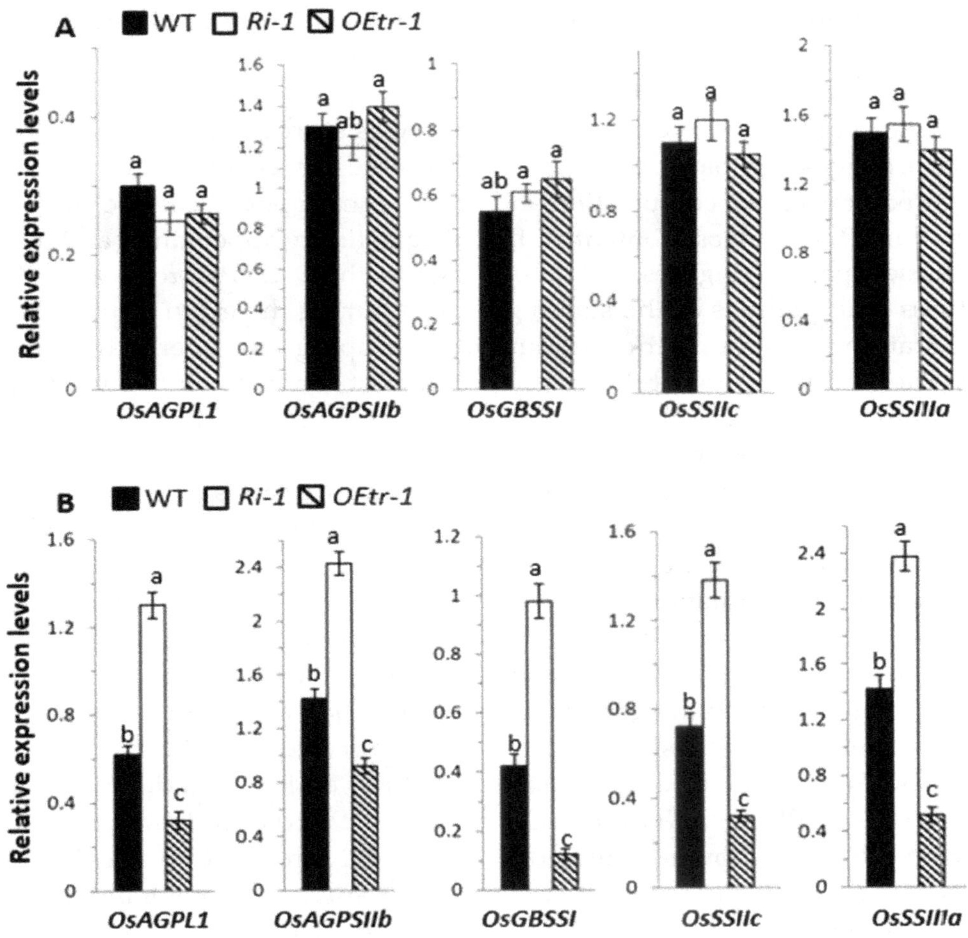

Figure 6. Expression of starch-biosynthesis-related genes during seed development in the WT, *Ri-1*, and *OEtr-1* plants. Total RNA was isolated from developing endosperms (**A**) 5 and (**B**) 12 DAF from the WT, *Ri-1*, and *OEtr-1* plants, and subjected to RT-PCR analysis. The relative expression levels of each gene were normalized to the expression level of the internal control *OsActin*. Different letters above the bars indicate significant differences, identified by performing ANOVAs ($p < 0.01$). Each value is the mean ± SD of three independent measurements. *OsAGPLI*: ADP-glucose pyrophosphorylase large subunit I; *OsAGPSIIb*: ADP-glucose pyrophosphorylase small subunit IIb; *OsGBSSI:* granule-bound starch synthase I; *OsSSIIc:* starch synthase IIc; *OsSSIIIa:* starch synthase IIIa. Primer sets and gene accession numbers are listed in Table S1.

3. Discussion

The amylose content of endosperms is an important parameter determining the eating quality of rice, which is negatively related with stickiness but positively related to rice grain hardness [34,35]. The key enzymes (genes) involved in the starch (amylopectin and amylose) biosynthesis during rice grain filling are well known. However, the control mechanisms by which the amylose and amylopectin biosynthesis is orchestrated are still not fully understood. Our results indicate for the first time that the protein kinase *OsCDPK1* is functionally negatively correlated with the amylose content, endosperm transparency, and seed size in developing rice seed.

The *OEtr-1* and *Ri-1* seeds had some distinct features compared with the WT seeds. For example, the *OEtr-1* grains were smaller, had lower amylose contents, and had more floury endosperms than the WT grains, and the *Ri*-1 grains were larger, had higher amylose contents, and had more transparent endosperms than the WT grains (Figures 1–4). This indicates that the *OsCDPK1* function is closely associated with the rice endosperm starch properties. The *OsCDPK1::GUS* expression profile in the developing rice grains gradually increased immediately after flowering and reached a maximum between 7 and 14 DAF (Figure 5). The *OsCDPK1::GUS* expression timing in the developing rice seeds was similar to that found in a study by Ohdan et al. [36], in which 27 different genes involved in starch biosynthesis were examined during rice-seed development. In that study, all the genes had been differentially expressed before 15 DAF. In this study, the GUS staining was found throughout the endosperm at 12 and 14 DAF of developing seeds, and was more intensified in the interior region at 12 DAF (Figure S4), suggesting that *OsCDPK1* first affects the expression of starch biosynthesis-related genes and later affects the starch composition and endosperm appearance. The starchy endosperm in the *OEtr-1* grains had a low amylose content and an opaque floury appearance, and the starch granules were small and loosely packed (Figures 1–3), indicating that the *OsCDPK1* roles were closely associated with the structures and qualities of the starch granules during the grain filling process. Moreover, *OsCDPK1::GUS* staining was found throughout the developing endosperm at 7–16 DAF (Figure 5B and Figure S4). These results suggest that the expression of some starch-biosynthesis-related genes in the endosperm cells may be affected by *OsCDPK1*, followed by changing the amylose content and resulting in the opaque endosperm in the *OEtr-1* grains.

Several mutations of rice genes involved in starch biosynthesis have been found to alter the structures and properties of the starch produced. For example, the *SSIIIa* mutation (*flo5*) was found to increase the amylose content, alter the amylopectin structure, and cause the endosperm to have a white core [20]. The *waxy* mutant (a mutation in *GBSSI*) produced an amylose-free, floury endosperm [11,12]. The *amylose-extender* mutation in *BEIIb* altered the fine structure of amylopectin and gave a floury endosperm [12]. A mutation in *BEI* (starch-branching enzyme I) caused the amylopectin fine structure to change but did not appear to affect the endosperm appearance [10]. Our data show that the expression of some starch-biosynthesis-related genes was affected by *OsCDPK1*, similar to the results of previous studies. During the middle phase (12 DAF) of endosperm development, examples of genes affected were *OsAGPLI*, *OsAGPSIIb*, *OsGBSSI*, *OsSSIIc*, and *OsSSIIIa*, which were significantly up-regulated in the *Ri-1* and down-regulated in *OEtr-1* (Figure 6). Changes in the expression of these genes caused the *OEtr-1* endosperm to have a lower amylose content and a floury appearance, whereas the *Ri-1* endosperm had a high amylose content and was more transparent (Figures 1, 2 and 4 and Tables 1 and 2). Calcium ions (Ca^{2+}) are secondary messengers in plant cells, with plant CDPK being a sensor to relay calcium signals via binding with calcium via the calcium-binding domains. We therefore suggest that *OsCDPK1* acts as an upstream regulator that is closely associated with starch biosynthesis. Some regulators have been found to regulate the expression of genes encoding key starch-biosynthesis enzymes, and null mutations in these regulators usually alter the amylopectin fine structure, the starch composition, the starch granule morphology and size, and the endosperm appearance. For instance, some starch-biosynthesis-related genes were found to be affected in five independent mutants, *flo2* (a tetratricopeptide repeat motif protein) [17], *flo4* (*pyruvate orthophosphate dikinase B*) [19], *flo7* (an unknown protein) [22], *osbzip58* (a bZIP transcription factor) [37], and *osbt1* (an ADP-glucose transporter) [38]. All these effects decreased the amylose content and changed the amylopectin composition. The endosperm appearance was affected differently in different parts; e.g., *flo2* had endosperm with a floury kernel, *flo4* and *osbzip58* and *osbt1* gave white-core phenotypes, and chalkiness was only found in the peripheral endosperm of *flo7*. Similarly, the low amylose content of the *OEtr-1* endosperm was expressed in a floury morphology (Figures 1 and 4). The results of previous studies and this study together indicate that a low-amylose content of rice endosperm gives a chalky or floury phenotype, suggesting that the amount of amylose present is an important factor affecting the quality and appearance of the starchy endosperm. This raises the question of whether rice can be engineered to have an endosperm with a high amylose content and, therefore, a transparent appearance. Here, we have provided direct evidence that silencing *OsCDPK1* (*Ri-1*) increases the amylose content of the endosperm, making the starchy grain more transparent (Figures 1 and 4 and Tables 1 and 2). These results will be useful in developing rice-breeding strategies aimed at maintaining (or even improving) grain quality in rice to cope with global warming. The transgenic lines *OEtrs* and *Ris* could also be ideal materials for investigating the mechanisms that control rice seed size and starch biosynthesis.

Temperature also strongly affects amylose synthesis during rice grain development. In previous studies, a lower temperature increased the expression of the *waxy* gene and protein and increased the amylose content in developing rice endosperms [39]; a higher temperature had the opposite effects [39,40]. Moreover, during the rice grain-filling process, a temperature higher than the optimum usually causes impaired starch accumulation, resulting in loosely packed starch granules with small air spaces between them, giving high proportions of opaque chalky or floury grains. These effects decrease the market value because the rice will have poor milling qualities (being easily broken), poor cooking and eating qualities, and a poor appearance [41,42]. It has recently been found that a high temperature (33 °C for 12 h light, 28 °C for 12 h dark) also induced the expression of three α-amylase genes—*Amy1A*, *Amy3C*, and *Amy3D*—in developing endosperms, causing the grains to be chalky, probably because of starch degradation and the accumulation of soluble sugars in the endosperm [43]. In contrast, we found that the *Ri-1* endosperm was more transparent at both a low temperature (22 °C for 16 h light, 20 °C for 8 h dark) and a high temperature (31 °C for 16 h light, 28 °C for 8 h dark) than that of the WT plant during the grain-filling process (Figure 2 and Table 2). Our results will be useful in developing rice-breeding strategies aimed at maintaining (or even improving) grain quality in rice selected to cope with global warming. Our results also improve our understanding of the molecular mechanisms involved in amylose biosynthesis.

It has been found in several studies that chalky or floury endosperms might be caused by loosely packed small, round starch granules formed in developing rice seeds [9,19,20,38]. The scanning electron microscopy images indicate that the WT and *Ri-1* endosperms contained closely packed polyhedral starch granules but that the white core of the *OEtr-1* endosperm contained loosely packed small starch granules. The starch granules in the *OEtr-1* endosperm were polyhedral (like in the WT and *Ri-1* endosperm) rather than small and round as in most chalky or floury mutants (Figure 3B). It is, therefore, likely that *OsCDPK1* plays a role in starch biosynthesis and negatively affects the sizes but not the shapes of the starch granules.

We previously found that *OsCDPK1* inhibits the feedback of GA biosynthesis through down-regulating *GA3ox2* and *GA20ox1* [32]. In this study, we put forward a model in which *OsCDPK1* plays key roles in negatively controlling the grain size, amylose content, and endosperm appearance, and also affects the physicochemical properties of the starch (Figure 7). Milled *OsCDPK1*-gene-silenced *Ri-1* grains were larger than WT grains, and there were numerous densely packed polyhedral starch granules accompanied by a high amylose content and a transparent endosperm. In contrast, the *OEtr-1* grains were smaller and contained loosely packed small, but still polyhedral, starch granules. The *OEtr-1* grains had lower amylose content and opaque white-cored endosperms. Notably, the phenotypes of the grain size and the floury endosperm in *OEtr-1*, *-2*, and *-3* were consistent and were unaffected by temperature during the grain-filling process (Figures 1 and 2). *OsCDPK1* therefore plays pleiotropic roles in rice seed development. Our results indicate that *Ri-1* and *OEtr-1* could be ideal materials for investigating the mechanisms controlling rice seed size and starch biosynthesis, and could also be valuable reference samples for rice breeding aimed at simultaneously improving grain yield and quality.

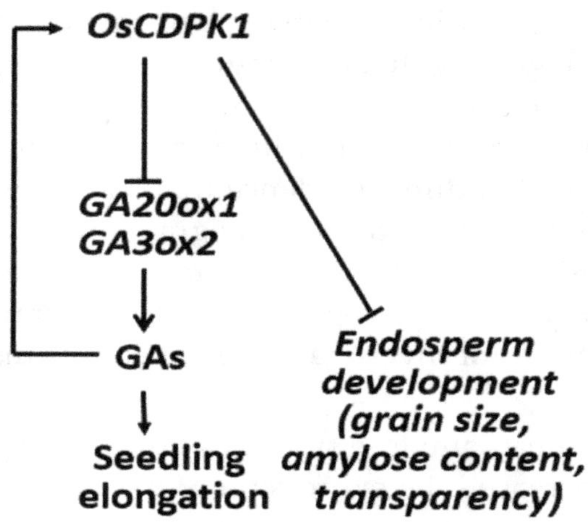

Figure 7. Proposed roles of *OsCDPK1* in the interconnecting GA biosynthesis and signaling pathways and endosperm developmental processes. The model is described in detail in the text. The arrows indicate activation and the blunt ends indicate inhibition.

4. Materials and Methods

4.1. Plant Materials

In our previous studies [32], 8 independent T1 transgenic lines that carried a single transgene in rice were subjected to ectopic overexpression of a constitutively active truncated form of *OsCDPK1* (*OEtr*) in rice seedlings, which all yielded a semi-dwarf phenotype that produced small seeds. By contrast, 5 independent T1 transgenic plants exhibiting a single copy of transgene were subject to *OsCDPK1* gene silencing (*Ri*) by RNA interference, which all gave a slender-like phenotype during seedling development and subsequently produced large seeds. According to the consistency of the phenotypic traits in *OEtr* lines of which all displayed semi-dwarf phenotype and produced small seeds. In *Ri* lines, which all gave a slender-like phenotype and produced large seeds [32]. Therefore, in this study, the homozygous transgenes of the T4 transgenic plants from *OEtr-1*, *OEtr-3*, and *OEtr-4*, and *Ri-1*, *Ri-2*, and *Ri-3* were selected to explore the roles of *OsCDPK1* in rice grain development.

4.2. Callus Induction

Rice *Oryza sativa* L. cv. Tainung 67 was used in the study. Immature seeds were de-hulled, sterilized with 2.4% NaOCl for 30 min, then washed thoroughly with sterile water. The seeds were then incubated on N6 agar medium [44] containing 10 μM 2,4-D to induce calli formation [45]. After about 30 days, the calli derived from the scutella were transferred to the fresh N6 agar medium containing 2,4-D for another 15 days and were subjected to *Agrobacterium*-mediated gene transformation [45].

4.3. Primers

The nucleotide sequences of all the primers used in the real-time polymerase chain reaction (qRT-PCR) amplification are shown in Table S1.

4.4. Construction of OsCDPK1::GUS Expression Vectors

The *OsCDPK1::GUS* expression vector was constructed by amplifying a 2007 bp DNA fragment containing the *OsCDPK1* promoter and its 5′-untranslated region (which contained a 607 bp intron)

(Figure 1A) by PCR using the forward primer *OsCDPK1-5P* (5′-ATACTGCAGTGGTCTTATT AGGTAAGGCC-3′) and the reverse primer *OsCDPK1-3B* (5′-ATAGGATCC TCCAAGAACTCCT TATGCAA-3′). The DNA fragment was cleaved using *Pst*I and *Bam*HI, and then cloned into vector *pBX-2* as described previously [45]. The *OsCDPK1::GUS* construct was linearized by digesting it with *Pst*I, then inserted into the *Pst*I site of the *pSMY1H* binary vector [45], then it was subjected to *Agrobacterium*-mediated gene transformation.

4.5. Plant Transformation

Recombinant binary plasmids were introduced into *Agrobacterium tumefaciens* strain EHA101 by electroporation, and rice calli were transformed as described previously [45].

4.6. Histochemical Staining of GUS Activity in Developing Rice Grains

Rice spikelets collected before flowering and 1–18 DAF were subjected to GUS activity staining to assess *OsCDPK1* gene expression profiles in developing rice grains. The lemma and palea were partially or completely removed from each spikelet (developing seed) before the staining process, and the spikelets were then incubated in a 1-mM 5-bromo-4-chloro-3-indolyl β-D-glucuronide solution (in 100 mM sodium phosphate containing 10 mM EDTA, 0.5 mM potassium ferrocyanide, 0.5 mM potassium ferricyanide, and 0.1% Triton X-100, at pH 7.0) at 37 °C in the dark for 4 h. The stained spikelets or immature grains were then preserved in 70% ethanol and rinsed with water before being photographed.

4.7. Quantitative RT-PCR

Developing seeds were collected from wild-type (WT), *Ri-1*, and *OEtr-1* at 5 and 12 DAF, respectively. Total RNA was isolated from developing endosperms by using TRIzol reagent (Invitrogen, Carlsbad, CA, USA), and DNA contamination was then removed using a TURBO DNA-free kit (Ambion, Foster City, CA, USA). A 5-µg aliquot of the total RNA was used to synthesize first strand cDNA using M-MuLV reverse transcriptase (New England Biolabs, Ipswich, MA, USA) and oligo (dT) primer. Quantitative RT-PCR was performed using an Eco Real-Time PCR System (Illumina, San Diego, CA, USA) following the manufacturer's instructions. Gene-specific primer sets (Table S1) localized at the 3′-untranslated regions for each gene examined were selected to allow assessment of the extent to which the starch-biosynthesis-related genes in WT, *OEtr-1*, and *Ri-1* were expressed. The relative expression levels were normalized to expression in the internal control, *OsActin 1*. All reactions were performed in triplicate.

4.8. Analysis of the Gelatinization Properties of the Starch

A 20-mg aliquot of rice endosperm powder derived from de-embryonic seeds was mixed with 1 mL of urea solution at a concentration of between 0 and 9 M, and the mixture was shaken vigorously for 24 h at room temperature [12]. All the tested mixtures were then centrifuged at 25,000× *g* for 60 min (Eppendorf, model 5427R, North Ryde, Australia) at a same time, and the volume of the gelatinized starch sediment was measured by the volume scale on the Eppendorf tube.

4.9. Apparent Amylose Content Analysis

A 20-mg aliquot of rice endosperm powder was gelatinized by adding 2 mL of 1 N NaOH and incubating the mixture at 25 °C for 24 h. Then, 4 mL of 1 N CH_3COOH was added, the mixture was mixed well, and 4 mL H_2O was added. A 0.8-mL aliquot of the solution was mixed with 0.2 mL I_2/KI (0.2%/2%), then 4 mL H_2O was added. The apparent amylose content was measured using the colorimetric method described by [33]. Absorbance at 620 nm was measured, and the apparent amylose content was determined by comparing the absorbance to a calibration curve prepared using potato amylose standards.

4.10. Scanning Electron Microscopy

Dehusked rice grains were cut transversely and analyzed using a scanning electron microscope (Quanta 200; FEI, Hillsboro, OR, USA) following the manufacturer's instructions.

5. Conclusions

Our results demonstrated that *OsCDPK1* plays key roles in negatively controlling the grain size, amylose content, and endosperm appearance, and also affects the physicochemical properties of the starch. Milled *OsCDPK1*-gene-silenced *Ri-1* grains were larger than WT grains, and there were numerous densely packed polyhedral starch granules accompanied by a high amylose content and a transparent endosperm. In contrast, the *OEtr-1* grains were smaller and contained loosely packed small, but still polyhedral, starch granules. The *OEtr-1* grains had lower amylose content and opaque white-cored endosperms. Moreover, the grain phenotypes in *OEtrs* were unaffected by temperature during the grain-filling process. *OsCDPK1* therefore plays pleiotropic roles in rice reproductive developmental processes, in a negative sense. Our results indicate that *Ri-1* and *OEtr-1* could be ideal materials for investigating the mechanisms that control rice seed size and starch biosynthesis, and for rice breeding to improve grain yield and quality.

Author Contributions: Conceptualization: S.-L.H.; Data curation: J.-Z.J. and S.-L.H.; Investigation: J.-Z.J., C.-H.K., B.-H.C. and M.-K.C.; Methodology: C.-S.L.; Supervision: S.-L.H.; Writing—original draft: S.-L.H.

References

1. Rabiei, B.; Valizadeh, M.; Ghareyazie, B.; Moghaddam, M.; Ali, A.J. Identification of QTLs for rice grain size and shape of Iranian cultivars using SSR markers. *Euphytica* **2004**, *137*, 325–332. [CrossRef]
2. Bao, J.; Jin, L.; Xiao, P.; Shen, S.; Sun, M.; Corke, H. Starch physicochemical properties and their associations with microsatellite alleles of starch synthesizing genes in a rice RIL population. *J. Agric. Food Chem.* **2008**, *56*, 1589–1594. [CrossRef] [PubMed]
3. Fitzgerald, M.A.; Lisle, A.J.; Martin, M. Chalky and translucent rice grains differ in starch composition and structure and cooking properties. *Cereal Chem.* **2000**, *77*, 627–632.
4. Jiang, H.; Dian, W.; Wu, P. Effect of high temperature on fine structure of amylopectin in rice endosperm by reducing the activity of the starch branching enzyme. *Phytochemistry* **2003**, *63*, 53–59. [CrossRef]
5. Morita, S.; Shiratsuchi, H.; Takanashi, J.; Fujita, K. Effect of high temperature on grain ripening in rice plants: Analysis of the effects of high night and high day temperatures applied to the panicle and other parts of the plant. *Jpn. J. Crop Sci.* **2004**, *73*, 77–83. [CrossRef]
6. Counce, P.A.; Bryant, R.J.; Bergman, C.J.; Bautista, R.C.; Wang, Y.J.; Siebenmorgen, T.J.; Moldenhauer, K.A.; Meullenet, J.F.C. Rice milling quality, grain dimensions, and starch branching as affected by high night temperatures. *Cereal Chem.* **2005**, *82*, 645–648. [CrossRef]
7. Patindol, J.; Wang, Y.J. Fine structures and physicochemical properties of starches from chalky and translucent rice kernels. *J. Agric. Food Chem.* **2003**, *51*, 2777–2784. [CrossRef] [PubMed]
8. Yamakawa, H.; Hirose, T.; Kuroda, M.; Yamaguchi, T. Comprehensive expression profiling of rice grain filling-related genes under high temperature using DNA microarray. *Plant Physiol.* **2007**, *144*, 258–277. [CrossRef] [PubMed]
9. Fu, F.F.; Xue, H.W. Coexpression analysis identifies rice starch regulator1, a rice AP2/EREBP family transcription factor, as a novel rice starch biosynthesis regulator. *Plant Physiol.* **2010**, *154*, 927–938. [CrossRef] [PubMed]
10. Mizuno, K.; Kimura, K.; Arai, Y.; Kawasaki, T.; Shimada, H.; Baba, T. Starch branching enzymes from immature rice seeds. *J. Biochem.* **1992**, *112*, 643–651. [CrossRef] [PubMed]

11. Smith, A.M.; Denyer, K.; Martin, C. The synthesis of the starch granule. *Annu. Rev. Plant Physiol. Plant Mol. Biol.* **1997**, *48*, 67–87. [CrossRef] [PubMed]
12. Nishi, A.; Nakamura, Y.; Tanaka, N.; Satoh, H. Biochemical and genetic analysis of the effects of amylose-extender mutation in rice endosperm. *Plant Physiol.* **2001**, *127*, 459–472. [CrossRef] [PubMed]
13. Singh, B.K.; Preiss, J. Starch branching enzymes from maize. *Plant Physiol.* **1985**, *79*, 34–40. [CrossRef] [PubMed]
14. Nakamura, Y.; Takeichi, T.; Kawaguchi, K.; Yamanouchi, H. Purification of two forms of starch branching enzyme (Q-enzyme) from developing rice endosperm. *Physiol. Plant.* **1992**, *84*, 329–335. [CrossRef]
15. Satoh, H.; Omura, T. New endosperm mutations induced by chemical mutagens in rice, *Oryza sativa* L. *Jpn. J. Breed.* **1981**, *31*, 316–326. [CrossRef]
16. Kaushik, R.P.; Khush, G.S. Genetic analysis of endosperm mutants in rice *Oryza sativa* L. *Theor. Appl. Genet.* **1991**, *83*, 146–152. [CrossRef] [PubMed]
17. She, K.C.; Kusano, H.; Koizumi, K.; Yamakawa, H.; Hakatae, M.; Imamura, T.; Fukuda, M.; Naito, N.; Tsurumaki, Y.; Yaeshima, M.; et al. A novel factor FLOURY ENDOSPERM 2 is involved in regulation of rice grain size and starch quality. *Plant Cell* **2010**, *22*, 3280–3294. [CrossRef] [PubMed]
18. Nishio, T.; Iida, S. Mutant having a low content of 16-kDa allergenic protein in rice (*Oryza sativa* L.). *Theor. Appl. Genet.* **1993**, *86*, 317–321. [CrossRef] [PubMed]
19. Kang, H.G.; Park, S.; Matsuoka, M.; An, G. White-core endosperm floury endosperm-4 in rice is generated by knockout mutations in the C_4-type pyruvate orthophosphate dikinase gene (*OsPPDKB*). *Plant J.* **2005**, *42*, 901–911. [CrossRef] [PubMed]
20. Ryoo, N.; Yu, C.; Park, C.S.; Baik, M.Y.; Park, I.M.; Cho, M.H.; Bhoo, S.H.; An, G.; Hahn, T.R.; Jeon, J.S. Knockout of a starch synthase gene *OsSSIIIa/Flo5* causes white-core floury endosperm in rice (*Oryza sativa* L.). *Plant Cell Rep.* **2007**, *26*, 1083–1095. [CrossRef] [PubMed]
21. Peng, C.; Wang, Y.H.; Liu, F.; Ren, Y.L.; Zhou, K.N.; Lv, J.; Zheng, M.; Zhao, S.L.; Zhang, L.; Wang, C.M.; et al. *FLOURY ENDOSPERM6* encodes a CBM48 domain-containing protein involved in compound granule formation and starch synthesis in rice endosperm. *Plant J.* **2014**, *77*, 917–930. [CrossRef] [PubMed]
22. Zhang, L.; Ren, Y.L.; Lu, B.Y.; Yang, C.Y.; Feng, Z.M.; Liu, Z.; Chen, J.; Ma, W.W.; Wang, Y.H.; Liu, Y.S.; et al. *FLOURY ENDOSPERM7* encodes a regulator of starch synthesis and amyloplast development essential for peripheral endosperm development in rice. *J. Exp. Bot.* **2016**, *67*, 633–647. [CrossRef] [PubMed]
23. Sanders, D.; Pelloux, J.; Brownlee, C.; Harper, J.F. Calcium at the crossroads of signaling. *Plant Cell* **2002**, *14*, S401–S417. [CrossRef] [PubMed]
24. Harper, J.F.; Breton, G.; Harmon, A. Decoding Ca^{2+} signals through plant protein kinases. *Annu. Rev. Plant Biol.* **2004**, *55*, 263–288. [CrossRef] [PubMed]
25. Mori, I.C.; Murata, Y.; Yang, Y.; Munemasa, S.; Wang, Y.F.; Andreoli, S.; Tiriac, H.; Alonso, J.M.; Harper, J.F.; Ecker, J.R.; et al. CDPKs CPK6 and CPK3 function in ABA regulation of guard cell S-type anion- and Ca^{2+}-permeable channels and stomatal closure. *PLoS Biol.* **2006**, *4*, e327. [CrossRef] [PubMed]
26. Sheen, J. Ca^{2+}-dependent protein kinases and stress signal transduction in plants. *Science* **1996**, *274*, 1900–1902. [CrossRef] [PubMed]
27. Ludwig, A.A.; Romeis, T.; Jones, J.D. CDPK-mediated signalling pathways: Specificity and cross-talk. *J. Exp. Bot.* **2004**, *55*, 181–188. [CrossRef] [PubMed]
28. Romeis, T.; Ludwig, A.A.; Martin, R.; Jones, J.D.G. Calcium-dependent protein kinases play an essential role in a plant defence response. *EMBO J.* **2001**, *20*, 5556–5567. [CrossRef] [PubMed]
29. Abbasi, F.; Onodera, H.; Toki, S.; Tanaka, H.; Komatsu, S. OsCDPK13, a calcium-dependent protein kinase gene from rice, is induced by cold and gibberellin in rice leaf sheath. *Plant Mol. Biol.* **2004**, *55*, 541–552. [CrossRef] [PubMed]
30. Wan, B.; Lin, Y.; Mou, T. Expression of rice Ca^{2+}-dependent protein kinases (CDPKs) genes under different environmental stresses. *FEBS Lett.* **2007**, *581*, 1179–1189. [CrossRef] [PubMed]
31. Asano, T.; Kunieda, N.; Omura, Y.; Ibe, H.; Kawasaki, T.; Takano, M.; Sato, M.; Furuhashi, H.; Mujin, T.; Takaiwa, F.; et al. Rice SPK, a calmodulin-like domain protein kinase, is required for storage product accumulation during seed development: Phosphorylation of sucrose synthase is a possible factor. *Plant Cell* **2002**, *14*, 619–628. [CrossRef] [PubMed]

32. Ho, S.L.; Huang, L.F.; Lu, C.A.; He, S.L.; Wang, C.C.; Yu, S.P.; Chen, J.; Yu, S.M. Sugar starvation- and GA-inducible calcium-dependent protein kinase 1 feedback regulates GA biosynthesis and activates a 14-3-3 protein to confer drought tolerance in rice seedlings. *Plant Mol. Biol.* **2013**, *81*, 347–361. [CrossRef] [PubMed]
33. Juliano, B.O. A simplified assay for milled-rice amylose. *Cereal Sci. Today* **1971**, *16*, 334–340.
34. Cameron, D.K.; Wang, Y.J. A better understanding of factors that affect the hardness and stickiness of long-grain rice. *Cereal Chem.* **2005**, *82*, 113–119. [CrossRef]
35. Li, H.; Prakash, S.; Nicholson, T.M.; Fitzgerald, M.A.; Gilbert, R.G. The importance of amylose and amylopectin fine structure for textural properties of cooked rice grains. *Food Chem.* **2016**, *196*, 702–711. [CrossRef] [PubMed]
36. Ohdan, T.; Francisco, P.B., Jr.; Sawada, T.; Hirose, T.; Terao, T.; Satoh, H.; Nakamura, Y. Expression profiling of genes involved in starch synthesis in sink and source organs of rice. *J. Exp. Bot.* **2005**, *56*, 3229–3244. [CrossRef] [PubMed]
37. Wang, J.C.; Xu, H.; Zhu, Y.; Liu, Q.Q.; Ca, X.L. OsbZIP58, a basic leucine zipper transcription factor, regulates starch biosynthesis in rice endosperm. *J. Exp. Bot.* **2013**, *64*, 3453–3466. [CrossRef] [PubMed]
38. Li, S.; Wei, X.; Ren, Y.; Qiu, J.; Jiao, G.; Guo, X.; Tang, S.; Wan, J.; Hu, P. *OsBT1* encodes an ADP-glucose transporter involved in starch synthesis and compound granule formation in rice endosperm. *Sci. Rep.* **2017**, *7*, 40124. [CrossRef] [PubMed]
39. Sano, Y.; Maekawa, M.; Kikuchi, H. Temperature effects on the Wx protein level and amylose content in the endosperm of rice. *J. Hered.* **1985**, *6*, 221–222. [CrossRef]
40. Asaoka, M.; Okuno, K.; Sugimoto, Y.; Kawakami, J.; Fuwa, H. Effect of environmental temperature during development of rice plants on some properties of endosperm starch. *Starch* **1984**, *36*, 189–193. [CrossRef]
41. Lyman, N.B.; Jagadish, S.V.K.; Nalley, L.L.; Dixon, B.L.; Siebenmorgen, T. Neglecting rice milling yield and quality underestimates economic losses from high-temperature stress. *PLoS ONE* **2013**, *8*, e72157. [CrossRef] [PubMed]
42. Shi, W.; Yin, X.; Struik, P.C.; Solis, C.; Xie, F.; Schmidt, R.C.; Huang, M.; Zou, Y.; Ye, C.; Jagadish, S.V.K. High day- and night-time temperatures affect grain growth dynamics in contrasting rice genotypes. *J. Exp. Bot.* **2017**, *68*, 5233–5245. [CrossRef] [PubMed]
43. Nakata, M.; Fukamatsu, Y.; Miyashita, T.; Hakata, M.; Kimura, R.; Nakata, Y.; Kuroda, M.; Yamaguchi, T.; Yamakawa, H. High temperature-induced expression of rice α-amylases in developing endosperm produces chalky grains. *Front. Plant Sci.* **2017**, *8*, 2089. [CrossRef] [PubMed]
44. Chu, C.C.; Wang, C.C.; Sun, C.S.; Hsu, C.; Yin, K.C.; Chu, C.Y.; Bi, F.Y. Establishment of an efficient medium for another culture of rice through comparative experiments on the nitrogen sources. *Sci. Sin.* **1975**, *5*, 659–668.
45. Ho, S.L.; Tong, W.F.; Yu, S.M. Multiple mode regulation of a cysteine proteinase gene expression in rice. *Plant Physiol.* **2000**, *122*, 57–66. [CrossRef] [PubMed]

3

Floral Scent Emission from Nectaries in the Adaxial Side of the Innermost and Middle Petals in *Chimonanthus praecox*

Zhineng Li [1,2,3,†], Yingjie Jiang [1,2,3,†], Daofeng Liu [1,2,3], Jing Ma [1,2,3], Jing Li [1,2,3], Mingyang Li [1,2,3] and Shunzhao Sui [1,2,3,*]

1 College of Horticulture and Landscape Achitecture, Southwest University, Chongqing 400715, China; znli@swu.edul.cn (Z.L.); sherneole@163.com (Y.J.); liu19830222@163.com (D.L.); majing427@swu.edu.cn (J.M.); acejing@126.com (J.L.); limy@swu.edu.cn (M.L.)

2 Key Laboratory of Horticulture Science for Southern Mountains Regions, Ministry of Education, Chongqing 400715, China

3 Chongqing Engineering Research Center for Floriculture, Chongqing 400715, China

* Correspondence: sszcq@swu.edu.cn

† These authors contributed equally to this work.

Abstract: Wintersweet (*Chimonanthus praecox*) is a well-known traditional fragrant plant and a winter-flowering deciduous shrub that originated in China. The five different developmental stages of wintersweet, namely, flower-bud period (FB), displayed petal stage (DP), open flower stage (OF), later blooming period (LB), and wilting period (WP) were studied using a scanning electron microscope (SEM) to determine the distribution characteristics of aroma-emitting nectaries. Results showed that the floral scent was probably emitted from nectaries distributed on the adaxial side of the innermost and middle petals, but almost none on the abaxial side. The nectaries in different developmental periods on the petals differ in numbers, sizes, and characteristics. Although the distribution of nectaries on different rounds of petals showed a diverse pattern at the same developmental periods, that of the nectaries on the same round of petals showed some of regularity. The nectary is concentrated on the adaxial side of the petals, especially in the region near the axis of the lower part of the petals. Based on transcriptional sequence and phylogenetic analysis, we report one nectary development related gene *CpCRC* (*CRABS CLAW*), and the other four YABBY family genes, *CpFIL* (*FILAMENTOUS FLOWER*), *CpYABBY2*, *CpYABBY5-1*, and *CpYABBY5-2* in *C. praecox* (accession no. MH718960-MH718964). Quantitative RT-PCR (qRT-PCR) results showed that the expression characteristics of these YABBY family genes were similar to those of 11 floral scent genes, namely, *CpSAMT*, *CpDMAPP*, *CpIPP*, *CpGPPS1*, *CpGPPS2*, *CpGPP*, *CpLIS*, *CpMYR1*, *CpFPPS*, *CpTER3*, and *CpTER5*. The expression levels of these genes were generally higher in the lower part of the petals than in the upper halves in different rounds of petals, the highest being in the innermost petals, but the lowest in the outer petals. Relative expression level of *CpFIL*, *CpCRC*, *CpYABBY5-1*, and *CpLIS* in the innermost and middle petals in OF stages is significant higher than that of in outer petals, respectively. SEM and qRT-PCR results in *C. praecox* showed that floral scent emission is related to the distribution of nectaries.

Keywords: *Chimonanthus praecox*; nectary; floral scent; gene expression

1. Introduction

Nectaries are glandular structures that secrete nectar, a carbohydrate-rich solution that is composed mainly of sugars and it generally serves as a reward for pollinators or for as protectors (e.g., ants) against herbivores, or, as a lure for animal prey in carnivorous plants [1]. Nectaries are most wide spread in angiosperms, particularly within flowers, and in ferns and Gnetales [2]. Arabidopsis *CRC* is expressed in the nectary throughout its development and plays a role in the specification and/or differentiation of the nectary [3]. *CRC* is also responsible for carpel growth and fusion, and floral meristem termination [2,4,5]. *CRC* encodes a putative transcription factor of the YABBY gene family, which also includes FIL, INO (INNER NO OUTER), YAB2 (YABBY2), YAB3, and YAB5 [6,7]. The YABBY family is characterized in *Arabidopsis* and rice [2,4,6,8–14]. INO expression occurs only in the abaxial domain of the ovule integument [4]. The "vegetative YABBYs" (FIL, YAB2, and YAB5) in angiosperm are exclusively expressed in leaf-homologous organs, both vegetative and floral, they are involved in leaf development, such as the leaf margin establishment that guides laminar growth and leaflet initiation; maintenance of leaf polarity; and, activation of leaf maturation processes and repression of shoot apical meristem genes [7]. Petunia *PhCRC1/2* expressed in developing nectaries and carpels, similar to *Arabidopsis CRC* expression [15,16]. No nectary glands develop in *crc* mutants [17]. Locations of nectaries are highly variable in broader taxonomic terms [18], although their locations within flowers are constant at the family level. Nectaries tend to be associated with the perianth in basal angiosperms [19], while they are usually associated with carpels and stamens in the eudicots. Fahn argued that nectaries position within flowers trends to shift from peripheral perianth positions in basal taxa to central positions that are associated with reproductive organs in more derived taxa [20].

Wintersweet (*C. praecox*) is a unique traditional deciduous woody flower that is popularly used in floral arrangement, bonsai growing, and landscaping in many countries because of its unique flowering time and distinctive fragrance in deep winter. Wintersweet is a potential spice material because of its volatile aromatic substances and can be used in perfumery, cosmetics, aromatic tea, aromatherapy, and the food industry [21–23].

More than 30 floral scent volatiles have been detected in *C. praecox* flowers, consisting almost exclusively of volatile benzenoids and terpenoids (monoterpenes and sesquiterpenes) [24,25]. Terpenoids play a leading role among these volatiles. These compounds in wintersweet have minimal molecular-genetic characterization, except for the homologous genes of *CpFPPS* and *CpSAMT* [22,26].

Here, we reported the distribution characteristics of nectaries in *C. praecox* and the expression profiles of nectary development related gene *CpCRC*, the other four *YABBY* family genes (*CpFIL*, *CpYABBY2*, *CpYABBY5-1*, and *CpYABBY5-2*) and floral scent related genes to help understand floral scent origination and the molecular regulation of nectary development in wintersweet and other plants of scented flowers.

2. Results

2.1. Distribution Characteristics of Nectary on Petals of Different Stages

SEM analysis was performed to determine the distribution characteristics of nectary on petals in different stages of petals in different developmental stages of floral meristem (FB, DP, OF, LB, and WP), and receptacle, pistil, and stamen (Figure 1A,F,K,R,W). The nectaries were mainly located in the adaxial side of the innermost and middle petals (red asterisk shown, Figure 1B–E,G–J,L–O,S–V,X–Z2). The adaxial/abaxial side of the outer petals and the abaxial side of innermost/middle petals of different stages had no nectary distribution. Figure 1P,Q show the adaxial and abaxial sides of the outer and middle petals in the OF stages (in green box), respectively, which there is no nectary detected at all. The numbers of nectaries changed in the FB to WP during flower senescence. The numbers of nectaries were equal or slightly higher in the middle petal than that in the innermost petal under the same magnification (×400, Figure 1). Morphological difference in different stages of nectaries are shown in Figure 1. With the development of floral meristem, the length-width ratio of nectaries became small and evaginated (×1800, Figure 1). The substance of floral scent was found beside the nectaries (red arrow shown, Figure 1S,Z1,Z2). The cell size increased from FB and DP to OF with the development of floral meristem. This phenomenon is similar to that in LB and WP. Larger number of nectaries was observed in FB or DP than in OF at the same magnification (×400) because of the different cell sizes in different development stages (Figure 1B,D,G,I,L,N). The number of nectaries in OF was higher than that in LB or WP (Figure 1L,N,S,U,X,Z1), with similar cell size in three different development stages.

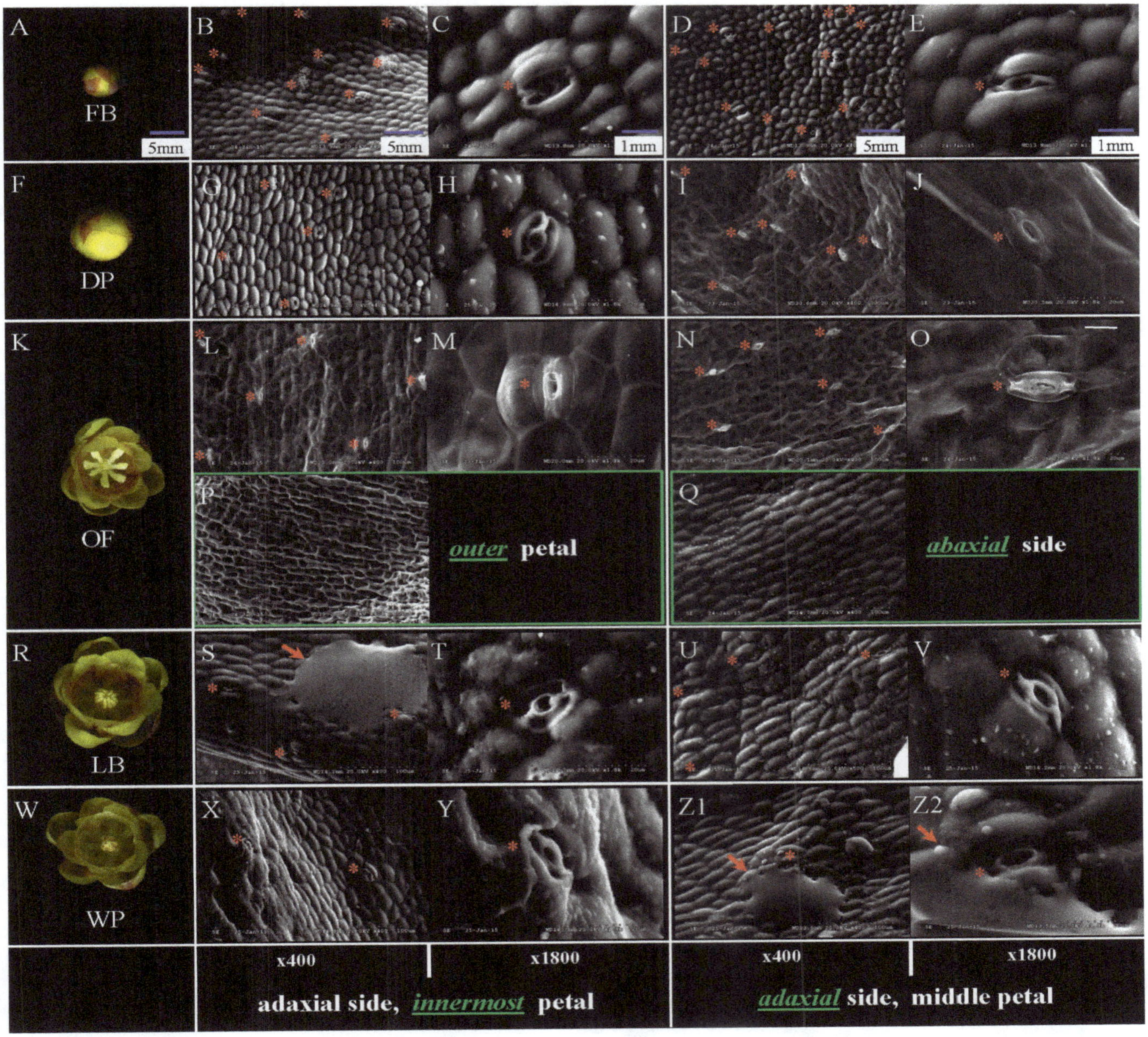

Figure 1. Morphology of the surface of petal glands on different stages in the *Chimonanthus praecox* by SEM. (**A,F,K,R,W**)floral meristem of five different developmental stages (FB, DP, OF, LB, and WP); (**B,C;G,H;L,M;S,T;X,Y**) nectaries on adaxial side of innermost petals in five different developmental stages. (**D,E;I,J;N,O;U,V;Z1,Z2**) nectaries on the adaxial side of middle petals in five different developmental stages. (**B,D,G,I,L,N,S,U,X,Z1**) surface of the glandular tissue, showing the nectary stomata (red asterisk). (**C,E,H,J,M,O,T,V,Y,Z2**) close-up of the nectary stomata. (**P,Q**) (in green box) no nectary distribution on adaxial side of outer petal and abaxial side of middle petals in OF stages. Scale bar is the same in (**A,F,K,R,W**); (**B,D,G,I,L,N,P,Q,S,U,X,Z1**); and (**C,E,H,J,M,O,T,V,Y,Z2**), respectively. Red arrow shows the substance of floral scent. The magnification was ×400 (left) and ×1800 (right), respectively.

No nectary was detected on the receptacle, perianth, stamen, and pistilin all five stages (FB, DP, OF, LB, and WP) (just show results in OF stage, Figure 2). Concentrated nectaries were found in the region near the axis of the lower part of petals, but almost none in the upper and edge of petals (Figure 3 just show the stages of FB and OF). Nectaries are not uniformly distributed in the petals (Figures 1B,D,G,I,L,N,S,U,X,Z1 and 3). These distribution characteristics of nectaries were perhaps related to the *YABBY* gene family, which controls the build of nectaries development and dorsiventral polarity.

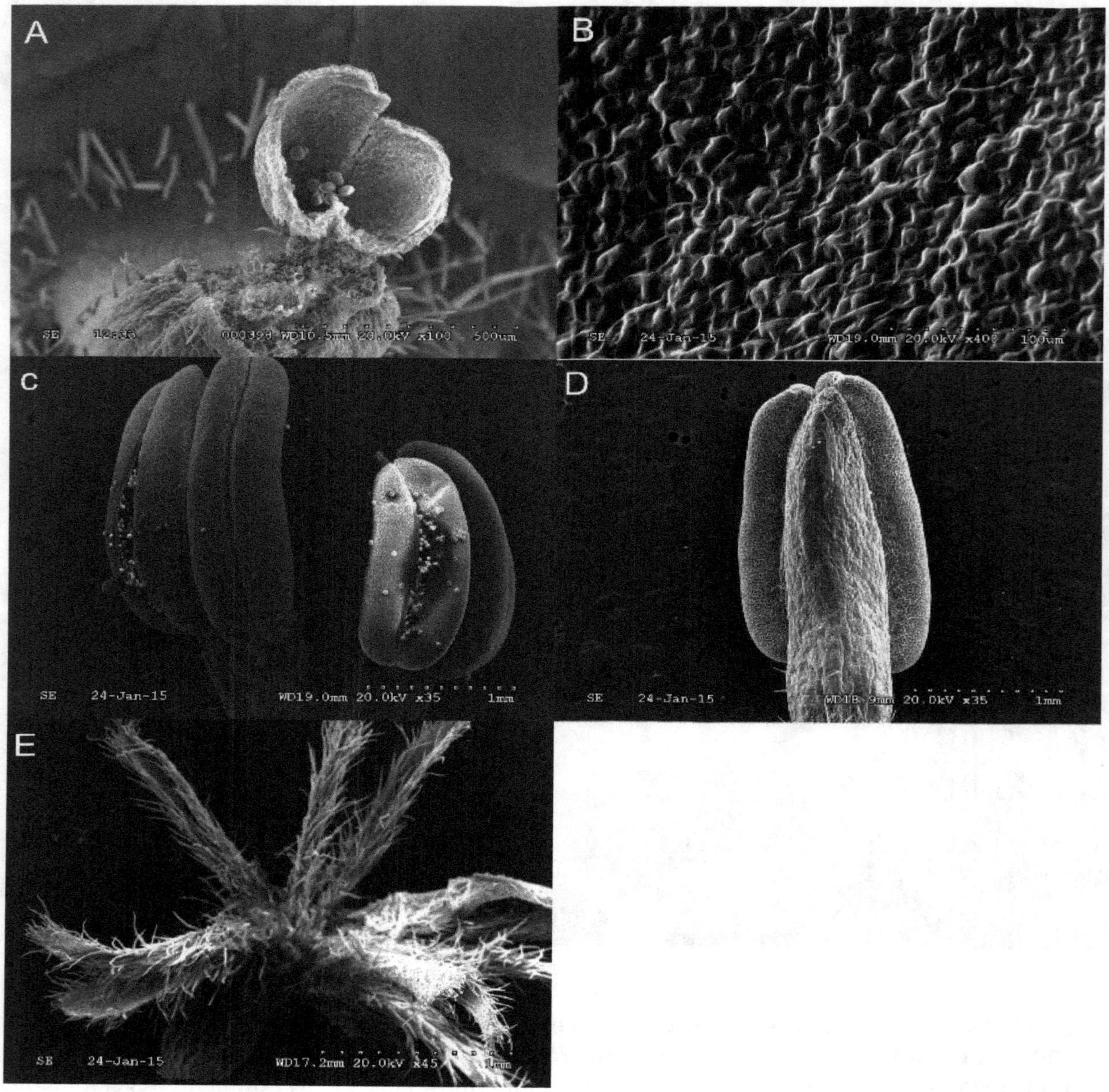

Figure 2. No nectaries distribution of the stamen and pistil in OF in the *Chimonanthus praecox* by SEM. (**A**), receptacle; (**B**), perianth; (**C**), front of stamen; (**D**), back of stamen; and, (**E**), pistil.

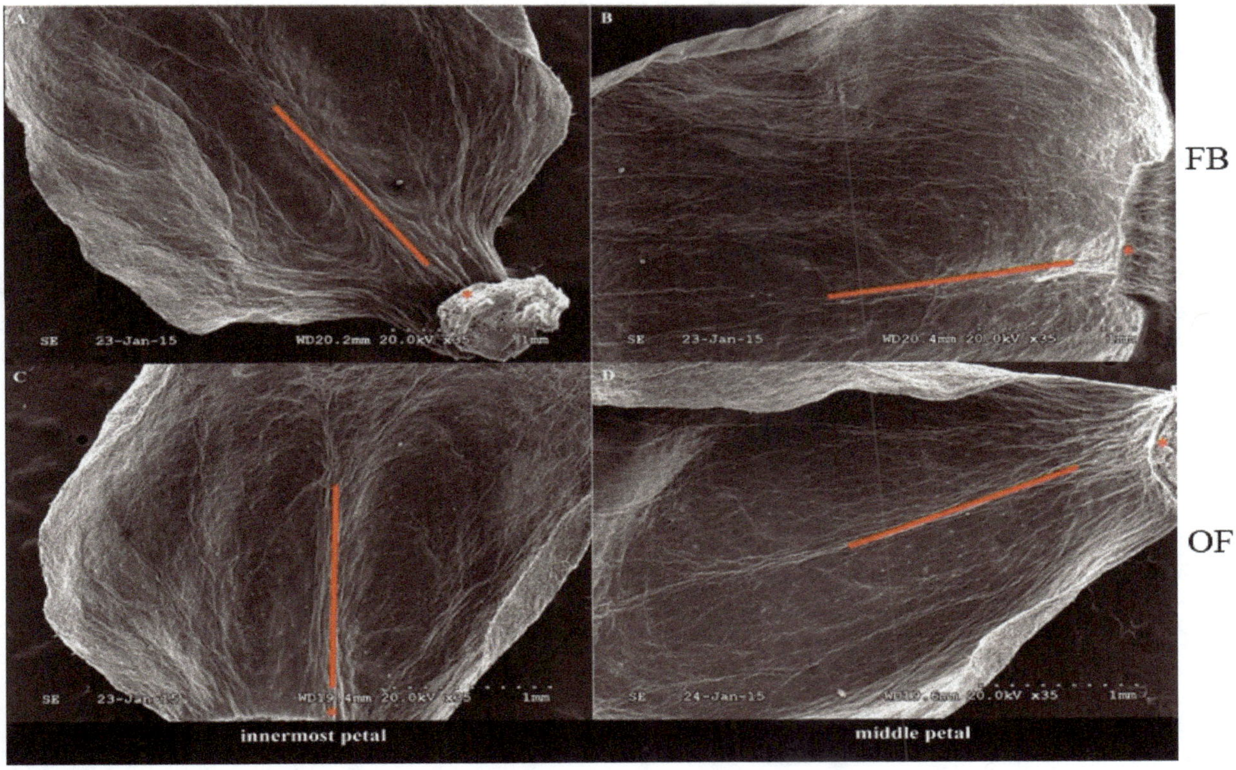

Figure 3. Distribution of the nectaries in FB and OF in the *Chimonanthus praecox* by SEM. Little white dot shown the nectaries concentrated near the axis of petals. Red line represented the axis of the petal. Red asterisks show the bottom of the petal.

2.2. Sequence Alignment and Phylogenetic Analysis

Partial or complete *CpFIL*, *CpCRC*, *CpYABBY2*, *CpYABBY5-1*, and *CpYABBY5-2* cDNAs contain open reading frame of 636, 519, 546, 555, and 552bp, respectively. The predicted CpFIL, CpCRC, CpYABBY2, CpYABBY5-1, and CpYABBY5-2 proteins of 212, 173, 182, 185, and 184 amino acid residues contains a zinc-finger domain in the N-terminus and a YABBY domain in the C-terminus (Figure 3A). The putative CpFIL protein shares 89% and 82% similarity with MgFIL from *Magnolia grandiflora* and Ny.coFIL from *Nymphaea colorata*. CpCRC protein shows 67% and 59% identity with the products of NnCRC-1/2 in *Nelumbo nucifera* and AtCRC in *Arabidopsis*. CpYABBY2 protein shares 72% and 57% similarity with NnYAB2 from *N. nucifera* and AtYAB2 from *Arabidopsis*, respectively. The putative CpYABBY5-1 protein shared high similarity (81%) with CpYABBY5-2, and they both showed 89% and 93% identity with the products of CsYAB5 in *Chloranthus serratus*, and 72% and 69% with that of AtYABBY5 in *Arabidopsis* (Figure 4A and Figure S1; Table S1).

The sequences of the five YABBYs protein homologues were aligned with the respective FIL, CRC, YABBY2, YABBY5-1, and YABBY5-2 proteins of the multiple angiosperm taxa for phylogenetic analysis. The CpYAB2 homologue clustered with non-core NnYAB2 of *Nelumbo nucifera*, DlYAB2 of *Dimocarpus longan*, Am.trYAB2 from *Amborella trichopoda* and core AtYAB2 from *A. thaliana*. The two YABBY5 sequences from *C. praecox*, CpYAB5-1 and CpYAB5-2, formed a sister group to the basal eudicot, CsYAB5 of *Chloranthus serratus*, formed a clade YABBY5 with AtYAB5 of *A. thaliana*. The CpFIL sequence aligned near to none-core basal eudicots, MgFIL of *Magnolia grandiflora*, NjFIL of *Nuphar japonica*, *Nymphaea colorata* Ny.coFIL and *A. trichopoda* Am.trFIL, and formed FIL clade. Together with *A. thaliana* AtCRC, the CpCRC, *A. trichopoda* Am.trCRC, *Oryza sativa* OsCRC, EsCRC of *Epimedium sagittatum*, EcCRC of *Eschscholzia californica*, *N. nucifera* NnCRC-1, and NnCRC-2 formed a clade of CRC (Figure 4B).

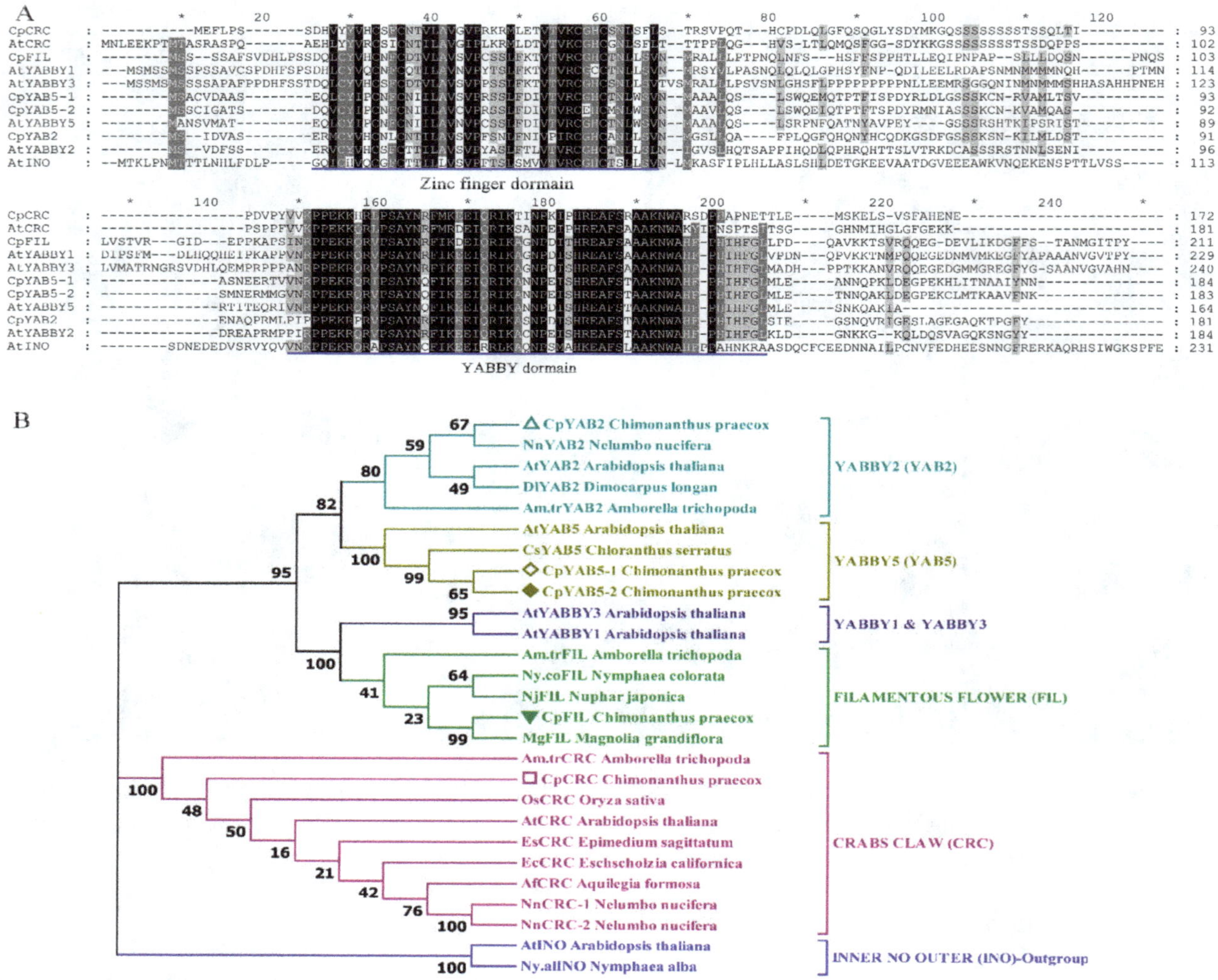

Figure 4. Sequence alignment and phylogenetic analysis of YABBY proteins. (**A**) Sequence alignment of YABBY proteins in *C. praecox* and *A. thaliana*. Conserved domains (Zinc finger domain and YABBY domain) are underlined in blue. Identical residues are highlighted in black and similar residues are highlighted in grey. Dotted line and asterisk represented the gap and the position of odd times of ten in protein sequence. (**B**) Phylogenetic analysis by Neighbor-joining (NJ) bootstrap analysis (1000 replications). AtINO (*A. thaliana*) and Ny.alINO (*Nymphaea alba*) as outgroup. The gene accession number is shown in Materials and Methods. Five different symbols in front of the protein name reprent the five YABBY protein of *Chimonanthus praecox* in this study.

2.3. Expression Analysis

In order to illustrate the correlation between nectary development and gene expression of *CpCRC* and the other four YABBY genes, also that of the nectary development and the floral scent, heatmap analysis with the RNA-Seq database in DP, OF, and WP stages [27] and qRT-PCR were conducted using cDNA derived from DP, OF, LB, and WP to determine the expression profile of five YABBY genes and 11 floral scent related genes in the flower buds of different developmental stages in *C. praecox* (Figure 1F,K,R,W). Almost no *CpDMAPP* was detected in DP, OF, LB, and WP. The relative expression of one nectary development related gene *CpCRC* and the other four YABBY family genes (*CpFIL*, *CpYABBY2*, *CpYABBY5-1*, and *CpYABBY5-2*) and seven floral scent genes (*CpSAMT*, *CpIPP*, *CpGPPS1*, *CpGPP*, *CpLIS*, *CpTER3*, and *CpTER5*) gradually increased in OF to LB and WP. The expression levels of *CpFIL*, *CpCRC*, *CpYABBY2*, *CpYABBY5-2*, and *CpTER3* were the highest, and those of *CpGPPS1* and *CpTER5* were the lowest in DP. The expression of *CpYABBY5-1*, *CpSAMT*, *CpIPP*, *CpGPP*, and *CpLIS* in DP was higher than that in OF but lower than that in WP. The highest and lowest expression levels of

CpGPPS2 and *CpMYR1* were in LB and OF, respectively. The expression of *CpFPPS* in DP was similar to that in LB, higher than that in OF, and lowest in WP (Figure 5).

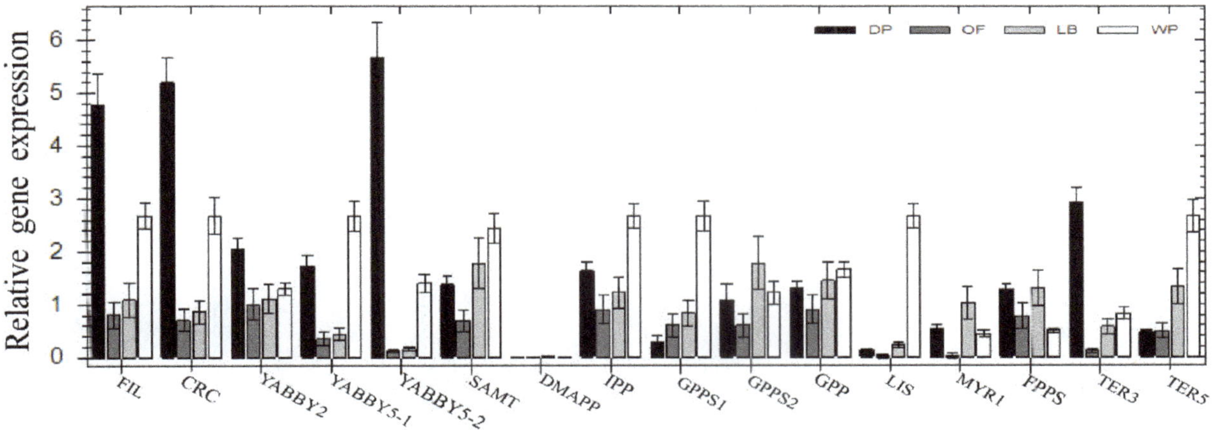

Figure 5. Quantitative real-time PCR analysis of different genes in four different developmental stages of DP, OF, LB, and WP. *Tublin* homologous gene of *Chimonanthus praecox* was used as an internal control.

According to the RNA-Seq database in DP, OF and WP stages, *CpFIL*, *CpCRC*, *CpYABBY2*, *CpYABBY5-1*, and *CpYABBY5-2* have similar expression pattern. The expression level in DP was higher than that in OF and WP and the lowest in WP (Figure 6).

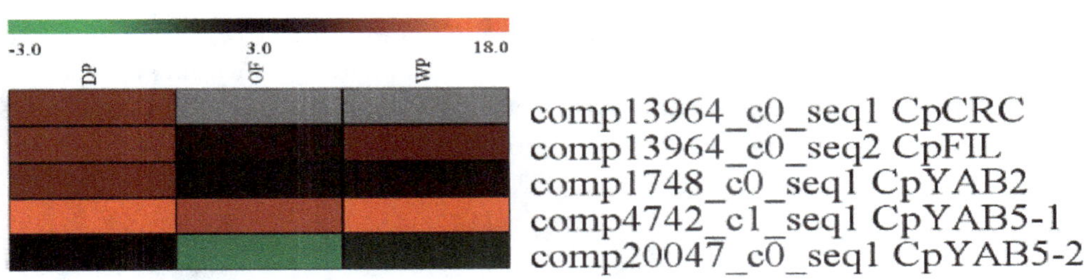

Figure 6. Heatmap analysis of 5 YABBY family gene expression in DP, OF, and WP stages.

To further clarify the correlation between the distribution characteristics of nectaries in three different round of petals and gene expression, the expression profile of five YABBY genes and five floral scent related genes in the innermost, middle, and outer petals in OF stages (Figure 1K) was detected using qRT-PCR. The expression level of 5 YABBY genes and five floral scent-related genes gradually decreased in the innermost to the middle and outer petals. The relative expression levels of *CpFIL*, *CpCRC*, *CpYABBY2*, *CpYABBY5-1*, and *CpYABBY5-2* in the innermost petals were 1.78- to 5.38-fold, and 3.24- to 12.57-fold higher than that in the middle and the outer petals, respectively. The relative expression levels of floral scent genes (*CpIPP*, *CpGPPS1*, *CpGPP*, *CpLIS*, and *CpTER5*) in the innermost petals were approximately 1.65- to 10.25-fold higher than those in the middle petals and 1.76- to 78.85-fold higher than those in the outer petals (Figure 7). They all have significant difference between the relative expression level of *CpFIL*, *CpCRC*, *CpYABBY5-1*, and *CpLIS* in middle petals and that of in outer petals (Figure 7).

Based on SEM results, for the sake of the relationship between the distridution characteristics of nectaries in the same round of petals and the gene expression, qRT-PCR was conducted using cDNA derived from the upper and lower halves in middle petals from DP, OF, and WP (Figure 1F,K,W) to further detect the expression profile of one nectary development related gene *CpCRC*, four YABBY family genes (*CpFIL*, *CpYABBY2*, *CpYABBY5-1/2*) and two floral scent genes (*CpIPP* and *CpGPPS1*). The relative expression of these five YABBY family genes and *CpIPP* had a similar expression pattern. The expression levels in the lower half of middle petals were higher than those in the upper halves in DP and OF stages, including the *CpYABBY5-1* and *CpYABBY5-2* in the WP stage and *CpGPPS1* in the

DP stage. However, the relative expression levels of *CpFIL*, *CpCRC*, *CpYABBY2*, *CpIPP*, and *CpGPPS1* in the lower half of middle petals were lower than those in the upper halves in WP stage and were similar to those of *CpGPPS1* in the OF stage (Figure 8).

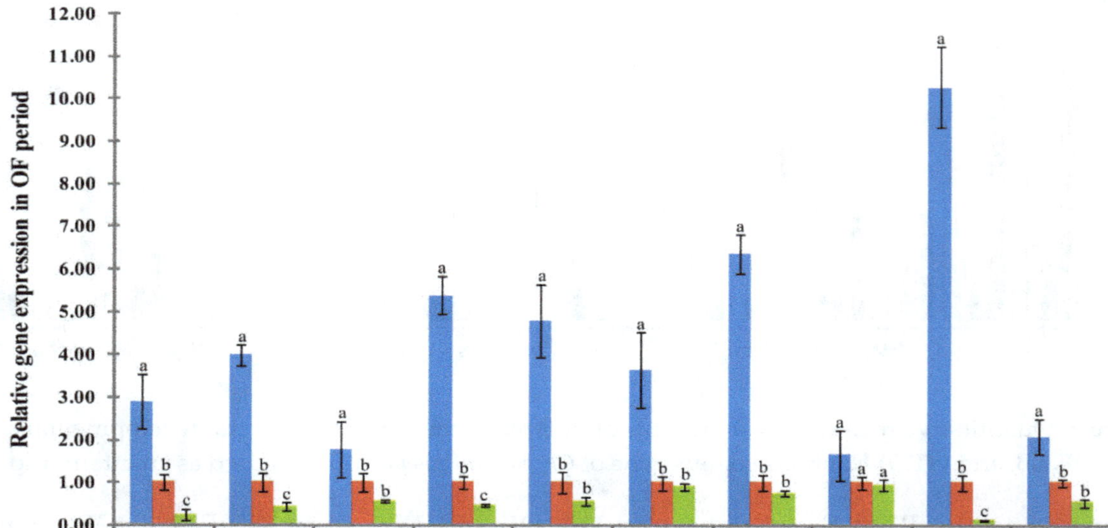

	FIL	CRC	YABBY2	YABBY5-1	YABBY5-2	IPP	GPPS1	GPP	LIS	TERS
inner.	2.89	3.98	1.78	5.38	4.76	3.64	6.35	1.65	10.25	2.07
mid.	1.00	1.00	1.00	1.00	1.00	1.00	1.00	1.00	1.00	1.00
out.	0.23	0.44	0.55	0.46	0.55	0.92	0.75	0.94	0.13	0.54

Figure 7. Quantitative real-time PCR analysis of different genes in three different rounds of petals in OF stages. Inner., mid., and out. represent innermost, middle, and outer petals, respectively. *Tublin* homologous gene of *C. praecox* was used as aninternal control. (Notes: *t*-test used for significant difference analysis; data is the means of relative expression; a, b, c show $p < 0.05$ significant level).

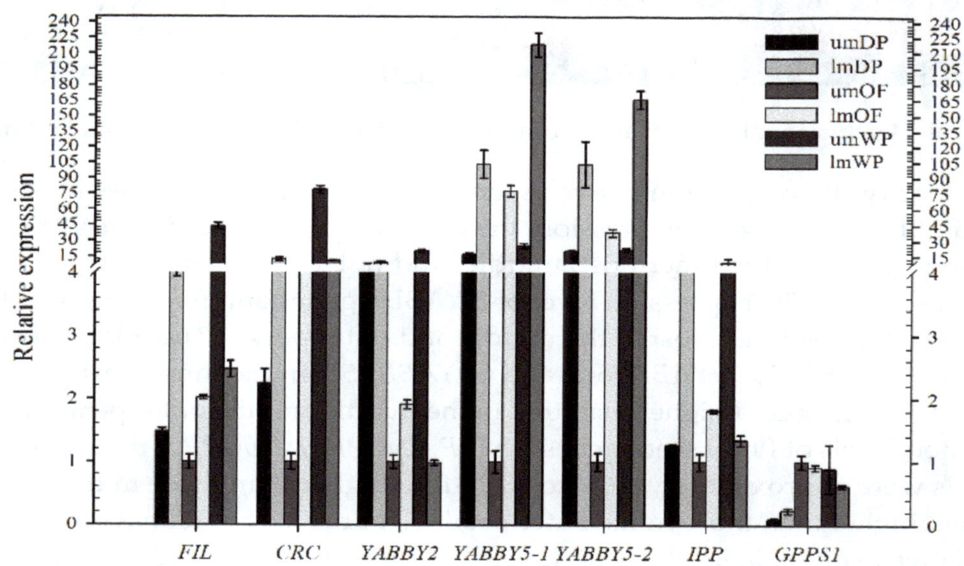

Figure 8. Quantitative real-time PCR analysis of different genes in the upper and lower halves of middle petals in DP, OF, and WP stages. *Tublin* homologous gene of *C. praecox* was used as an internal control. "um" and "lm" represent upper and lower half of middle petals, respectively.

3. Discussion

Results of SEM analysis in *C. praecox* show that nectaries were distributed on the adaxial side of the innermost and middle petals but not on the abaxial side. No nectary was detected in all five stages (FB, DP, OF, LB, and WP) on the outer petals, including in the receptacle, perianth, stamen, and pistil. The surface morphology of the innermost and middle glands of *C. praecox* is similar to that of the inner petal glands of *Alphonsea glandulosa* and *Petunia* [16,28]. The surface of the nectar glands is different from the surrounding epidermis, and nectar stomata are found across the surface of the glandular tissues [28]. The nectar stomata are raised slightly above the epidermis with an aperture for nectar secretion [29]. Although the locations of nectaries within flowers vary highly in terms of broader taxonomic terms, their locations are constant at the family level [30]. Nectaries are usually associated with carpels and stamens in eudicots, but are related to perianth in basal angiosperms [19]. *C. praecox* belongs to Calycanthaceae, Laurales, Magnoliids, and is clustered to Magnoliales, Piperales, and Canellales, which are close to Chloranthales, Austrobaileyales, Nymphaeales, and Amborellales [31]. In Magnoliidae, *C. praecox* has no nectary distribution on its receptacle, perianth, stamen, or pistil, but has some on the adaxial side of the innermost and middle petals; this finding partly supports that of a previous study, nectaries position within flowers trends to shift from peripheral perianth in basal taxa to central reproductive organs in more derived taxa [19,20].

Monoterpenes, such as myrcene, geraniol, linalool and sesquiterpene compounds, are the main aroma components of *C. praecox* [32,33]. Therefore, the concentration of universal precursor of monoterpene (GPP) and its substrate IPP can indirectly reflect the aroma production of *C. praecox*. qPCR analysis of *CpIPP* and *CpGPPS* gene in different parts of petals can indirectly determine the location of aroma substances. The expression levels of the nectary development related genes *CpCRC*, the other four YABBY family genes (*CpFIL*, *CpYABBY2*, and *CpYABBY5-1/2*) and five floral scent genes (*CpIPP*, *CpGPPS1*, *CpGPP*, *CpLIS*, and *CpTER5*) in the innermost petals of *C. praecox* were higher than those in the middle and outer petals, but they were the lowest in the outer petals. The *CpLIS* expression was increased seven-fold at the OF stage, which is responsible for α-linalool biosynthesis [27]; and α-linalool accounts for 36% of the total quantity of volatile compounds has been reported in wintersweet flowers [34]. The expression results were consistent with the characteristics of nectary distribution based on SEM analysis (Figures 1 and 7).

The expression pattern of five YABBY genes in *C. praecox* by qPCR was in accordance with the RNA-Seq in DP, OF, and WP stages (Figures 5 and 6). The expression levels of *CpIPP* and *CpGPPS* were significantly different in the different halves of the petals; those in the upper halves were significantly lower than those in the lower halves during the first two periods. This result is consistent with that of SEM analysis (Figure 1), which stated that the nectaries were mainly distributed in the lower half part of the petals near the axis and were rarely distributed on the edge and upper half part of the petals. Nectaries are not uniformly distributed in the petals, that is why the numbers of nectaries were equal or slightly higher in the middle petal than that in the innermost ($\times$400, Figure 1). The expression characteristics of *CpFIL*, *CpCRC*, *CpYABBY2*, and *CpYABBY5-1/2* were generally similar to those of *CpIPP* and *CpGPPS*.

At least one YABBY gene family member *CpCRC* was expressed in all asymmetric above-ground organs in a polarity, suggesting that this gene is involved in establishing dorsiventral polarity in all of these organs. The YABBY gene family controls the build of dorsiventral (abaxial/adaxial) polarity [4,10,35]. Therefore, we proposed that the floral scent mainly originates from the nectaries that are distributed neither on the abaxial side of the innermost and middle petals nor on the outer petals, but on the adaxial side of the innermost and middle petals. This unbalanced distribution of the nectaries is caused by dorsiventrality differentiation, one of the most important polarities in the development of lateral organs in plants.

4. Materials and Methods

4.1. Plant Material

C. praecox plants of 21-years old were grown in the campus of Southwestern University (106°43′ E, 29°83′ N, Beibei District, Chongqing City, China) under natural photoperiod. Flower development was divided into the following five stages: FB is the stage wherein the flower bud is closed, and the petals are yellow; DP wherein the petals unroll; OF wherein the petals reach full opening, and the stamens bent toward the adaxial side of innermost petals and away from the pistils at a right angle; LB that occurs after two days of OF, where the stamens commence to move to enclose the pistils; and WP wherein the flower is pollinated, and the petals and stamens start to wilt. Floral tissue samples, such as receptacle, sepals, petals, stamens, and pistils were obtained from five different stages. Some of the petals were divided into upper and lower halves. All plant materials were harvested then fixed with FAA buffer or frozen in liquid nitrogen and stored at −80 °C for RNA extraction.

4.2. Scanning Electron Microscope (SEM)

Fresh petals of *C. praecox* were soaked for an hour in pre-cold 2% glutaraldehyde solution and then were rinsed three to four times with 0.1 M phosphate buffer (pH 7.2) for 1 h. The buffer was discarded, and ethanol dehydration was conducted in a step-by-step gradient. Ethanol concentrations were 30%, 50%, 70%, 80%, 90%, and 100% for 25 min each. The alcohol was washed, and 1:1 mixture of isoamyl acetate to ethanol was added. Then, the solution was added with pure isoamyl acetate, soaked for 10–20 min for each step, stirred properly, and air dried before the electron microscope observation.

4.3. Sequence Alignment and Phylogenetic Analysis

The sequences of one nectary development related gene *CRC*, four other YABBY family genes (*FIL, YABBY2, YABBY5-1/2*) and 11 floral scent genes in *C. praecox* were selected from the Illumina deep sequencing [27]. Blastn of these genes were obtained and named as *CpFIL, CpCRC, CpYABBY2, CpYABBY5-1, CpYABBY5-2, CpSAMT, CpDMAPP, CpIPP, CpGPPS1, CpGPPS2, CpGPP, CpLIS, CpMYR1, CpFPPS, CpTER3*, and *CpTER5.* The sequences included in the analysis were downloaded from the NCBI GenBank (http://www.ncbi.nlm.nih.gov). The amino acid sequences of the YABBY family were aligned using ClustalX 1.83 [36]. Neighbor-joining (NJ) bootstrap analysis (1000 replications) with Poisson correction for the amino acids was performed using MEGA 4 [37]. Sequence data for analysis can be found in the GenBank/EMBL databases under the following accession numbers: *CpFIL, CpCRC, CpYABBY2, CpYABBY5-1* and *CpYABBY5-2* from *C. praecox*; *AtINO, AtCRC, AtYABBY1, AtYABBY2, AtYABBY3*, and *AtYABBY5* (AAF23754, NP_177078, NP_566037, AF136539, AF136540, NM_179749) from *A. thaliana*; *Am.trCRC, Am.trFIL*, and *Am.trYAB2* (AJ877257, AB168113, AB126654) from *Amborella trichopoda*; *AfCRC* (AY854797) from *Aquilegia formosa*; *CsYAB5* (BAF65259) from *Chloranthus serratus*; *DlYAB2* (ACN59438) from *Dimocarpus longan*; *EsCRC* (GH62810) from *Epimedium sagittatum*; *EcCRC* (CAQ17052) from *Eschscholzia californica*; *NnYAB2, NnCRC-1*, and *NnCRC-2* (XP_010247861, XM_010259669, XM_010259670) from *Nelumbo nucifera*; *NjFIL* (BAD83708) from *Nuphar japonica*; *Ny.alINO* (AB092980) from *Nymphaea alba*; *Ny.coFIL* (BAF65258) from *Nymphaea colorata*; *MgFIL* (BAF65261) from *Magnolia grandiflora*; and *OsCRC* (AAR84663) from *Oryza sativa*.

4.4. Gene Expression Analysis

Tissues sampled for gene expression analysis include flower buds of four different developmental stages (DP, OF, LB, and WP), three different rounds (innermost, middle, and outer) of petals in OF stages, the upper and lower halves of middle petals in DP, OF, and WP stages. Total RNA for the expression analysis was extracted using RNAprep pure kit (Tiangen, Beijing, China) according to the manufacturer's instructions. Exactly 3 µg of RQ1 RNase-Free DNase (Promega, Madison, WI, USA) pre-treated total RNA was reverse transcribed according to the instructions of the Primescript RT reagent kit (Takara, Tokyo, Japan). qRT-PCR was performed to determine the expression pattern of one nectary development related gene*CpCRC*, four YABBY family genes (*CpFIL, CpYABBY2, CpYABBY5-1*, and *CpYABBY5-2*) and 11 floral scent genes, such as *CpSAMT, CpDMAPP, CpIPP, CpGPPS1, CpGPPS2, CpGPP, CpLIS, CpMYR1, CpFPPS, CpTER3*, and *CpTER5*.

The primers for qRT-PCR are listed in Table 1. Reactions were performed with the Sso Fast Eva Green Supermix (Bio-Rad, Hercules, CA, USA) and analyzed using Bio-Rad CFX96 (Bio-Rad CFX Manager Software Version 1.6). Thermocycler conditions were 95 °C for 30 s, followed by 40 cycles of

95 °C for 5 s and 60 °C for 5 s. qRT-PCR products were amplified using 5 µL 2× Sso Fast Eva Green Supermix, 0.5 µL RT reaction mixture, 0.5 µL of forward and reverse primer (10 µmol/µL) each, and RNase Free dH_2O to a final volume of 10 µL. Relative amounts of transcripts were calculated using the comparative CT method ($2^{-\Delta\Delta Ct}$), and the values were normalized. The house-keeping gene *CpTublin* of *C. praecox* was used as internal control. Data are shown as mean values ± standard deviation (SD) from three replicates for each sample. Significant difference was carried out by *t*-test ($p < 0.05$).

The expression patterns of the five YABBY genes were estimated by FPKM values and were visualized using MultiExperiment Viewer (Broad Institute of MIT and Harvard University, Boston, MA, USA [38]).

Table 1. Primer for real-time PCR.

Gene Name	Forward Primer Sequence	Reverse Primer Sequence
Actin	AGGCTAAGATTCAAGACAAGG	TTGGTCGCAGCTGATTGCTG
CpFIL	AATCCCGACATAACCCACAGAGAG	TCCTGTTGGCGCACGCTAGTT
CpCRC	CCTCCCGTCACCTTACAAACTACAG	CTGCTACAAGGAACACTGACCGC
CpYABBY2	CCATTGTCAAGATAAAGGTAGCGATT	CTGGTGGTGGTATAGGTAGCATTCG
CpYABBY5-1	TCTCCCTCTCTATTTATCCTCGTTT	GTAAAAGGCTAAAGCAGGATCATG
CpYABBY5-2	TTTTGAACACTGGAAACTTCGTCTT	GATGCAGCTCGACATCTCACTATCT
SAMT	ACCATTTTCACATCATTGCCAGAC	CTTCCTCTTTTACCATCAAGTGCTG
DMAPP	ATCGGAGAAGAAAGTGAGCGAGAGT	GCCGTGTATCGAAGCAGCAGT
IPP	CAGACCATCTCTTTCTCCCACTTTC	GGTCGGAGAGAAGGTGGTAGAGGTA
GPPS1	GTTAGCCAACTTTCCATACCATTTC	GAGTGACAACATCATCAAAGAAGGG
GPPS2	ATGAAGATGATTAGATTTCGAGTCCAAG	ATAACCAATTTACAACCCCTGACCC
GPP	TCTACAGAAAATGGGAGAAAACGAT	TATCTGTTTCTGTCACCAAATCCAC
LIS	GGCCAAAGTTAATGAAGTGAGATCC	CGTATATGCCATCGTTGCTGCC
MYR1	TTTCACAAAAATTGCCTTCAACCTT	CAAGGTGATGGAGAACTAAAACAAAAC
FPPS	TCTTTGTCCAGTTCTTCCAGCGTT	ATCAGTGAAATCAAAGGCGGAATCT
TER3	AGAGTTGAATTGCACAGGGTGATAG	GCAGTGGATGTTGTTGATCAGCTC
TER5	CTCTCCCTCAGTCTCTTCTCCCTTT	ATCTCCATGCAACATTGGCTACAG

Author Contributions: Methodology, Z.L. and Y.J.; Software, Z.L.; Validation, Z.L. and J.L.; Formal Analysis, Y.J. and Z.L.; Resources, D.L.; Data Curation, J.M.; Writing-Original Draft Preparation, Y.J. and Z.L.; Writing-Review & Editing, Z.L.; Visualization, Z.L. and Y.J.; Supervision, Z.L.; Project Administration, S.S. and M.L.; Funding Acquisition, Z.L., S.S. and J.M.

Abbreviations

SEM	scanning electron microscope
qRT-PCR	quantitative Reverse Transcript-Polymerase Chain Reaction
CRC	*CRABS CLAW*
FIL	*FILAMENTOUS FLOWER*
INO	*INNER NO OUTER*
YAB	*YABBY*
SAMT	*S*-adenosyl-*L*-methionine: salicylic acid carboxyl methyltransferase
DMAPP	dimethylallyl pyrophosphate
IPP	isopentenyl pyrophosphate
GPPS	geranyl diphosphate synthase
GPP	geranyl pyrophosphate
LIS	*S*-linalool synthase
MYR1	myrcenesynthase
FPPS	farnesyl pyrophosphonate synthase
TER	α-terpineol synthase

References

1. Schmid, R. Reproductive versus extra-reproductive nectaries-historical perspective and terminological recommendations. *Bot. Rev.* **1988**, *54*, 179–227. [CrossRef]

2. Lee, J.Y.; Baum, S.F.; Alvarez, J.; Patel, A.; Chitwood, D.H.; Bowman, J.L. Activation of CRABS CLAW in the Nectaries and Carpels of *Arabidopsis*. *Plant Cell* **2005**, *17*, 25–36. [CrossRef] [PubMed]
3. Baum, S.F.; Eshed, Y.; Bowman, J.L. The *Arabidopsis* nectary is an ABC-independent floral structure. *Development* **2001**, *128*, 4657–4667. [PubMed]
4. Bowman, J.L. The *YABBY* gene family and abaxial cell fate. *Curr. Opin. Plant Biol.* **2000**, *3*, 17–22. [CrossRef]
5. Gross, T.; Broholm, S.; Becker, A. *CRABS CLAW* Acts as a Bifunctional Transcription Factor in Flower Development. *Front. Plant Sci.* **2018**, *9*, 835. [CrossRef] [PubMed]
6. Goldshmidt, A.; Alvarez, J.P.; Bowman, J.L.; Eshed, Y. Signals derived from *YABBY* gene activities in organ primordia regulate growth and partitioning of *Arabidopsis* shoot apical meristems. *Plant Cell* **2008**, *20*, 1217–1230. [CrossRef] [PubMed]
7. Bartholmes, C.; Hidalgo, O.; Gleissberg, S. Evolution of the *YABBY* gene family with emphasis on the basal eudicot *Eschscholzia californica* (Papaveraceae). *Plant Biol.* **2012**, *14*, 11–23. [CrossRef] [PubMed]
8. Siegfried, K.R.; Eshed, Y.; Baum, S.F.; Otsuga, D.; Drews, G.N.; Bowman, J.L. Members of the *YABBY* gene family specify abaxial cell fate in *Arabidopsis*. *Development* **1999**, *126*, 4117–4128. [PubMed]
9. Eshed, Y.; Baum, S.F.; Bowman, J.L. Distinct mechanisms promote polarity establishment in carpels of *Arabidopsis*. *Cell* **1999**, *99*, 199–209. [CrossRef]
10. Yamaguchi, T.; Nagasawa, N.; Kawasaki, S.; Matsuoka, M.; Nagato, Y.; Hirano, H.Y. The *YABBY* gene *DROOPING LEAF* regulates carpel specification and midrib development in *Oryza sativa*. *Plant Cell* **2004**, *16*, 500–509. [CrossRef] [PubMed]
11. Liu, H.L.; Xu, Y.Y.; Xu, Z.H.; Chong, K. A rice *YABBY* gene, *OsYABBY4*, preferentially expresses in developing vascular tissue. *Dev. Genes Evol.* **2007**, *217*, 629–637. [CrossRef] [PubMed]
12. Ohmori, Y.; Abiko, M.; Horibata, A.; Hirano, H.Y. A transposon, Ping, is integrated into intron 4 of the *DROOPING LEAF* gene of rice, weakly reducing its expression and causing a mild drooping leaf phenotype. *Plant Cell Physiol.* **2008**, *49*, 1176–1184. [CrossRef] [PubMed]
13. Ohmori, Y.; Toriba, T.; Nakamura, H.; Ichikawa, H.; Hirano, H.Y. Temporal and spatial regulation of *DROOPING LEAF* gene expression that promotes midrib formation in rice. *Plant J.* **2011**, *65*, 77–86. [CrossRef] [PubMed]
14. Tanaka, W.; Toriba, T.; Ohmori, Y.; Yoshida, A.; Kawai, A.; Mayama-Tsuchida, T.; Ichikawa, H.; Mitsuda, N.; Ohme-Takagi, M.; Hirano, H.Y. The *YABBY* Gene *TONGARI-BOUSHI1* Is Involved in Lateral Organ Development and Maintenance of Meristem Organization in the Rice Spikelet. *Plant Cell* **2012**, *24*, 80–95. [CrossRef] [PubMed]
15. Lee, J.Y.; Baum, S.F.; Oh, S.H.; Jiang, C.Z.; Chen, J.C.; Bowman, J.L. Recruitment of *CRABS CLAW* to promote nectary development within the eudicot clade. *Development* **2005**, *132*, 5021–5032. [CrossRef] [PubMed]
16. Morel, P.; Heijmans, K.; Ament, K.; Chopy, M.; Trehin, C.; Chambrier, P.; Rodrigues Bento, S.; Bimbo, A.; Vandenbussche, M. The Floral C-Lineage Genes Trigger Nectary Development in *Petunia* and *Arabidopsis*. *Plant Cell* **2018**, *30*, 2020–2037. [CrossRef] [PubMed]
17. Bowman, J.L.; Smyth, D.R. *CRABS CLAW*, a gene that regulates carpel and nectary development in *Arabidopsis*, encodes a novel protein with zinc finger and helix-loop-helix domains. *Development* **1999**, *126*, 2387–2396. [PubMed]
18. Brown, W.H. The bearing of nectaries on the phylogeny of flowering plants. *Proc. Am. Phil. Soc.* **1938**, *79*, 549–595.
19. Endress, P.K. The Flowers in Extant Basal Angiosperms and Inferences on Ancestral Flowers. *Int. J. Plant Sci.* **2001**, *162*, 1111–1140. [CrossRef]
20. Fahn, A. The topography of the nectary in the flower and its phylogenetic trend. *Phytomorphology* **1953**, *3*, 424–426.
21. Kozomara, B.; Vinterhalter, B.; Radojevic, L.; Vinterhalter, D. In vitro propagation of *Chimonanthus praecox* (L.), a winter flowering ornamental shrub. *In Vitro Cell Dev. Biol. Plant* **2008**, *44*, 142–147. [CrossRef]
22. Xiang, L.; Zhao, K.; Chen, L. Molecular cloning and expression of *Chimonanthus praecox* farnesyl pyrophosphate synthase gene and its possible involvement in the biosynthesis of floral volatile sesquiterpenoids. *Plant Physiol. Biochem.* **2010**, *48*, 845–850. [CrossRef] [PubMed]
23. Zhao, K.G.; Zhou, M.Q.; Chen, L.Q.; Zhang, D.L.; Robert, G.W. Genetic diversity and discrimination of *Chimonanthus praecox* (L.) link germplasm using ISSR and RAPD markers. *Hortscience* **2007**, *42*, 1144–1148.

24. Vainstein, A.; Lewinsohn, E.; Pichersky, E.; Weiss, D. Floral Fragrance. New Inroads into an Old Commodity. *Plant Physiol.* **2001**, *127*, 1383–1389. [CrossRef] [PubMed]
25. Kram, B.W.; Carter, C.J. *Arabidopsis thaliana* as a model for functional nectary analysis. *Sex. Plant Reprod.* **2009**, *22*, 235–246. [CrossRef] [PubMed]
26. Ma, L.; Li, H.F.; Peng, C.C.; Chen, Z.Z.; Long, Z.F. Cloning of *SAMT* gene cDNA from *Chimonanthus praecox* and its expression in *Escherichia coli*. *Agric. Sci. Technol.* **2012**, *13*, 82–87.
27. Liu, D.; Sui, S.; Ma, J.; Li, Z.; Guo, Y.; Luo, D.; Yang, J.; Li, M. Transcriptomic analysis of flower development in wintersweet (*Chimonanthus praecox*). *PLoS ONE* **2014**, *9*, e86976. [CrossRef] [PubMed]
28. Xue, B.; Shao, Y.Y.; Saunders, R.M.; Tan, Y.H. *Alphonsea glandulosa* (Annonaceae), a New Species from Yunnan, China. *PLoS ONE* **2017**, *12*, e0170107. [CrossRef] [PubMed]
29. Nepi, M. Nectary structure and ultrastructure. In *Nectaries and Nectar*; Nicolson, S.W., Nepi, M., Pacini, E., Eds.; Springer: Dordrecht, The Netherlands, 2007; pp. 129–166.
30. Elias, T.S. Extrafloral nectaries: Their structure and distribution. In *The Biology of Nectaries*; Bentley, B., Elias, T.S., Eds.; Columbia University Press: New York, NY, USA, 1983; pp. 174–203.
31. Bremer, B.; Bremer, K.; Chase, M.W.; Fay, M.F.; Reveal, J.L.; Soltis, D.E.; Soltis, P.S.; Stevens, P.F.; Anderberg, A.A.; Moore, M.J.; et al. An update of the Angiosperm Phylogeny Group classification for the orders and families of flowering plants: APG III. *Bot. J. Linn. Soc.* **2009**, *161*, 105–121.
32. Deng, C.; Song, G.; Hu, Y. Rapid determination of volatile compounds emitted from *Chimonanthus praecox* flowers by HS-SPME-GC-MS. *Z. Naturforsch. C* **2004**, *59*, 636–640. [CrossRef] [PubMed]
33. Miller, E.R.; Taylor, G.W.; Eskew, M.H. The volatile oil of calycanthus floridus. *J. Am. Chem. Soc.* **1914**, *36*, 2182–2185. [CrossRef]
34. Azuma, H.T.M.; Asakawa, Y. Floral Scent Chemistry and Stamen Movement of *Chimonanthus praecox* (L.) Link (Calycanthaceae). *Acta Phytotaxon. Geobot.* **2005**, *56*, 197–201.
35. Juarez, M.T.; Twigg, R.W.; Timmermans, M.C. Specification of adaxial cell fate during maize leaf development. *Development* **2004**, *131*, 4533–4544. [CrossRef] [PubMed]
36. Thompson, J.D.; Gibson, T.J.; Plewniak, F.; Jeanmougin, F.; Higgins, D.G. The CLUSTAL_X windows interface: Flexible strategies for multiple sequence alignment aided by quality analysis tools. *Nucleic Acids Res.* **1997**, *25*, 4876–4882. [CrossRef] [PubMed]
37. Tamura, K.; Dudley, J.; Nei, M.; Kumar, S. MEGA4: Molecular Evolutionary Genetics Analysis (MEGA) software version 4.0. *Mol. Biol. Evol.* **2007**, *24*, 1596–1599. [CrossRef] [PubMed]
38. Yeung, K.Y.; Fraley, C.; Murua, A.; Raftery, A.E.; Ruzzo, W.L. Model-based clustering and data transformations for gene expression data. *Bioinformatics* **2001**, *17*, 977–987. [CrossRef] [PubMed]

4

Expression Characteristics and Functional Analysis of the *ScWRKY3* Gene from Sugarcane

Ling Wang [1], Feng Liu [1], Xu Zhang [1], Wenju Wang [1], Tingting Sun [1], Yufeng Chen [1], Mingjian Dai [1], Shengxiao Yu [1], Liping Xu [1,2], Yachun Su [1,2,*] and Youxiong Que [1,2,*]

1 Key Laboratory of Sugarcane Biology and Genetic Breeding, Ministry of Agriculture, Fujian Agriculture and Forestry University, Fuzhou 350002, China; lingw2017@126.com (L.W.); 18359162091@163.com (F.L.); zahngxuqq7@126.com (X.Z.); wwj1470665850@163.com (W.W.); sunting3221@163.com (T.S.); CYF9410@163.com (Y.C.); 18906071567@163.com (M.D.); dbzq666666@163.com (S.Y.); xlpmail@126.com (L.X.)

2 Key Laboratory of Ministry of Education for Genetics, Breeding and Multiple Utilization of Crops, College of Crop Science, Fujian Agriculture and Forestry University, Fuzhou 350002, China

* Correspondence: syc2009mail@163.com (Y.S.); queyouxiong@126.com (Y.Q.)

Abstract: The plant-specific WRKY transcriptional regulatory factors have been proven to play vital roles in plant growth, development, and responses to biotic and abiotic stresses. However, there are few studies on the *WRKY* gene family in sugarcane (*Saccharum* spp.). In the present study, the characterization of a new subgroup, IIc WRKY protein ScWRKY3, from a *Saccharum* hybrid cultivar is reported. The ScWRKY3 protein was localized in the nucleus of *Nicotiana benthamiana* leaves and showed no transcriptional activation activity and no toxic effects on the yeast strain Y2HGold. An interaction between ScWRKY3 and a reported sugarcane protein ScWRKY4, was confirmed in the nucleus. The *ScWRKY3* gene had the highest expression level in sugarcane stem pith. The transcript of *ScWRKY3* was stable in the smut-resistant *Saccharum* hybrid cultivar Yacheng05-179, while it was down-regulated in the smut-susceptible *Saccharum* hybrid cultivar ROC22 during inoculation with the smut pathogen (*Sporisorium scitamineum*) at 0–72 h. *ScWRKY3* was remarkably up-regulated by sodium chloride (NaCl), polyethylene glycol (PEG), and plant hormone abscisic acid (ABA), but it was down-regulated by salicylic acid (SA) and methyl jasmonate (MeJA). Moreover, transient overexpression of the *ScWRKY3* gene in *N. benthamiana* indicated a negative regulation during challenges with the fungal pathogen *Fusarium solani* var. *coeruleum* or the bacterial pathogen *Ralstonia solanacearum* in *N. benthamiana*. The findings of the present study should accelerate future research on the identification and functional characterization of the WRKY family in sugarcane.

Keywords: sugarcane; WRKY; subcellular localization; gene expression pattern; protein-protein interaction; transient overexpression

---

1. Introduction

Plant growth and development are vulnerable to several external environmental challenges, such as drought, high salinity, cold, and pathogens. There are complex metabolic regulation mechanisms in plants which enhance their resistance to a wide range of stresses through physiological changes largely controlled at the molecular level [1]. The transcription factors (TFs) in plant cells interact with specific DNA sequences in target gene promoters to activate or inhibit transcription and expression of target genes, thereby regulating the expression of these genes. This modulation causes adaptation to the effects and damage from various stresses [2].

As one of the largest plant-specific families of TFs, WRKY has been proven to be widely implicated in responses to biotic and abiotic stresses [3,4]. WRKY TFs are also a vital part of the signaling pathway network of plants, regulating physiological and biochemical processes [5]. The WRKY proteins were named based on a DNA-binding WRKY domain, which contains approximately 60 amino acid residues. This domain features a WRKYGQK sequence at its N-terminal end together with a $Cx_{45}Cx_{22\text{-}23}HxH$ (C_2H_2-type) or $Cx_7Cx_{23}HxC$ (C_2HC-type) zinc finger-like motif at the C-terminal [1,5,6]. Although the DNA-binding domain is highly conserved, the overall structure of the WRKY proteins is highly diverse and can be divided into three groups (I, II, and III) including five subgroups (IIa-IIe) in group II. These groupings are categorized according to the number of WRKY domains and are also based on features of the zinc finger-like motif [6,7]. Previous studies have indicated that WRKYs with similar roles usually have related functions. For example, WRKYs in groups I and III are involved in epidermal development, senescence, and abiotic stress, while WRKYs in group II are related to low phosphorus stress, disease resistance, secondary root formation, and abiotic stress, with a few exceptions observed [8].

Currently, *WRKY* genes have been identified in various plant species. There is a total of 72 *WRKYs* in the model dicot *Arabidopsis thaliana* [5]. In monocots, there are 103 *WRKYs* in *Oryza sativa* [9], 116 *WRKYs* in *Zea mays* [10], 105 *WRKYs* in *Setaria italica* [11], 68 *WRKYs* in *Sorghum bicolor* [12], and 45 *WRKYs* in *Hordeum vulgare* [13]. It has been reported that 30 *WRKY* genes in *A. thaliana* were responsive to salt stress [14]. Also, 58 *WRKY* genes in *Z. mays* and 19 *WRKY* genes in *Phaseolus vulgaris* were related to drought stress [15,16]. Qiu et al. [17] indicated that ten of 13 candidate *WRKY* genes in rice can respond to sodium chloride (NaCl), polyethylene glycol (PEG), low temperature, or high temperature stress. Wu et al. [18] showed that eight of the 15 candidate *WRKY* genes in wheat responded to low temperature, NaCl, or PEG stress. Previous studies showed that 49 *A. thaliana* *WRKY* genes were induced by *Pseudomonas syringae* or salicylic acid (SA) [19]. Fifteen *WRKYs* were induced by *Magnaporthe grisea*, and 12 of them were simultaneously induced by *Xanthomonas oryzae* pv. *oryzae* [9]. Several other reports also demonstrated that *WRKYs* can positively or negatively regulate the responses of plants to external biotic or abiotic stresses [20–22]. Moreover, numerous studies have reported that *WRKYs* are widely involved in a complicated signal transduction network, which may work together with upstream or downstream components, or may interact with other WRKY proteins during physiological processes or in response to various biotic and abiotic stimuli [1,23,24]. These reports provide the foundation for studying the tolerance mechanism of plant *WRKY* genes to environmental stress.

Sugarcane (*Saccharum* spp.) is not only the foremost sugar-producing crop, but also the one that has potential as a bioenergy resource [25]. The study on signal transduction in sugarcane growth, development, and its responses to the external environment, especially the functional analysis of TFs, is of great significance for sugarcane molecular breeding. As reported, there were 26 WRKY-like proteins discovered in a publicly available sugarcane expressed sequence tag (EST) database via an *in silico* study, and their phylogenetic relationships were determined [26]. Beyond that, only two other sugarcane group IIc WRKY proteins, Sc-WRKY (GenBank Accession No. GQ246458.1) [27] and ScWRKY4 (GenBank Accession No. MG852087.1) [28], have been isolated from *Saccharum* hybrid cultivar FN22 and *Saccharum* hybrid cultivar ROC22 respectively and characterized by molecular techniques. *Sc-WRKY* and *ScWRKY4* were both shown to be related to tolerance enhancement to PEG and NaCl stresses [27,28]. Under biotic treatment, *Sc-WRKY* may play a positive role in response to smut pathogen [27], while *ScWRKY4* may be negatively or probably not involved in this regulation [28], suggesting the functional differentiation of group IIc ScWRKYs in smut pathogen resistance. In this study, a new group IIc *WRKY* gene family member, *ScWRKY3* (GenBank Accession No. MK034706), was screened from our previous sugarcane transcriptome data [29]. The sequence characteristics of ScWRKY3 and its subcellular localization, transcriptional activation activity, and its protein-protein interaction with ScWRKY4 were analyzed. The expression profiles of *ScWRKY3* in sugarcane tissues in response to various stresses were assessed, as well as the effects that occurred in *Nicotiana benthamiana* leaves after challenging with the bacterial pathogen *Ralstonia solanacearum* and the fungal pathogen *Fusarium solani* var. *coeruleum*.

2. Results

2.1. Bioinformatics Analysis of ScWRKY3 Gene

There was no nucleic acid sequence difference or amino acid sequence difference in *ScWRKY3* between smut-susceptible *Saccharum* hybrid cultivar ROC22 and smut-resistant *Saccharum* hybrid cultivar Yacheng05-179 (Figure S1). As shown in Figure 1, the *ScWRKY3* gene has a cDNA length of 910 bp containing an open reading frame (ORF) from position 162 to 872, and its encoded amino acid residues contain a conserved WRKY domain from position 166 to 223. Bioinformatics analysis revealed that the ScWRKY3 protein has a molecular weight of 25.98 kDa (Table S1). The theoretical isoelectric point (pI), grand average of hydrophobicity (GRAVY), and instability index (II) of ScWRKY3 were 8.58, -0.49, and 56.08 (Table S1), respectively, suggesting that ScWRKY3 might be an unstable basic hydrophilic protein. Secondary structure prediction showed that ScWRKY3 is mainly composed of random coil (69.07%), alpha-helix (18.22%), and extended strand (12.70%) portions (Figure S2). In addition, the ScWRKY3 protein was predicted to have no signal peptide or transmembrane domain (Figure S3). Euk-mPLoc 2.0 software [30] showed that ScWRKY3 has the highest probability of localization in the nucleus (Figure S4).

```
1    ACCACCACTAACCCCAAAGCATCATCGTCATCGTCAACCTTATCGTGCACAGTTATTGACACCAAAAGGAGAGAGATGCATTCCGATCCA
91   TCCTGCGAATTTTCTGCCCCGGCCGCCCAACCAAGATCCACACTTTGATCATCAATCGATCGCTAGCGCAT
162  ATGCAGGCATATATGGAGGGAGGCCAGTTGAGTGCTTGCCTTCCTAGCTTCCTTGTGCCGGATCACTACGCCGGCTTCCCCCTTCCTCTC
      M  Q  A  Y  M  E  G  G  Q  L  S  A  C  L  P  S  F  L  V  P  D  H  Y  A  G  F  P  L  P  L
252  CCGCTACAACTTCCTAGCCAACCAAACAACAAGCTTTTCCAGATGCCGTTTGTAGTTAACCAGGAAGAGACAGAGAACCATGGCGGCATG
      P  L  Q  L  P  S  Q  P  N  N  K  L  F  Q  M  P  F  V  V  N  Q  E  E  T  E  N  H  G  G  M
342  CTCTCCTCCGACCATTGTGGTGGACTATACCCGCTGCCGGCACTGCCCTTCGGCAGCTGCTCCGGCGCCGCCGCCGCAACAGCATGCGGT
      L  S  S  D  H  C  G  G  L  Y  P  L  P  A  L  P  F  G  S  C  S  G  A  A  A  A  T  A  C  G
432  GGGAAGCCGACGGCCGGTTTCATGCCCAGTGCTATTGGCGCTGAGGAGGTCTGCACCTCGGTGGCTACTAAACTAGGTTGCAACGAGAGT
      G  K  P  T  A  G  F  M  P  S  A  I  G  A  E  E  V  C  T  S  V  A  T  K  L  G  C  N  E  S
522  AATAGCACATGGTGGAAGGGCTCAGCAGCTACGACTGCGGAGAGAGGGAAGATGAAGGTGAGGAGGAAGATGAGGGAACCGAGGTTTTGC
      N  S  T  W  W  K  G  S  A  A  T  T  A  E  R  G  K  M  K  V  R  R  K  M  R  E  P  R  F  C
612  TTCCAGACCAGAAGCGACGTGGATGTACTGGATGATGGCTACAAGTGGAGGAAGTACGGGCAGAAGGTTGTCAAGAACAGCCTCCATCCA
      F  Q  T  R  S  D  V  D  V  L  D  D  G  Y  K  W  R  K  Y  G  Q  K  V  V  K  N  S  L  H  P
702  AGGAGCTATTTCCGGTGCACTCACAGCAACTGCCGCGTGAAGAAACGGGTGGAGCGGCTGTCGACGGACTGCCGCATGGTGATGACCACG
      R  S  Y  F  R  C  T  H  S  N  C  R  V  K  K  R  V  E  R  L  S  T  D  C  R  M  V  M  T  T
792  TACGAGGGCCGCCACACGCACTCTCCCTGCAGCGACGACGCTTCCTCCGCCGACCACACCGATTGCTTCAGCTCCTTCTGAATCTATCTA
      Y  E  G  R  H  T  H  S  P  C  S  D  D  A  S  S  A  D  H  T  D  C  F  S  S  F  *
882  TCCCACGACCGCATACATTGACGACGCAG
```

Figure 1. Nucleotide acid sequences and deduced amino acid sequences of the sugarcane *ScWRKY3* gene obtained by PCR amplification. The sequence of the WRKY motif (WRKYGQK) is highlighted in the red box, and that of the C_2H_2 domain ($C_{X4}C_{X23}H_XH$) in the black box. The upstream sequences to start codon ATG (marked in red font) is 5′ untranslated region (UTR) and the downstream sequences to stop codon TGA (marked in red font) is 3′UTR of *ScWRKY3*. *: stop codon.

Amino acid sequence alignment (Figure S5) indicated that the similarity of ScWRKY3 to *S. bicolor* SbWRKY57 (XP_002452824.2), *Miscanthus lutarioriparius* MlWRKY12 (AGQ46321.1), *Z. mays* ZmWRKY51 (XP_020393361.1), *S. italica* SiWRKY12 (XP_004953301.1), *O. sativa* OsWRKY12 (XP_015624962.1) (all these accession numbers in brackets are from GenBank), sugarcane ScWRKY4 and Sc-WRKY were 93%, 93%, 87%, 87%, 66%, 53% and 24%, respectively. A conserved WRKY domain (WRKYGQK) and a conserved zinc-finger motif ($C_{X4}C_{X23}H_XH$) at the C-terminus were found (Figure 2). The phylogenetic tree of sugarcane ScWRKY3, ScWRKY4, Sc-WRKY and WRKYs from other plant species demonstrated that WRKY proteins could be divided into three groups with no obvious distinction between monocots and dicots. ScWRKY3 was classified into group IIc, along with AtWRKY13, OsWRKY22, AtWRKY57, TaWRKY10, Sc-WRKY, and ScWRKY4 (Figure 3). MEME software prediction showed that all WRKYs except TaWRKY46 and OsWRKY46 contained motif 1 (WRKY domain) and motif 2 (zinc-finger domain). In addition, motif 3 (WRKY domain) and motif 4 (unknown domain) were detected in group I WRKYs. Some group IIc WRKYs, for example ScWRKY3, ScWRKY4, AtWRKY13, OsWRKY23 and AtWRKY57, contained motif 4. The WRKYs in group IId had their unique motif 5 (unknown domain) (Figure 3). On the whole, the phylogenetic analysis showed that most WRKYs within the same group generally had a similar structure.

2.2. Subcellular Localization

The recombinant vector pMDC83-*ScWRKY3-GFP* was generated to investigate the subcellular distribution of ScWRKY3. We used 4′,6-diamidino-2-phenylindole (DAPI) staining as a nuclear marker. As shown in Figure 4, the green fluorescence of the control (*35S::GFP*) in *N. benthamiana* was distributed through the whole cell, including the plasma membrane, nucleus, and cytoplasm, while the fusion protein of ScWRKY3::GFP was only found in the nucleus, which was consistent with the software prediction.

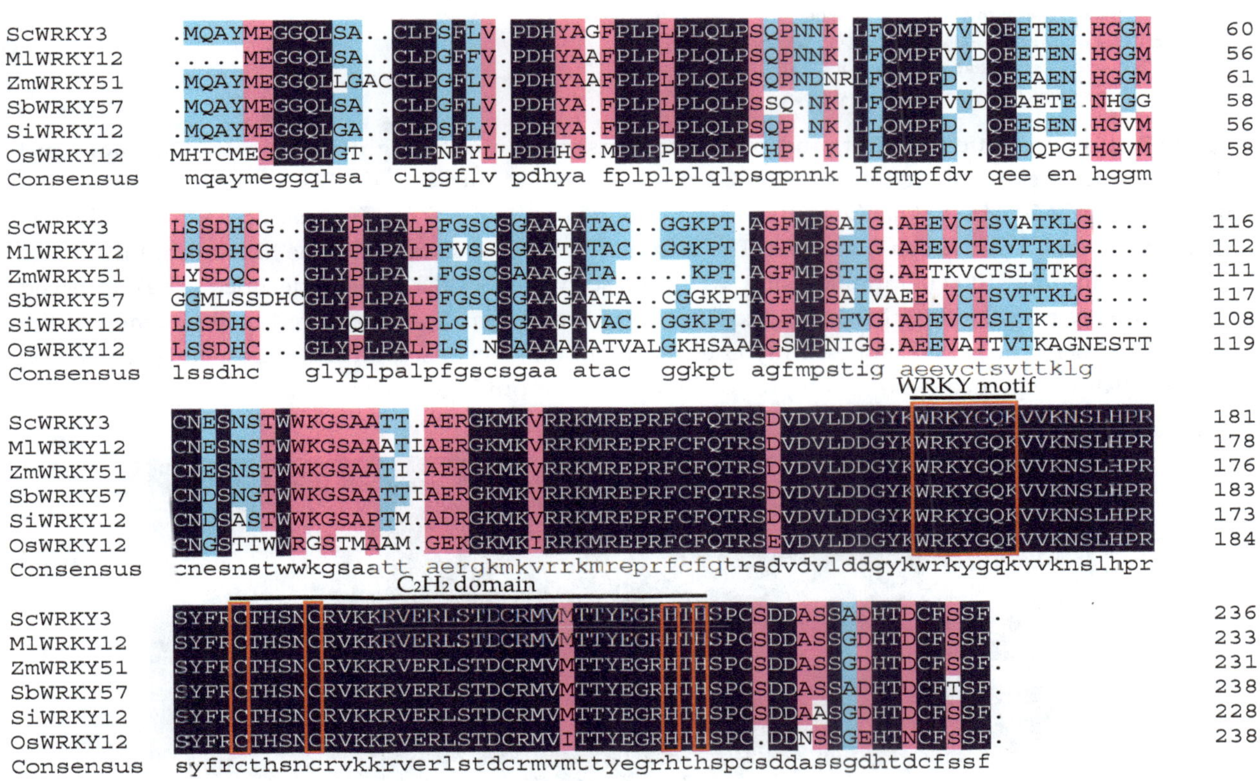

Figure 2. Amino acid sequence alignment of ScWRKY3 and WRKYs from other plant species by DNAMAN (version 6.0.3.99, Lynnon Biosoft) software. The amino acid sequences of *Miscanthus lutarioriparius* MlWRKY12 (AGQ46321.1), *Zea mays* ZmWRKY51 (XP_020393361.1), *Sorghum bicolor* SbWRKY57 (XP_002452824.2), *Setaria italica* SiWRKY12 (XP_004953301.1), and *Oryza sativa* OsWRKY12 (XP_015624962.1) are from GenBank. The black, pink, blue, and white colors indicate the homology level of conservation of the amino acid residues in the alignment at 100, ≥75, ≥50, and <50%, respectively. The sequences of the WRKY motif (WRKYGQK) and the C_2H_2 domain ($C_{X4}C_{X23}H_XH$) are highlighted by the red rectangle.

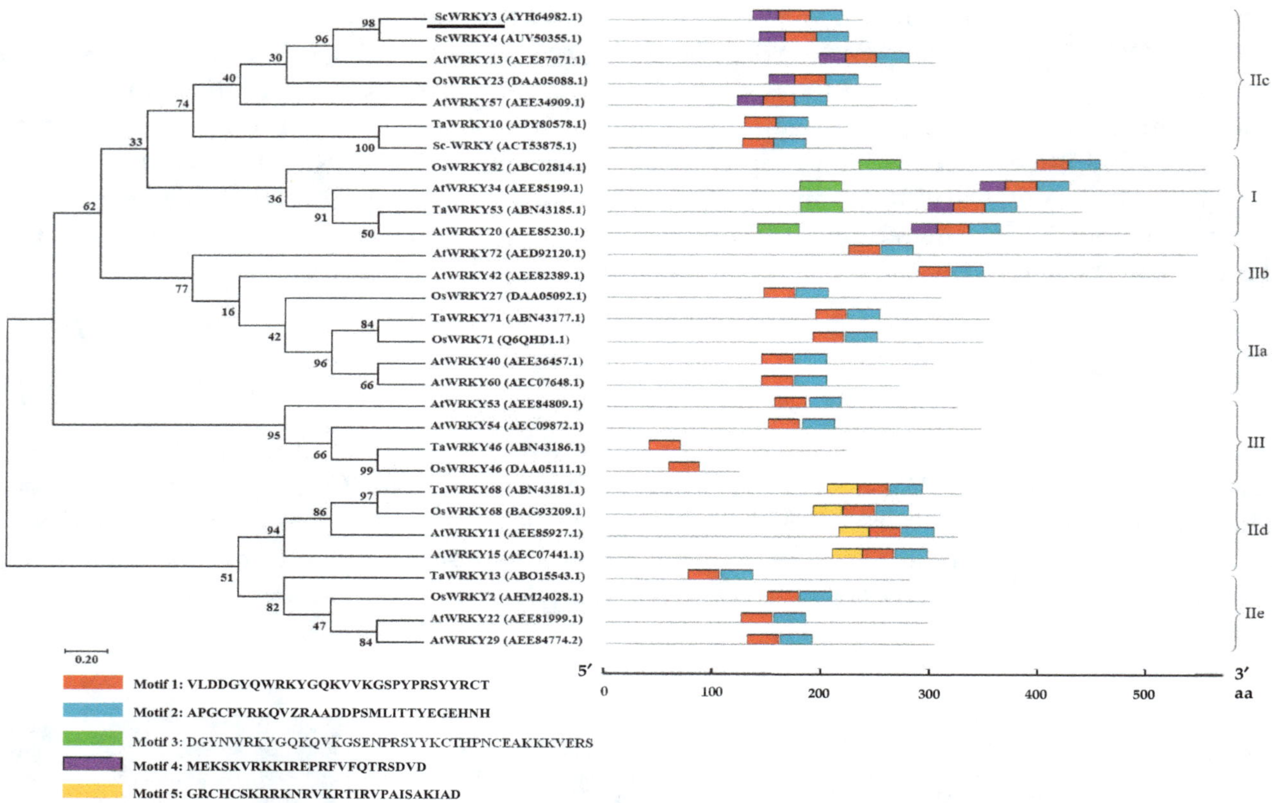

Figure 3. Phylogenetic tree (left) and predicted conserved motifs (right) of ScWRKY3 protein and WRKYs from various plant species. The GenBank accession number of WRKY proteins follows the protein name. Ml, *Miscanthus lutarioriparius*; Sb, *Sorghum bicolor*; Zm, *Zea mays*; Si, *Setaria italic*; Os, *Oryza sativa*; and At, *Arabidopsis thaliana*. The unrooted tree is constructed by the Maximum Likelihood with bootstrapping (1000 iterations) using MEGA7.0 software. ScWRKY3 is underlined. The conserved domains were predicted by MEME Suite 5.0.2 software. The different-colored boxes named at the bottom represent conserved motifs. Gray lines represent the nonconserved sequences, and the position of each WRKY sequence is exhibited proportionally. The motif logo is shown in Figure S6.

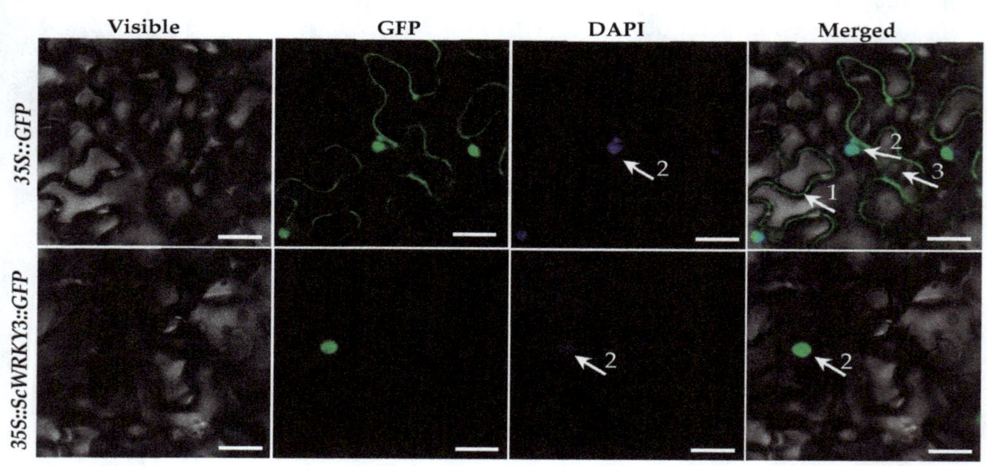

Figure 4. Subcellular localizations of *35S::GFP* and *35S::ScWRKY3::GFP* in *Nicotiana benthamiana* leaves. The epidermal cells of *N. benthamiana* are used for capturing images of visible light, green fluorescence, blue fluorescence, and visible light merged with green and blue fluorescence. White arrows 1, 2, and 3 indicate plasma membrane, nucleus, and cytoplasm, respectively. Scale bar = 50 μm. *35S::GFP*, the *Agrobacterium tumefaciens* strain carrying the empty vector pMDC83-*GFP*. *35S::ScWRKY3::GFP*, the *A. tumefaciens* strain carrying the recombinant vector pMDC83-*ScWRKY3-GFP*. DAPI, 4′,6-diamidino-2-phenylindole.

2.3. *Transcription Activation Activity of ScWRKY3*

The Y2H Gold-GAL4 yeast two-hybrid system was used to detect the transcriptional activation activity of ScWRKY3. As shown in Figure 5, yeast cells transformed with either the positive control pGADT7+pGBKT7-p53, the negative control pGBKT7-p53, or the recombinant plasmid pGBKT7-*ScWRKY3* all grew well in SDO (SD/-Trp, SD minimal medium without tryptophan) medium plates, while only the positive control turned blue in SDO/X (SD/-Trp/X-α-Gal, SDO plates with X-α-D-galactosidase) medium plates. These results indicated that all the plasmids were successfully transfected into yeast strain Y2HGold. The GAL4-BD combined with ScWRKY3 protein can successfully express tryptophan but cannot activate the *MEL1* gene in the presence of X-α-gal. After aureobasidin A (AbA) resistance screening, the yeast cells transformed with pGBKT7-*ScWRKY3* and with the negative control did not activate the two reporter genes, *AUR1-C* and *MEL1*. However, the positive control did survive, and its X-α-gal detection system showed a blue color, indicating that the ScWRKY3 protein does not possess transcriptional activation activity. This protein showed no toxicity to the yeast strain Y2HGold. This result implies that the bait protein of ScWRKY3 can be used for yeast two-hybrid screening.

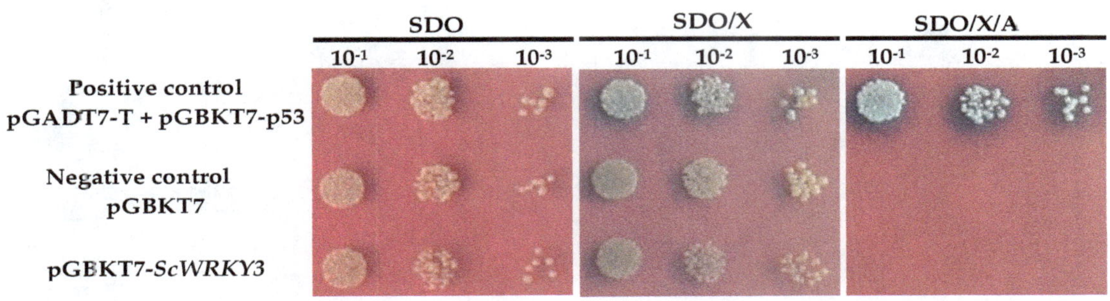

Figure 5. Testing of the ScWRKY3 transactivation activity assay. SDO (SD/-Trp), synthetic dropout medium without tryptophan; SDO/X (SD/-Trp/X-α-Gal), synthetic dropout medium without tryptophan, but plus 5-bromo-4-chloro-3-indoxyl-α-D-galactopyranoside; SDO/X/A (SD/-Trp/X-α-Gal/AbA), synthetic dropout medium without tryptophan, but plus 5-bromo-4-chloro-3-indoxyl-α-D-galactopyranoside and aureobasidin A.

2.4. *Interaction Between ScWRKY3 and ScWRKY4*

As shown in Figure 6, all plasmid combinations grew normally on DDO (SD/-Leu/-Trp) plates. However, when transferred to QDO (SD/-Ade/-His/-Leu/-Trp) and QDO/X/A (SD/-Ade/-His/-Leu/-Trp/X-α-Gal/AbA) plates, only the AD-ScWRKY4+BD-ScWRKY3 combination and the positive control pGADT7-T+pGBKT7-p53 continued to grow and turned blue with X-α-gal detection. This indicates that ScWRKY4 may function downstream of ScWRKY3. Additionally, we have further proved the above results by bimolecular fluorescence complementation (BiFC). When ScWRKY3 was fused to pUC-SPYNE (this fusion was named ScWRKY3-YFP^{N}), and ScWRKY4 was fused to pUC-SPYCE (this fusion was named ScWRKY4-YFP^{C}), a fluorescent complex was formed and was visualized in the nucleus of *N. benthamiana* leaf cells.

While, when ScWRKY3 was fused to pUC-SPYCE (this fusion was named ScWRKY3-YFP^{C}), and ScWRKY4 was fused to pUC-SPYNE (this fusion was named ScWRKY4-YFP^{N}), no fluorescent complex was formed in *N. benthamiana* leaf cells. The results were consistent with those of yeast two-hybrid system and showed that there was an interaction between ScWRKY3 and ScWRKY4, and the specific protein complex was located in the nucleus.

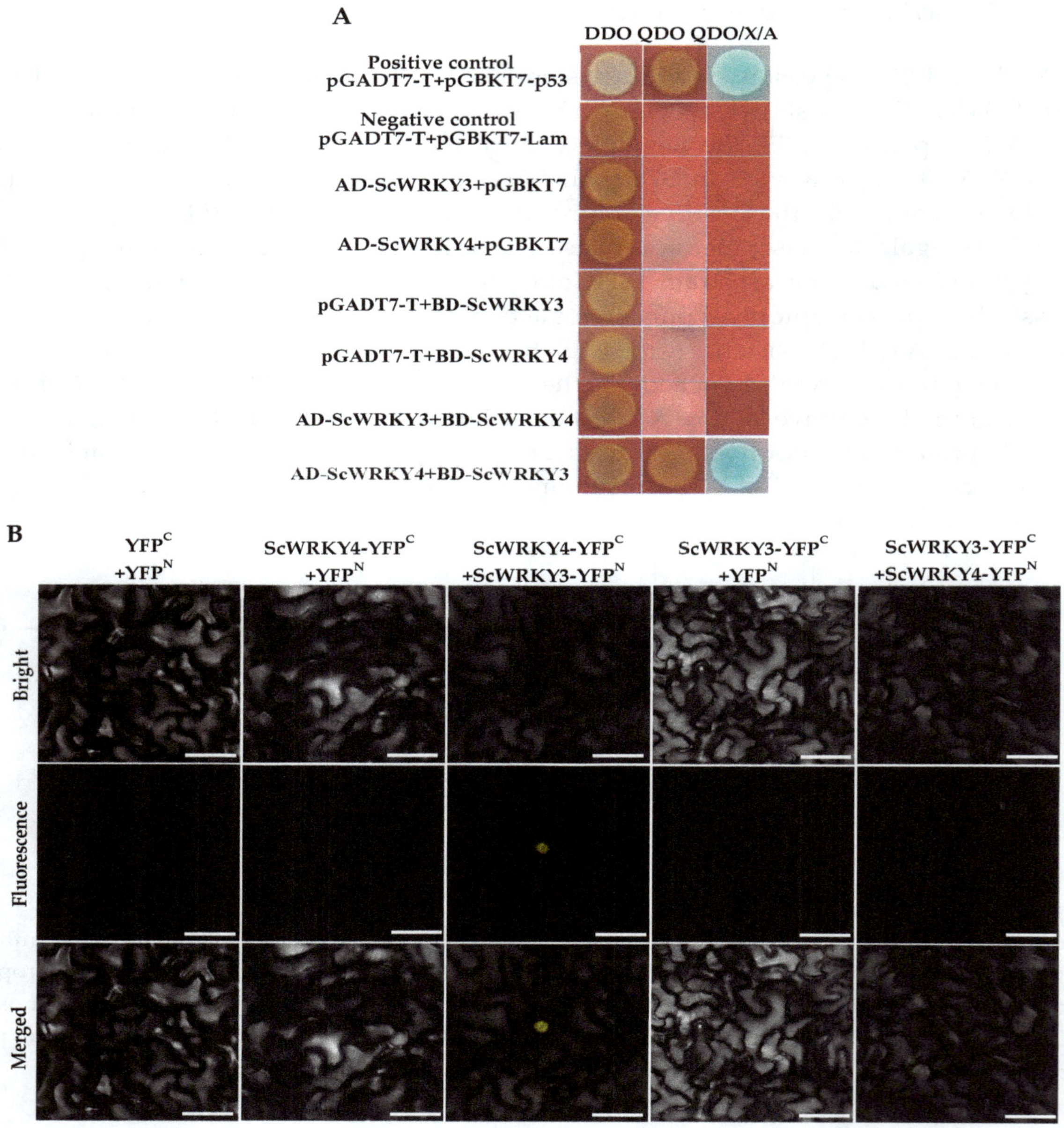

Figure 6. Interaction between ScWRKY3 and ScWRKY4 in yeast and in *Nicotiana benthamiana* leaves. (**A**) The interaction between ScWRKY3 and ScWRKY4 was verified using a yeast two-hybrid system. A variety of BD and AD vectors were combined and transformed into GoldY2H yeast. Left to Right: Transformations were grown and screened on DDO (SD/-Leu/-Trp, SD medium without leucine and tryptophan), QDO (SD/-Ade/-His/-Leu/-Trp, SD medium without adenine, histidine, leucine, or tryptophan), and QDO/X/A (SD/-Ade/-His/-Leu/-Trp/X-α-Gal/AbA, QDO medium with X-α-D-Galactosidase and aureobasidin medium. (**B**) The bimolecular fluorescence complementation (BiFC) assay for the location determination of the interaction between ScWRKY3 and ScWRKY4. Scale bar = 50 μm. YFP^C, YFP^N, ScWRKY4-YFP^C, ScWRKY4-YFP^N, ScWRKY3-YFP^C, and ScWRKY3-YFP^N represent the plasmids pUC-SPYCE and pUC-SPYNE and the recombinant plasmids ScWRKY4-pUC-SPYCE, ScWRKY4-pUC-SPYNE, ScWRKY3-pUC-SPYCE and ScWRKY3-pUC-SPYNE, respectively.

2.5. Gene Expression Patterns of ScWRKY3 in Response to Various Stress Conditions

The expression patterns of the *ScWRKY3* gene in sugarcane tissues and under various stresses were investigated using real-time fluorescent quantitative PCR (qRT-PCR). The results indicated that *ScWRKY3* was constitutively expressed in different sugarcane tissues, with the highest expression level in stem epidermis. It remained at lower expression levels in other tissues (root, bud, leaf, and stem pith) (Figure 7A). During the treatments with NaCl and PEG, the transcript of *ScWRKY3* in ROC22 was remarkably up-regulated by 3.28-fold at 24 h and 38.57-fold at 3 h, respectively, and remained unchanged at other time points (Figure 7B). Moreover, the expression of *ScWRKY3* in ROC22 was markedly down-regulated under both SA and methyl jasmonate (MeJA) treatments, but it was up-regulated under ABA stress with a 1.73-fold higher level than in the control (Figure 7C). These results indicated that the *ScWRKY3* gene might have positive responses to ABA, PEG, and NaCl stimuli but a negative response to SA and MeJA. After infection by the smut pathogen, the expression level of *ScWRKY3* was almost unchanged after 72 h in the smut-resistant cultivar Yacheng05-179 and during the period of 48–72 h in the smut-susceptible cultivar ROC22, while it was significantly down-regulated (0.76-fold) at 24 h in ROC22 (Figure 7D). This suggested that *ScWRKY3* may play a role in the smut pathogen response.

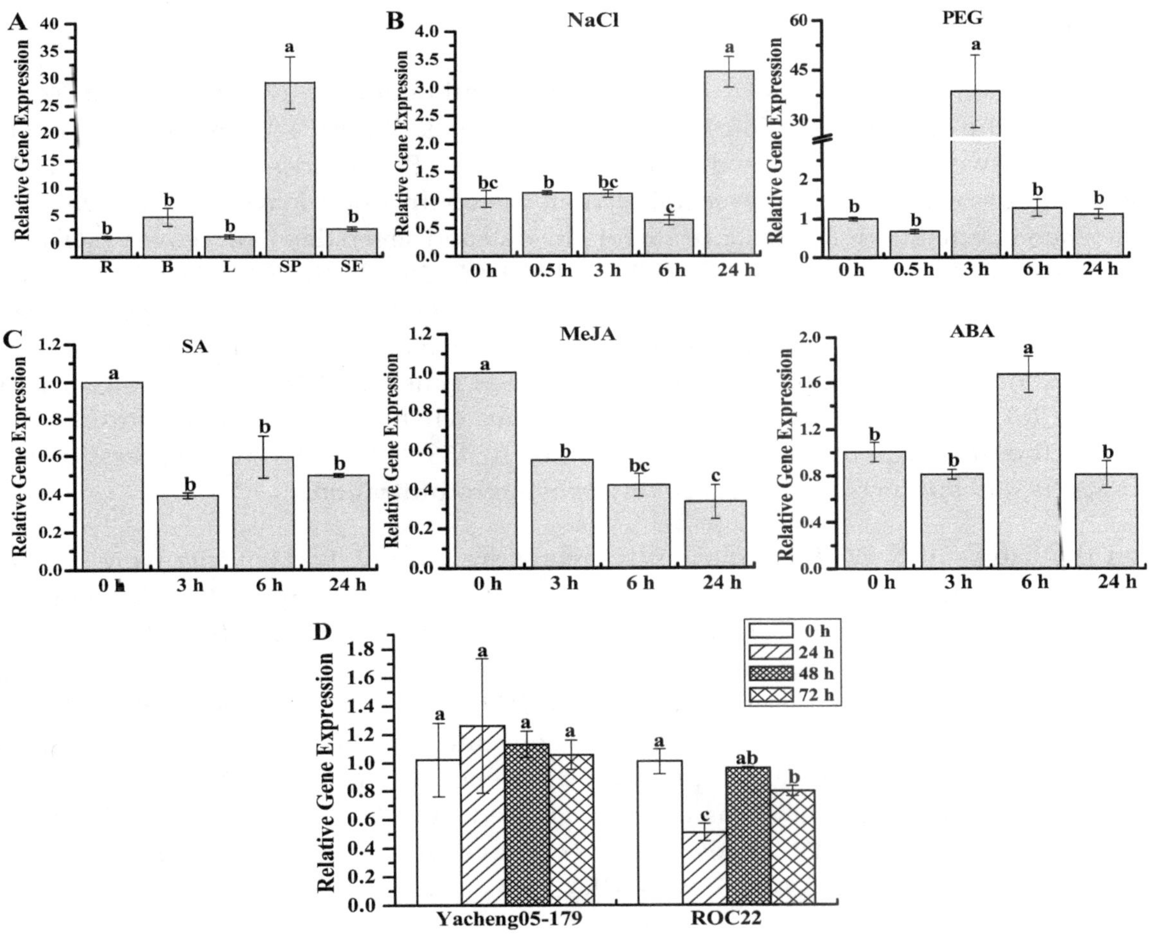

Figure 7. Gene expression assay of *ScWRKY3*. (**A**) Tissue-specific expression analysis of *ScWRKY3* in different 10-month-old ROC22 tissues by qRT-PCR. The tissues (root, bud, leaf, stem pith, and stem epidermis) are represented by R, B, L, SP, and SE, respectively; (**B**) Gene expression patterns of *ScWRKY3* in 4-month-old ROC22 plantlets under abiotic stress. NaCl, sodium chloride (simulating salt stress) (250 mM); PEG, polyethylene glycol (simulating drought treatment) (25.0%); (**C**) Gene expression patterns of *ScWRKY3* in 4-month-old ROC22 plantlets under plant hormone stress. SA, salicylic acid (5 mM); MeJA, methyl jasmonate (25 μM); ABA, abscisic acid (100 μM); (**D**) Gene expression patterns of the *ScWRKY3* gene after infection with smut pathogen. Yacheng05-179 is a smut-resistant *Saccharum* hybrid cultivar, and ROC22 is a smut-susceptible *Saccharum* hybrid cultivar. Data are normalized to the glyceraldehyde-3-phosphate dehydrogenase (*GAPDH*) expression level. All data points are means ± standard error ($n = 3$). Bars superscripted by different lowercase letters indicate significant differences, as determined by Duncan's new multiple range test (p-value < 0.05).

2.6. Transient Overexpression of ScWRKY3 in N. benthamiana Leaves

The *ScWRKY3* gene was inserted into the plant overexpression vector, and was transformed into *N. benthamiana* leaves by the *Agrobacterium tumefaciens*-mediated method to analyze whether the target gene could induce a plant immune response. The transcripts of *ScWRKY3* in *N. benthamiana* leaves were detected using a semi-quantitative PCR technique (Figure 8A). The phenotypic observation after injection for one day was shown in Figure 8B, and no significant difference in superficial characteristics was demonstrated between the experimental group and the control group. However, qRT-PCR results demonstrated that six immunity-associated marker genes, including the hypersensitive response (HR) marker genes, *NtHSR203* and *NtHSR515*, the SA pathway related gene *NtPR1*, the JA pathway associated gene *NtPR3*, and two ethylene synthesis-dependent genes, *NtEFE26* and *NtAccdeaminase*, were all up-regulated with a higher fold change range from 1.47 to 14.16 than the control (Figure 8C). These results suggest that transiently overexpressed *ScWRKY3* may take part in the immune response in *N. benthamiana* leaves.

To detect the effect of *ScWRKY3* in response to pathogen, the *N. benthamiana* leaves, were transformed with the control in the left half blade and with *35S::ScWRKY3* in the right half blade for one day. Then they were inoculated by the bacterial pathogen *R. solanacearum*. As shown in Figure 8D, there was a slight symptomatic difference between the control half-leaves and the *35S::ScWRKY3* half-leaves when injected with *R. solanacearum* for one day and seven days. Moreover, qRT-PCR results revealed that the HR marker genes *NtHSR201* and *NtHSR515* and the SA-related gene *NtNPR1* all showed significantly lower expression in *35S::ScWRKY3*-overexpressing *N. benthamiana* leaves after one day and seven days of *R. solanacearum* inoculation when compared to the control. No remarkable transcript difference or down-regulation of the SA-related gene *NtPR-1a/c* or the JA-associated genes *NtPR2* and *NtPR3* was observed in *35S::ScWRKY3* leaves when compared with controls. Compared to the control, the transcript abundance of *NtHSR203* in *35S::ScWRKY3*-overexpressing leaves was decreased at one day but increased at seven days post-agroinfiltration.

When the leaves of *ScWRKY3*-transiently-overexpressing *N. benthamiana* were inoculated by the fungal pathogen *F. solani* var. *coeruleum* for one day and seven days, a heavier wilting disease symptom was observed in the *N. benthamiana* leaves containing *35S::ScWRKY3* than in the control (Figure 8F). Additionally, in comparison with the control, *NtHSR201*, *NtPR3*, and *NtAccdeaminase* showed significantly higher expression, while *NtHSR515* and *NtNPR1* presented obviously lower expression in *35S::ScWRKY3*-overexpressing leaves at one day or seven days after *F. solani* var. *coeruleum* infection. No statistically significant expression difference in *NtPR-1a/c*, *NtPR2*, or *NtEFE26* was found at one day, while *NtPR-1a/c* was down-regulated, *NtPR2* was up-regulated, and *NtEFE26* was unchanged in *35S::ScWRKY3* leaves after seven days with *F. solani* var. *coeruleum* inoculation. The expression level of *NtHSR203* in *35S::ScWRKY3* was lower at one day but higher at seven days after inoculation than in the control (Figure 8G).

These results demonstrated that in comparison with the transiently overexpressing pEarleyGate 203 vector, *ScWRKY3* transient overexpression in *N. benthamiana* leaves significantly decreased the transcript abundance of *NtHSR515*, *NtPR1*, and *NtPR-1a/c* after *R. solanacearum* or *F. solani* var. *coeruleum* infection. It was anticipated that *ScWRKY3* can negatively regulate the HR marker genes or SA signaling pathway-mediated genes to reduce the tolerance of *N. benthamiana* to pathogens.

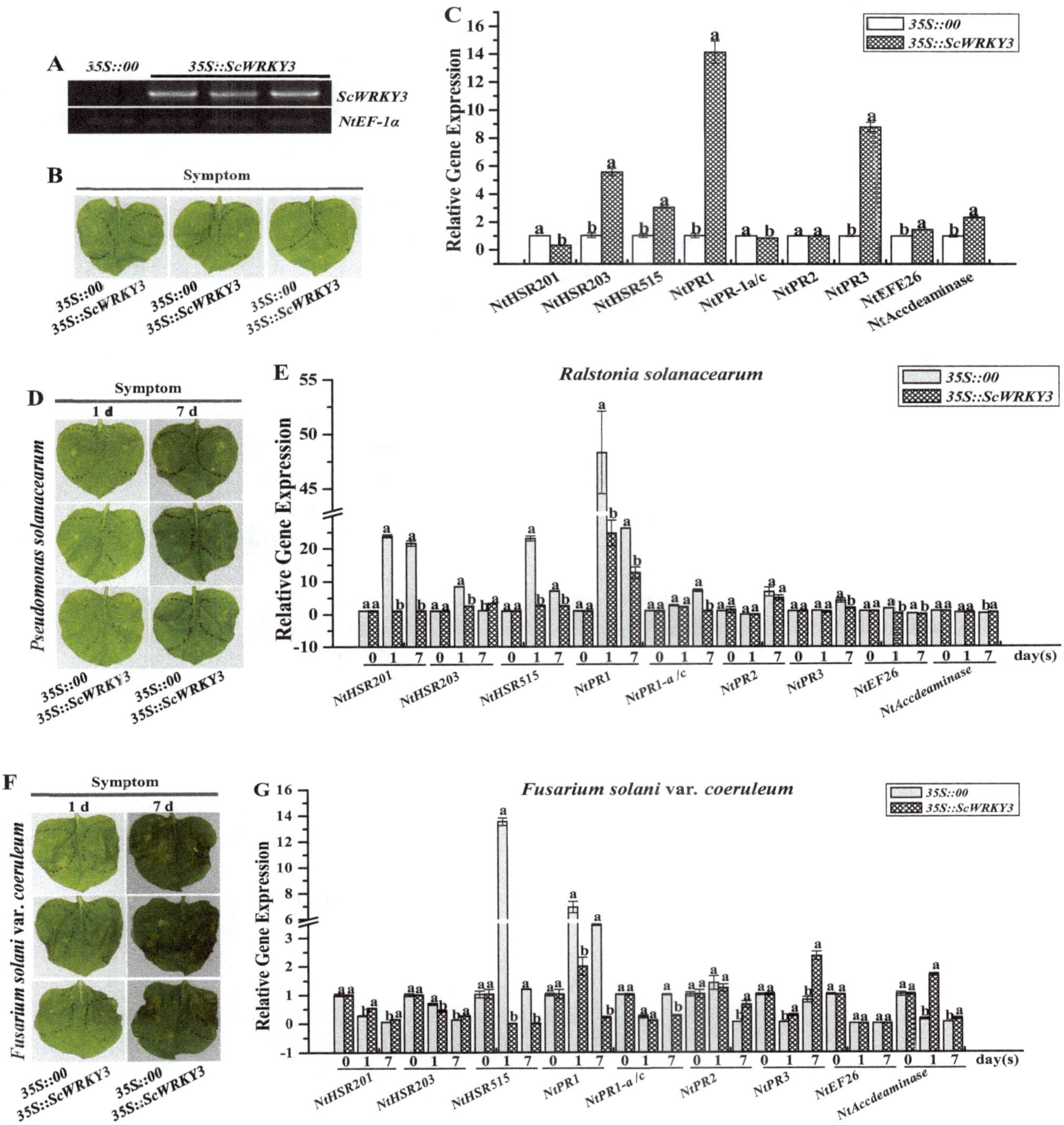

Figure 8. Effects of transient overexpression of *ScWRKY3* in *Nicotiana benthamiana* leaves. (**A**) Semi-quantitative PCR analysis of *ScWRKY3* in *N. benthamiana* leaves after one day of infiltration by *Agrobacterium* strain GV3101 carrying pEarleyGate 203-*ScWRKY3* (*35S::ScWRKY3*) and the empty vector pEarleyGate 203 (*35S::00*). (**B**) Phenotype of *N. benthamiana* leaves after one day of agroinfiltration. (**C**) The transcript level of nine immunity-associated marker genes in the *N. benthamiana* leaves after one day of agroinfiltration. (**D**,**F**) Disease symptoms of *N. benthamiana* post-inoculation with *Ralstonia solanacearum* and *Fusarium solani* var. *coeruleum* are observed after one day and seven days of agroinfiltration. (**E**,**G**) The transcripts of nine immunity-associated marker genes in the *N. benthamiana* leaves after inoculation with *R. solanacearum* or *F. solani* var. *coeruleum* for one day and seven days. Data are normalized to the *NtEF-1α* expression level. All data points are means ± standard error ($n = 3$). Bars superscripted by different lowercase letters indicate significant differences, as determined by Duncan's new multiple range test (p value < 0.05).

3. Discussion

As one of the largest groups of TFs, the WRKY proteins have been found in a wide range of plant species since the initial *WRKY* cDNA was isolated from sweet potato [31]. Although sugarcane is an important bioenergy and cash crop [25], there are only three reports about WRKYs in sugarcane [26–28]. In this study, a novel sugarcane *ScWRKY3* gene was isolated and identified. As reported, there is a functional similarity of WRKYs in the same or phylogenetically closely related group [2,3]. Phylogenetic tree analysis indicated that ScWRKY3 is a member of the group IIc WRKY proteins, along with Sc-WRKY [27] and ScWRKY4 [28] (Figure 3). This is helpful for further functional comparative studies on the same WRKY family members in sugarcane. The structure of TFs is usually composed of four functional domains, namely, the DNA binding domain, the transcriptional activation or repression domain, the oligomerization sites, and the nuclear localization signals [32]. These four components are the core regions that perform the functions of TFs or interact with the *cis*-acting elements in the promoter regions of various stress related genes [33]. In our study, ScWRKY3 protein had one WRKY domain, which is a DNA binding domain containing 60 amino acids. WRKY domain was mainly composed of motif 1 and motif 2 (Figure 3) and can bind specifically to the DNA sequence motif (T)(T)TGAC(C/T) which is known as the W-box and existed in many promoters of plant defense-related genes [6]. As showed by Wei et al. [10], subgroup IId WRKYs possess two basic amino acid sequences, including a RCHCSK[RK][RK]K[LN]R motif, which may function as a nuclear localization signal, and a KRxIxVPAISxKxAD motif. Similarly, the motif 5 (Figure 3) also contains these amino acid sequences. While further work is required to clarify the function of the other unknown motifs, such as the predicted motif 4 in Figure 3. Previous studies showed that the functions of the two WRKY domains in the group I WRKYs are different. The WRKY domain at the C-terminal can bind to their target DNA, while another WRKY domain at the N-terminal may be as the site where proteins interact with each other [8,12,15].

Subcellular localization analysis is valuable for determining the functions of proteins. The present study showed that the fusion protein of ScWRKY3::GFP was detected in the nucleus of *N. benthamiana* leaf cells (Figure 4), which was consistent with the software prediction results and previous studies on other plant WRKYs [28,34–36]. This indicated that ScWRKY3 may play a role as a nuclear-localized protein to regulate cellular processes.

Transcriptional activity analysis is important for the functional analysis of TFs [37]. Since the GAL4 yeast two-hybrid system was first discovered [38], this method has been increasingly used to study the interactions between WRKY proteins [39]. In this study, the full-length ScWRKY3 cDNA showed no auto-activation (Figure 5), so it could be used as the bait to screen interacting proteins in a yeast two-hybrid system. Post-translational modifications or interactions with cofactors are needed for ScWRKY3 protein to fulfill its function. Screening and identifying WRKY interacting proteins is important to reveal the role of WRKY in plant signal transduction [40,41]. It has been reported that WRKYs have the activities of self-regulation and mutual regulation, and they can form functional homo- or heterodimers among some WRKY proteins or interact with other functional proteins to play roles [4,5]. WRKY6 and WRKY22, which both belong to group II WRKYs in *A. thaliana*, interact with MPKl0 and MPK3/MPK6, respectively [42,43]. Previous studies also proved that AtWRKY30, AtWRKY53, AtWRKY54, and AtWRKY70, which all belong to group III of the WRKY proteins, have interactive effects in yeast [39]. Similarly, yeast two-hybrid and BiFC results showed that a fluorescent complex from ScWRKY3 and ScWRKY4 was formed (Figure 6B). This was visualized in the nucleus in *N. benthamiana* leaf cells, which indicated that there may be an interacting relationship between ScWRKY3 and ScWRKY4. In *Arabidopsis*, previous study indicated that WRKYs in the group IIb or group III could interact with themselves and with group IIa WRKYs, while group IId WRKYs could only interact with group IIa WRKY members [41]. Groups IIc ZmWRKY25 and ZmWRKY47 had interactions and may be involved in the response to drought stress by interacting with other WRKYs [15]. Besides, ZmWRKY39 was down-regulated under light drought stress and its phylogenetically closely related protein ZmWRKY106 showed a positive response to this stress, while they had up-regulated co-expression interaction under drought stress [15]. However, the nature of the interaction as well as the biological implications of this process between ScWRKY3 and ScWRKY4 requires further study.

WRKY genes are expressed differentially in different tissues of plants, which demonstrates that *WRKY* genes can be expressed in different physiological conditions and in different types of cells, and may regulate a series of life activities including growth, development, and morphological composition [35,40,44]. Twenty-eight *CiWRKYs* were detected in the roots, stems, and leaves of wild *Caragana intermedia*, and different *CiWRKYs* showed differential expression in various tissues. For example, *CiWRKY69–1* had the highest expression level in roots, while *CiWRKY40–1* and *CiWRKY30* were mainly expressed in leaves [35]. Among the 37 *A. thaliana WRKY* genes reported by Bakshi et al. [45], 12 were specifically expressed in the mature zone of root cells, suggesting that these *WRKY* genes may be involved in the regulation of root cell maturation in *A. thaliana*. *TaWRKY44*, a *WRKY* gene of *Triticum aestivum*, was differentially expressed in all organs examined, including root, stem, leaf, pistil, and stamen, with the highest expression level in the leaves and the lowest expression level in the pistils [40]. It is known that *ScWRKY4* is constitutively expressed in the root, bud, leaf, stem pith, and stem epidermis of sugarcane, with the highest expression level in the stem epidermis [28]. This was similar to the findings on tissue-specific expression of *ScWRKY3* in the present study. Can the coexpression of *ScWRKY3* and *ScWRKY4* in the same tissue be tied together with their BiFC interaction results (Figure 6B)? Can they co-localize in the same organelle in the same tissue? These remain to be validated by future research, for example, only if we obtain the promoters specific to *ScWRKY3* and *ScWRKY4* respectively, can we check that if they co-localize in the same organelle in the same tissue. WRKY TFs are critical for signal transduction, plant growth and stress responses [44]. Salt and drought, the two representative abiotic stresses, adversely affect the growth and development of plants, but the response can be resisted by activating the ABA signal transduction pathway to induce the expression of a series of stress-responsive genes [46]. *AtWRKY1* played a negative role in ABA-mediated drought resistance, and the *AtWRKY1* knockout mutant could enhance the drought tolerance of *A. thaliana* [47]. In the *OsWRKY11* knockout mutant, drought responsive genes were induced to enhance the drought tolerance of rice [36]. The present study showed that the expression of *ScWRKY3* was increased by PEG, NaCl, and exogenous ABA. Previous studies also determined that *Sc-WRKY* and *ScWRKY4* showed up-regulated expression levels under PEG and NaCl treatments [27,28]. Furthermore, under ABA stress, the transcript of *ScWRKY4* was remarkably up-regulated by 1.59-, 2.87-, and 1.26-fold at 0.5 h, 6 h, and 24 h higher than the control, respectively [28]. These results suggest that *ScWRKY3* and *ScWRKY4* may participate in sugarcane resistance to drought and salt stresses which may be mediated through ABA signaling. Similar to the expression characteristics seen in group II *WRKYs* of other plants, the expression level of *AtWRKY57* was remarkably up-regulated in drought conditions. This occurred through the direct activation of the expression of an important functional gene (*AtNCED3*) in the ABA synthesis pathway, which enhanced the tolerance of *A. thaliana* to drought stress [48].

Plants have evolved at least two sets of biochemical defenses to protect themselves against external challenges [49]. One defense response, caused by infection of pathogenic bacteria, initiates a localized hypersensitive reaction to confine the injured site and to prevent further infection of pathogens. This is known as systemic acquired resistance (SAR) [50]. The other response is mainly activating the expression of defense genes through various signal molecules to exhibit resistance in plants. The signal transduction pathways related to this defense response may be mediated by SA, JA, or ABA [51,52]. Previous studies indicated that the *WRKY* genes of group IIc are related to plant immunity. For instance, *AtWRKY28* and *AtWRKY75* can be induced by infection with *Sclerotinia sclerotiorum* in connection with SA- and JA/ET-mediated defense signaling pathways [53]. Overexpression of rice *WRKY89* increased the expression level of SA and enhanced the resistance to rice blast fungus [54]. *AtWRKY57* is negatively associated with resistance to *Botrytis cinerea* in *A. thaliana* by regulating the expression of JA pathway-related genes [49]. *OsWRKY13* enhances the resistance of rice to *X. oryzae* pv. *oryzae* and *M. grisea* [55,56]. In the present study, *ScWRKY3* was down-regulated by smut pathogen in the smut-susceptible cultivar ROC22 at 24 h, while it was almost unchanged in the smut-resistant cultivar Yacheng05-179. Conversely, Wang et al. [28] showed that the expression of *ScWRKY4* was quite stable in ROC22 but was down-regulated in Yacheng05-179 under the stress of smut pathogen. Liu et al. [27] found that the expression of *Sc-WRKY* was remarkably up-regulated at 24 and 60 h after infection by smut pathogen treatment in sugarcane FN22. Moreover, *ScWRKY3* was down-regulated under SA and MeJA treatments, which was opposite to the expression patterns of *ScWRKY4* and *Sc-WRKY* [27,28]. The results revealed that *ScWRKY3* might be a negative regulatory gene in the sugarcane response to smut pathogen. This can be further proved by antimicrobial test in more stable overexpression and knockout plants. In *Arabidopsis*, *AtWRKY25* gene was proved to play a negative regulatory role in the SA-mediated defense response to *Pseudomonas syringae* [57] but a positive regulatory role in response to NaCl stress [58]. Yokotani et al. [59] demonstrated that overexpression of *OsWRKY76* in rice plants suppressed the induction of defense related genes after inoculation with blast fungus (*Magnaporthe oryzae*) but up-regulated the expressions of abiotic stress-associated genes using microarray analysis. It is therefore possible for *WRKY* genes to play opposite roles in biotic resistance and abiotic tolerance. In the future, genetic transformations can be done to further our understanding on the responses of *ScWRKY* genes to biotic and abiotic stresses. In addition, whether there are sequence differences in the promoter regions of the *ScWRKY* genes from different sugarcane cultivars which may cause differences in gene expression patterns for the biotic or abiotic stress need further investigation.

As Liu et al. [60] proved, overexpression of the *Gossypium hirsutum GhWRKY25* in *N. benthamiana* is involved in the regulation of expression of multiple defense-associated marker genes, including the SA-, ET-, and JA-mediated genes, to decrease the resistance to the fungal pathogen *B. cinerea*. *GhWRKY40* has been revealed to be inducible by stress from the bacterial pathogen *R. solanacearum*, and *GhWRKY40* expression was up-regulated by SA, MeJA, and ET [61]. When *GhWRKY40* was transiently overexpressed in *N. benthamiana* leaves, most of the resistance-associated genes, including the SA-, ET-, JA-, and HR-responsive genes, were down-regulated after infection with *R. solanacearum*, which indicated that overexpression of *GhWRKY40* reduces the tolerance to *R. solanacearum* [61]. *A. thaliana* WRKY27 negatively regulated resistance genes during infection with the pathogen *R. solanacearum*, and the symptom development in *R. solanacearum* appeared earlier than in the mutant wrky27-1 plants, which lacks the function of WRKY27 [62]. In this study, most of the immunity-associated marker genes, including the HR-, SA-, JA-, and ET-related genes, were up-regulated when *ScWRKY3* was transiently overexpressed in *N. benthamiana*, suggesting that *ScWRKY3* may play a role in the plant immune response. After infection with the bacterial pathogen *R. solanacearum*, most of the detected immunity-associated marker genes, including the HR marker genes *NtHSR201* and *NtHSR515*, the SA-related genes *NtPR-1a/c* and *NtNPR1*, and the JA-associated gene *NtPR3*, were lower in the *35S::ScWRKY3* overexpressing leaves than in the control, revealing that *ScWRKY3* may play a negative role in the response to the bacterial pathogen *R. solanacearum*. After *F. solani* var. *coeruleum* infection, the wilting disease symptoms were greater in *35S::ScWRKY3* leaves than in the control. The qRT-PCR results showed that the transcript abundance of JA- and ET-related genes was remarkably higher, but the SA-related genes were evidently reduced in *35S::ScWRKY3* leaves compared to their levels in the control. These results indicated that there was a crosstalk between JA-/ET-related genes and SA-related genes in the response to the fungal pathogen *F. solani* var. *coeruleum* when *ScWRKY3* was overexpressed in *N. benthamiana*.

4. Materials and Methods

4.1. Plant Materials and Treatments

In China, ROC22 has been the main sugarcane cultivar grown for the past 20 years and encompasses approximately 60% of the total sugarcane cultivated area. ROC22 is a *Saccharum* hybrid cultivar which is susceptible to smut disease and results in a poor ratoon performance.

Yacheng05-179, an intergeneric hybrid (BC2) with smut resistant properties, is generated from *S. officinarum* × *S. arundinaceum*. In this study, ROC22 and Yacheng05-179 were used as plant materials and collected from the Key Laboratory of Sugarcane Biology and Genetic Breeding, Ministry of Agriculture, Fuzhou, China.

To analyze the tissue-specific expression level of the target gene, nine healthy and uniform 10-month-old ROC22 plants were randomly selected from one field. The white root, bud, +1 leaf, stem pith, and stem epidermis were immediately frozen in liquid nitrogen and kept at -80 °C until extraction of total RNA. Each sample contained three biological replicates.

For biotic treatment, the robust and healthy stems of 10-month-old ROC22 and Yacheng05-179 were harvested and soaked in water for germination at 32 °C. Then the two-bud setts of both sugarcane cultivars were inoculated with 0.5 μL suspensions of 5×10^6 smut spores/mL (plus 0.01% (*v/v*) Tween-20), while the control was inoculated with aseptic water in 0.01% (*v/v*) Tween-20 [63]. The treated samples were cultured at 28 ± 1 °C in a photoperiod of 16-h light and 8-h darkness. Three biological replicates were set, and five buds were randomly chosen at 0 h, 24 h, 48 h, and 72 h for each biological replicate, respectively.

For abiotic and hormone stimuli, healthy and uniform approximately 4-month-old ROC22 plantlets were transferred to water for one week and then treated with six different exogenous stresses. Two groups were separately cultured in aqueous solutions of 250 mM NaCl and 25% PEG 8000, and the leaves were sampled at 0, 0.5, 3, 6, and 24 h, respectively [63–65]. The other three groups were sprayed with 100 μM ABA, 5 mM SA in 0.01% (*v/v*) Tween-20 and 25 μM MeJA for 0, 3, 6, and 24 h, respectively [63–65]. Each treatment was prepared with three biological replicates that contained three plants. All collected samples were immediately frozen in liquid nitrogen and kept at −80 °C until use.

4.2. RNA Extraction and First-strand cDNA Synthesis

The total RNAs of all the samples were extracted with TRIzol® reagent (Invitrogen, Shanghai, China). The RNA quality was determined by 1.0% agarose gel electrophoresis and measured at wavelengths of 260 and 280 nm using a spectrophotometer (NanoVueplus, GE, USA). The residual DNA was removed by DNase I (Promega, Madison, WI, USA). The RevertAid First Strand cDNA Synthesis Kit (Fermentas, Shanghai, China) was used to synthesize the first-strand cDNA from ROC22 and Yacheng05-179 leaves which was treated as templates for cloning the target gene. Prime-Script™ RT Reagent Kit (Perfect Real Time) (TaKaRa Biotechnology, Dalian, China) was used to synthesize the first-strand cDNA of the other samples for expression profile analysis.

4.3. Cloning, Sequencing, and Bioinformatic Analysis of the ScWRKY3 Gene

A gene which codes for a predicted *WRKY* transcriptional regulator named *ScWRKY3* was screened from our previous transcriptome data of sugarcane infected by smut fungus [29]. The specific amplification primers (Table S2) were designed using National Center of Biotechnology Information (NCBI) online software (https://www.ncbi.nlm.nih.gov/tools/primer-blast/). The reverse transcription-polymerase chain reaction (RT-PCR) system contained 1.0 μL cDNA template, 1.0 μL each of the forward and reverse primers (10 μM), 2.5 μL 10× ExTaq buffer (Mg^{2+} plus), 2.0 μL dNTPs (2.5 mM), and 0.125 μL ExTaq enzyme (5.0 U/μL) (TaKaRa Biotechnology, Dalian, China), and 17.375 μL ddH_2O. The RT-PCR reaction conditions were as follows: 94 °C for 4 min; 35 cycles of 94 °C for 30 s, 58 °C for 30 s, and 72 °C for 1 min 30 s; and 72 °C for 10 min. The amplified fragment, which had been gel-purifed using a Gel Extraction Kit (Tiangen, Beijing, China), was linked to the pMD19-T vector (TaKaRa Biotechnology, Dalian, China) and transformed into *Escherichia coli* strain DH5α cells. The positive clones were selected for sequencing (Biosune, Fuzhou, China).

The sequence of the *ScWRKY3* gene was analyzed using the ORF Finder (https://www.ncbi.nlm.nih.gov/orffinder/) and a conserved domains program (http://www.ncbi.nlm.nih.gov/Structure/cdd/wrpsb.cgi) [66]. ProtParam (https://web.expasy.org/protparam/) [67] and NPS@ srever (https://npsa-prabi.ibcp.fr/cgi-bin/npsa_automat.pl?page=/NPSA/npsa_hnn.html) [68] were used for analyzing the primary structure and secondary structure of the ScWRKY3 protein, respectively. The online programs SignalP 4.1 Server (http://www.cbs.dtu.dk/services/SignalP/) [69,70], TMHMM Server v. 2.0 (http://www.cbs.dtu.dk/services/TMHMM/) [71], and Euk-mPLoc 2.0 Server (http://www.csbio.sjtu.edu.cn/bioinf/euk-multi-2/) [30] were used to predict the signal peptide, the transmembrane domain, and the subcellular localization of the target protein, respectively. The BLASTp program (https://blast.ncbi.nlm.nih.gov/Blast.cgi?PROGRAM=blastp&PAGE_TYPE=BlastSearch&LINK_LOC=blasthome) in NCBI was used to find the homologous amino acid sequences from other plants. The multiple alignment was performed using DNAMAN 6.0.3.99 software. Then the MEGA 7.0 software [72] with the Maximum Likelihood (ML) (1000 BootStrap) method was used to construct the unrooted phylogenetic tree of ScWRKY3 with sugarcane Sc-WRKY, ScWRKY4, and WRKY proteins from *Arabidopsis thaliana* [14] and other plants [13,73]. The online software MEME Suite 5.0.2 (http://meme.sdsc.edu/meme/intro.html) [73] was used to build the logo representations of the conservative domain and the rest of the alignment.

4.4. Subcellular Localization

The complete coding region of *ScWRKY3* without a stop codon was amplified using the primers *ScWRKY3*-Gate-F and *ScWRKY3*-Gate-R (Table S2), which were designed based on the sequences of *ScWRKY3* and the Gateway® donor vector of pDONR221. The gel-purified product was linked into pDONR221 using the Gateway BP Clonase™ II enzyme mix (Invitrogen, Carlsbad, CA, USA) and transformed into DH5α cells and sequenced (Biosune, Fuzhou, China). The Gateway LR Clonase™ II enzyme mix (Invitrogen) was used to ligate pDONR221-*ScWRKY3* into the subcellular localization vector pMDC83-*GFP* [74]. GV3101 cells, carrying the recombinant vector pMDC83-*ScWRKY3-GFP* or the pMDC83-*GFP* vector were inoculated into LB liquid medium supplemented with 35 μg/mL rifampicin and 50 μg/mL kanamycin, and shaken overnight in an incubator at 200 rpm and 28 °C. Subsequently, Murashige and Skoog (MS) liquid medium was used to dilute the cell density of the *Agrobacterium* solutions to an OD_{600} of 0.8. This was supplemented with 200 μM acetosyringone and cultured in the dark for 30 minutes. Then the *Agrobacterium* solutions were injected into the leaves of eight-leaf stage *N. benthamiana* using a 1.0 mL sterilized syringe [75,76]. After two days of infiltration, the treated leaves were collected and stained with 1.0 μg/mL DAPI solution in dark conditions for one h. The subcellular localization result was observed using a Leica Microsystems microscope (model Leica TCS SP8, Mannheim, Germany) with a 10 × lens, a chroma GFP filter set for EGFP (excitation at 488 nm), and a DAPI filter set for chromatin (excitation at 458 nm) [60].

4.5. Analysis of Transcriptional Activation of ScWRKY3 in Yeast Cells

To analyze the transcriptional activation of ScWRKY3, the Y2HGold-GAL4 yeast two hybrid system (containing four reporter genes, including *AUR1-C*, *HIS3*, *ADE2*, and *MEL1*) was used following the manufacturer's instructions for the Matchmaker Gold yeast two-hybrid system [49]. The *ScWRKY3* gene was PCR-amplified from pMD19-T-*ScWRKY3* using primers *ScWRKY3*-BD-F and *ScWRKY3*-BD-R (Table S2). The gel-purified product was double-digested with *Nde* I and *Bam*H I enzymes, as was the plasmid pGBKT7. Then the recombinant plasmid pGBKT7-*ScWRKY3* was constructed using T4 DNA ligase (5 U/μL) (Thermo Fisher, Shanghai, China). The pGBKT7 vector, containing the nutritional screening marker gene *TRP1*, was used as a negative control. Plasmids of pGBKT7-53+pGADT7-T have been proven to bind the 53 protein and the T protein in yeast cells. The hybrid vector can activate the reporter gene *AUR1-C* on a plate that contains the AbA antibiotics, so it was used as the positive control. The empty vector plasmid pGBKT7 and plasmids pGBKT7-53+pGADT7-T and pGBKT7-*ScWRKY3* were transformed into yeast strain Y2HGold following the manufacturer's protocol for Y2HGold Chemically Competent Cells (TaKaRa Biotechnology, Dalian, China). The positive colonies were screened from selective medium plates for transferring onto the SDO (SD/-Trp, SD minimal medium without tryptophan), the SDO/X (SD/-Trp/X-α-Gal, SDO plates with X-α-D-Galactosidase), and the SDO/X/A (SD/-Trp/X-α-Gal/AbA, SDO/X plates with aureobasidin A) plates, respectively. Then the transcriptional activation activities were calculated by observing and imaging the growth conditions of the yeast cells after incubating for 2–3 days in a 29 °C incubator.

4.6. Analysis of Interaction Between ScWRKY3 and ScWRKY4

Transcriptional activation analysis in this study and a previous study [28] showed that ScWRKY3 and ScWRKY4 (GenBank Accession No. AUV50355.1) did not possess transcriptional activation activity, and the bait protein has no toxic effect on the yeast strain Y2HGold. Hence, the interacting relationship between ScWRKY3 and ScWRKY4 was identified by a yeast two-hybrid system and BiFC analysis. In the yeast two-hybrid system, AD-ScWRKY3 or AD-ScWRKY4 was used as a prey vector, and BD-ScWRKY4 or BD-ScWRKY3 was used as a bait vector, respectively. A double-enzyme digestion method was used for bait vector and prey vector construction. The specific primers with corresponding restriction enzyme sites are shown in Table S2. pGADT7-T was used as prey control, and pGBKT7-p53 or pGBKT7-Lam was used as the positive or negative bait control, respectively. These combination constructs, including the positive control pGADT7-T + pGBKT7-p53, the negative control pGADT7-T + pGBKT7-Lam, AD-ScWRKY4 + pGBKT7, AD-ScWRKY3 + pGBKT7, pGADT7-T + BD-ScWRKY3, pGADT7-T + BD-ScWRKY4, AD-ScWRKY3 + BD-ScWRKY4, or AD-ScWRKY4 + BD-ScWRKY3, were co-transformed into yeast strain Y2HGold following the manufacturer's protocol for Y2HGold Chemically Competent Cells (TaKaRa Biotechnology, Dalian, China). Subsequently, the transformed yeast cells were selected using yeast selective medium DDO (SD/-Leu/-Trp) to detect whether all the plasmids were successfully transfected into the yeast strain Y2HGold. Then the interaction between ScWRKY3 and ScWRKY4 was detected using QDO (SD/-Ade/-His/-Leu/-Trp) and QDO/X/A (SD/-Ade/-His/-Leu/-Trp/X-α-Gal/AbA) medium [64]. For BiFC vector construction, we used the Gateway method [77]. The coding sequences of *ScWRKY3* and *ScWRKY4* were amplified and linked into the non-fluorescent fragment in the pUC-SPYNE or pUC-SPYCE vector through LR-recombination using Gateway primers (Table S2). The two cooperating plasmids were transformed into the *N. benthamiana* leaves using the *Agrobacterium*-mediated method [78,79]. After five days of infiltration, the presence of fluorescence from yellow fluorescent protein (YFP) was observed using Leica Microsystems (model Leica TCS SP8, Mannheim, Germany) with a 10 × lens and a YFP filter (excitation at 561 nm).

4.7. Expression Patterns of ScWRKY3 in Sugarcane Tissues under Various Stresses

For the expression pattern analysis of *ScWRKY3* in sugarcane tissues (root, bud, leaf, stem pith, and stem epidermis) and in response to various stresses (NaCl, PEG, SA, MeJA, and ABA), the qRT-PCR primers ScWRKY3-QF/R (Table S2) were designed using the Beacon Designer V8.14 software. Glyceraldehyde-3-phosphate dehydrogenase (*GAPDH*) (GenBank Accession Number: CA254672) was used as the reference gene (Table S2). An ABI 7500 Real-Time PCR System (Applied Biosystems, Foster City, CA, USA) and a SYBR Green PCR Master Mix Kit (Roche, Shanghai, China) were used for qRT-PCR analysis. The qRT-PCR reaction system was subjected to 50 °C for 2 min, 95 °C for 10 min, 95 °C for 15 s, and 59 °C for 1 min, for 40 cycles. A melting curve analysis was performed at 95 °C for 15 s, 60 °C for 1 min, 95 °C for 15 s, and 60 °C for 30 s. Each sample was set up for triplicate technical replicates, and sterile water was used as the negative control template. The $2^{-\Delta\Delta C_T}$ method [80], DPS 9.50 software, and Origin 8 software were adopted to calculate the relative expression of the target gene (Tables S3–S6), to analyze the significance level of the experimental data, and to structure the histogram, respectively. To reduce the effects of mechanical injury on the expression of the target gene in the inoculation test with the smut pathogen, the relative expression level of the *ScWRKY3* gene was determined by subtracting the expression of the sterile water at the corresponding time according to Su et al. [81].

4.8. Transient Expression of ScWRKY3 in N. benthamiana

The Gateway LR Clonase^{TM} II enzyme mix (Invitrogen) was used to ligate pDONR221-*ScWRKY3*, as mentioned above, into the overexpression vector of pEarleyGate 203 [82]. The plasmid of pEarleyGate 203-*ScWRKY3* was transformed from the Gateway LR reaction into *A. tumefaciens* strain GV3101, while the pEarleyGate 203 vector which was transformed into GV3101 alone was used as a control. GV3101 cells were shaken overnight in LB liquid medium supplemented with 35 μg/mL rifampicin and 50 μg/mL kanamycin at 200 rpm and 28 °C. The *Agrobacterium* solutions were collected and resuspended to OD_{600} = 0.8 using the MS liquid medium and supplemented with 200 μM acetosyringone. Then the *Agrobacterium* suspensions were injected into the lower epidermis of the eight-leaf stage of *N. benthamiana* leaves using a 1.0 mL sterilized syringe and cultured at 28 °C with a photoperiod of 16-h light and 8-h darkness [76]. Each group of the injected leaves was collected for RNA extraction to analyze the expression level of *ScWRKY3* in *N. benthamiana* by semi-quantitative PCR with the specific primer *ScWRKY3*-Gate-F/R (Table S2). The *NtEF-1α* (GenBank Accession No. D63396) gene was used as the reference gene. The semi-quantitative PCR program was set as: 94 °C, 4 min; 94 °C, 30 s; 65 °C, 30 s; 72 °C, 1 min plus 30 s; 35 cycles; and 72 °C, 10 min. Two important tobacco pathogens, including the bacterial pathogen of *R. solanacearum* and the fungal pathogen of *F. solani* var. *coeruleum*, were cultured overnight in potato dextrose water (PDW) liquid medium at 200 rpm and 28 °C. Then the two cultured pathogen cells were separately injected into the 1-day overexpressing *N. benthamiana* leaves after being diluted to OD_{600} = 0.6 using 10 mM magnesium chloride ($MgCl_2$) solution. All treated plants were maintained for one week at 28 °C with a photoperiod of 16-h light/8-h darkness to track the changes in leaf symptoms and to analyze the relative transcript level of nine tobacco immunity-associated marker genes, including the HR marker genes *NtHSR201*, *NtHSR203*, and *NtHSR515*; the SA-related genes *NtPR-1a/c* and *NtNPR1*; the JA-associated genes *NtPR2* and *NtPR3*; and the ET synthesis-dependent genes *NtEFE26* and *NtAccdeaminase* (Tables S2 and S7–S9) [83,84]. All the treatments were carried out in three replicates. For representational observation, *Agrobacterium* suspensions carrying the vector pEarleyGate 203 and the recombinant vector pEarleyGate 203-*ScWRKY3* were injected into the left and right side of each selected *N. benthamiana* leaf, respectively.

5. Conclusions

In this study, a novel *ScWRKY3* gene was isolated from sugarcane and functionally characterized. ScWRKY3 belongs to group IIc of the WRKY family as a nucleoprotein, with no auto-activation. It has an interaction with another group IIc sugarcane WRKY protein, ScWRKY4, however the interaction mechanism and its corresponding function need further investigation. ScWRKY3 may participate in sugarcane resistance to drought and salt stimuli, and this resistance may be mediated by ABA signaling pathways. The transcript abundance of *ScWRKY3* was stable in the smut-resistant cultivar Yacheng05-179, while it was down-regulated in the smut-susceptible cultivar ROC22 at 24 h, during inoculation with *S. scitamineum*. In addition, *ScWRKY3* showed a negative regulatory effect on the bacterial pathogen *R. solanacearum* and the fungal disease *F. solani* var. *coeruleum* in *35S::ScWRKY3*-overexpressing *N. benthamiana*. These results may be useful for the functional identification of the WRKY family in sugarcane and good for the interaction analysis of ScWRKY3 with other WRKY proteins or other functional proteins.

Supplementary Materials

Figure S1: Nucleic acid sequences alignment of *ScWRKY3* in ROC22 and Yacheng05-179; **Figure S2.** Secondary structure prediction of ScWRKY3; **Figure S3.** Signal peptide and transmembrane domain prediction of ScWRKY3; **Figure S4.** Subcellular localization prediction of ScWRKY3; **Figure S5.** Amino acid sequences alignment of ScWRKY3 and other WRKYs; **Figure S6.** The logo of predicted conserved motifs in the WRKYs; **Table S1.** Primary structure analysis of ScWRKY3; **Table S2.** Primers used in this study; **Table S3.** Raw calculations of tissue-specific expression of *ScWRKY3* in different 10-month-old ROC22 tissues by qRT-PCR; **Table S4.** Raw calculations of gene expression patterns of *ScWRKY3* in 4-month-old ROC22 plantlets under abiotic stress; **Table S5.** Raw calculations of gene expression of *ScWRKY3* in 4-month-old ROC22 plantlets under plant hormone stress; **Table S6.** Raw calculations of expression of the *ScWRKY3* gene after infection with smut pathogen; **Table S7.** Raw calculations of the transcript level of nine immunity-associated marker genes in the *Nicotiana benthamiana* leaves after one day of agroinfiltration; **Table S8.** Raw calculations of the transcripts of nine immunity-associated marker genes in the *Nicotiana benthamiana* leaves after inoculation with *Ralstonia solanacearum* for one day and seven days; **Table S9**. Raw calculations of the transcripts of nine immunity-associated marker genes in the *Nicotiana benthamiana* leaves after inoculation with *Fusarium solani* var. *coeruleum* for one day and seven days.

Author Contributions: L.W., Y.S., and Y.Q. conceived and designed the research. L.W. and X.Z. prepared the materials. L.W., F.L., X.Z., W.W., T.S., Y.C., M.D., and S.Y. conducted the experiments. L.W., F.L., and X.Z. analyzed the data. L.W. wrote the manuscript. Y.S., Y.Q., and L.P. helped to revise the manuscript. All authors read and approved the final manuscript.

References

1. Ulker, B.; Somssich, I.E. WRKY transcription factors: From DNA binding towards biological function. *Curr. Opin. Plant Biol.* **2004**, *7*, 491–498. [CrossRef] [PubMed]
2. Jiang, J.J.; Ma, S.H.; Ye, N.H.; Jiang, M.; Cao, J.S.; Zhang, J.H. WRKY transcription factors in plant responses to stresses. *J. Integr. Plant Biol.* **2017**, *59*, 86–101. [CrossRef] [PubMed]
3. Singh, K.B.; Foley, R.C.; Oñatesánchez, L. Transcription factors in plant defense and stress responses. *Curr. Opin. Plant Biol.* **2002**, *5*, 430–436. [CrossRef]
4. Chen, L.G.; Song, Y.; Li, S.J.; Li, S.J.; Zhang, L.P.; Zou, C.S.; Yu, D.Q. The role of WRKY transcription factors in plant abiotic stresses. *Biochim. Biophys. Acta* **2012**, *1819*, 120–128. [CrossRef] [PubMed]
5. Rushton, P.J.; Somssich, I.E.; Ringler, P.; Shen, Q.J. WRKY transcription factors. *Trends Plant Sci.* **2010**, *15*, 247–258. [CrossRef] [PubMed]
6. Eulgem, T.; Rushton, P.J.; Robatzek, S.; Somssich, I.E. The WRKY superfamily of plant transcription factors. *Trends Plant Sci.* **2000**, *5*, 199–206. [CrossRef]
7. Mohanta, T.K.; Park, Y.H.; Bae, H. Novel genomic and evolutionary insight of WRKY transcription factors in plant lineage. *Sci. Rep.* **2016**, *6*, 37309. [CrossRef] [PubMed]
8. Chen, F.; Hu, Y.; Vannozzi, A.; Wu, K.C.; Cai, H.Y.; Qin, Y.; Mullis, A.; Lin, Z.G.; Zhang, L.S. The WRKY transcription factor family in model plants and crops. *Plant Sci.* **2018**, *36*, 1–25. [CrossRef]
9. Ramamoorthy, R.; Jiang, S.Y.; Kumar, N.; Venkatesh, P.N.; Ramachandran, S. A comprehensive transcriptional profiling of the *WRKY* gene family in rice under various abiotic and phytohormone treatments. *Plant. Cell Physiol.* **2008**, *49*, 865–879. [CrossRef] [PubMed]
10. Wei, K.; Chen, J.; Chen, Y.F.; Wu, L.J.; Xie, D.X. Multiple-strategy analyses of ZmWRKY subgroups and functional exploration of *ZmWRKY* genes in pathogen responses. *Mol. Biosyst.* **2012**, *8*, 1940–1949. [CrossRef] [PubMed]
11. Muthamilarasan, M.; Bonthala, V.S.; Khandelwal, R.; Jaishankar, J.; Shweta, S.; Nawaz, K.; Prasad, M. Global analysis of WRKY transcription factor superfamily in *Setaria* identifies potential candidates involved in abiotic stress signaling. *Front. Plant Sci.* **2015**, *6*, 910. [CrossRef] [PubMed]
12. Pandey, S.P.; Somssich, I.E. The role of WRKY transcription factors in plant immunity. *Plant Physiol.* **2009**, *150*, 1648–1655. [CrossRef] [PubMed]
13. Mangelsen, E.; Kilian, J.; Berendzen, K.W.; Kolukisaoglu, U.H.; Harter, K.; Jansson, C.; Wanke, D. Phylogenetic and comparative gene expression analysis of barley (*Hordeum vulgare*) WRKY transcription factor family reveals putatively retained functions between monocots and dicots. *BMC Genomics* **2008**, *9*, 194. [CrossRef] [PubMed]
14. Wei, X.A.; Yao, W.J.; Jiang, T.B.; Zhou, B.R. Identification of *WRKY* gene in response to abiotic stress from WRKY transcirption factor gene family of *Arabidopsis thaliana*. *J. Northeast. For. Univ.* **2016**, *44*, 45–48.
15. Zhang, T.; Tan, D.; Zhang, L.; Zhang, X.; Han, Z. Phylogenetic analysis and drought-responsive expression profiles of the WRKY transcription factor family in maize. *Agric. Gene* **2017**, *3*, 99–108. [CrossRef]
16. Wu, J.; Chen, J.B.; Wang, L.F.; Wang, S.M. Genome-wide investigation of WRKY transcription factors involved in terminal drought stress response in common bean. *Front. Plant Sci.* **2017**, *8*, 380. [CrossRef] [PubMed]
17. Qiu, Y.Q.; Jing, S.J.; Fu, J.; Li, L.; Yu, D.Q. Cloning and analysis of expression profile of 13 *WRKY* genes in rice. *Chin. Sci. Bull.* **2004**, *49*, 2159–2168. [CrossRef]

18. Wu, H.L.; Ni, Z.F.; Yao, Y.Y.; Guo, G.G.; Sun, Q.X. Cloning and expression profiles of 15 genes encoding WRKY transcription factor in wheat (*Triticum aestivem* L.). *Prog. Nat. Sci. Mater. Int.* **2008**, *18*, 697–705. [CrossRef]
19. Dong, J.; Chen, C.H.; Chen, Z.X. Expression profiles of the *Arabidopsis WRKY* gene superfamily during plant defense response. *Plant Mol. Biol.* **2003**, *51*, 21–37. [CrossRef] [PubMed]
20. Kim, K.C.; Lai, Z.; Fan, B.; Chen, Z. *Arabidopsis* WRKY38 and WRKY62 transcription factors interact with histone deacetylase 19 in basal defense. *Plant Cell* **2008**, *20*, 2357–2371. [CrossRef] [PubMed]
21. Liu, D.L.; Leib, K.; Zhao, P.Y.; Kogel, K.H.; Langen, G. Phylogenetic analysis of barley WRKY proteins and characterization of HvWRKY1 and repressors of the pathogen-inducible gene *HvGER4c*. *Mol. Genet. Genomics* **2014**, *289*, 1331–1345. [CrossRef] [PubMed]
22. Peng, Y.; Bartley, L.E.; Chen, X.W.; Dardick, C.; Chern, M.; Ruan, R.; Canlas, P.E.; Ronald, P.C. *OsWRKY62* is a negative regulator of basal and *Xa21*-mediated defense against *Xanthomonas orvzae* pv. *orvzae* in rice. *Mol. Plant* **2008**, *1*, 446–458. [CrossRef] [PubMed]
23. Berri, S.; Abbruscato, P.; Faivre-Rampant, O.; Brasileiro, A.C.; Fumasoni, I.; Satoh, K.; Kikuchi, S.; Mizzi, L.; Morandini, P.; Pè, M.E.; et al. Characterization of WRKY co-regulatory networks in rice and *Arabidopsis*. *BMC Plant Biol.* **2009**, *9*, 120. [CrossRef] [PubMed]
24. Banerjee, A.; Roychoudhury, A. WRKY proteins: Signaling and regulation of expression during abiotic stress responses. *Sci. World J.* **2015**, 807560. [CrossRef] [PubMed]
25. Chen, R.K.; Xu, L.P.; Lin, Y.Q. *Modern Sugarcane Genetic Breeding*; China Agriculture Press: Beijing, China, 2011.
26. Lambais, M.R. *In silico* differential display of defense-related expressed sequence tags from sugarcane tissues infected with *Diazotrophic endophytes*. *Genet. Mol. Biol.* **2001**, *24*, 103–111. [CrossRef]
27. Liu, J.X.; Que, Y.X.; Guo, J.L.; Xu, L.P.; Wu, J.Y.; Chen, R.K. Molecular cloning and expression analysis of a WRKY transcription factor in sugarcane. *Afr. J. Biotechnol.* **2012**, *11*, 6434–6444.
28. Wang, L.; Liu, F.; Dai, M.J.; Sun, T.T.; Su, W.H.; Wang, C.F.; Zhang, X.; Mao, H.Y.; Su, Y.C.; Que, Y.X. Cloning and expression characteristic analysis of *ScWRKY4* gene in sugarcane. *Acta Agron. Sin.* **2018**, *44*, 1367–1379.
29. Que, Y.X.; Su, Y.C.; Guo, J.L.; Wu, Q.B.; Xu, L.P. A global view of transcriptome dynamics during *Sporisorium scitamineum* challenge in sugarcane by RNAseq. *PLoS ONE* **2014**, *9*, e106476. [CrossRef] [PubMed]
30. Chou, K.C.; Shen, H.B. A new method for predicting the subcellular localization of eukaryotic proteins with both single and multiple sites: Euk-mPLoc 2.0. *PLoS ONE* **2010**, *5*, e9931. [CrossRef] [PubMed]
31. Ishiguro, S.; Nakamura, K. Characterization of a cDNA encoding a novel DNA-binding protein, *SPF1*, that recognizes SP8 sequences in the 5′ upstream regions of genes coding for sporamin and beta-amylase from sweet potato. *Mol. Gen. Genet.* **1994**, *244*, 563–571. [CrossRef] [PubMed]
32. Eulgem, T.; Rushton, P.J.; Schmelzer, E.; Hahlbrock, K.; Somssich, I.E. Early nuclear events in plant defence signalling: Rapid gene activation by WRKY transcription factors. *Embo J.* **1999**, *18*, 4689–4699. [CrossRef] [PubMed]
33. Liu, L.S.; White, M.J.; Macrae, T. Transcription factors and their genes in higher plants. *FEBS J.* **2010**, *262*, 247–257.
34. Fan, Z.Q.; Tan, X.L.; Shan, W.; Kuang, J.F.; Lu, W.J.; Chen, J.Y. BrWRKY65, a WRKY transcription factor, is involved in regulating three leaf senescence-associated genes in Chinese flowering cabbage. *Int. J. Mol. Sci.* **2017**, *18*, 1228.
35. Wan, Y.Q.; Mao, M.Z.; Wan, D.L.; Yang, Q.; Yang, F.Y.; Mandlaa; Li, G.J.; Wang, R.G. Identification of the *WRKY* gene family and functional analysis of two genes in *Caragana intermedia*. *BMC Plant Biol.* **2018**, *18*, 31. [CrossRef] [PubMed]
36. Lee, H.; Cha, J.; Choi, C.; Choi, N.; Ji, H.S.; Park, S.R.; Lee, S.; Hwang, D.J. Rice *WRKY11* plays a role in pathogen defense and drought tolerance. *Rice* **2018**, *11*, 5. [CrossRef] [PubMed]
37. Wang, H.H.; Meng, J.; Peng, X.X.; Tang, X.K.; Zhou, P.L.; Xiang, J.H.; Deng, X.B. Rice *WRKY4* acts as a transcriptional activator mediating defense responses toward *Rhizoctonia solani*, the causing agent of rice sheath blight. *Plant Mol. Biol.* **2015**, *89*, 157–171. [CrossRef] [PubMed]
38. Fields, S.; Song, O. A novel genetic system to detect protein-protein interactions. *Nature* **1989**, *340*, 245. [CrossRef] [PubMed]
39. Sébastien, B.; Li, J.; Palva, E.T. WRKY54 and WRKY70 co-operate as negative regulators of leaf senescence in *Arabidopsis thaliana*. *J. Exp. Bot.* **2012**, *63*, 2667–2679.

40. Wang, X.T.; Zeng, J.; Li, Y.; Rong, X.L.; Sun, J.T.; Sun, T.; Li, M.; Wang, L.Z.; Feng, Y.; Chai, R.H.; et al. Expression of *TaWRKY44*, a wheat *WRKY* gene, in transgenic tobacco confers multiple abiotic stress tolerances. *Front. Plant Sci.* **2015**, *6*, 615. [CrossRef] [PubMed]
41. Chi, Y.C.; Yang, Y.; Zhou, Y.; Zhou, J.; Fan, B.F.; Yu, J.Q.; Chen, Z.X. Protein-protein interactions in the regulation of WRKY transcription factors. *Mol. Plant* **2013**, *6*, 287–300. [CrossRef] [PubMed]
42. Robatzek, S.; Somssich, I.E. Targets of *AtWRKY6* regulation during plant senescence and pathogen defense. *Genes Dev.* **2002**, *16*, 1139–1149. [CrossRef] [PubMed]
43. Popescu, S.C.; Popeseu, G.V.; Bachan, S.; Zhang, Z.; Gerstein, M.; Snyder, M.; Dinesh-Kumar, S.E. MAPK target networks in *Arabidopsis thaliana* revealed using functional protein microarrays. *Genes Dev.* **2009**, *23*, 80–92. [CrossRef] [PubMed]
44. Zhou, L.; Wang, N.N.; Kong, L.; Gong, S.Y.; Li, Y.; Li, X.B. Molecular characterization of 26 cotton *WRKY* genes that are expressed differentially in tissues and are induced in seedlings under high salinity and osmotic stress. *Plant Cell Tissue Org. Cult.* **2014**, *119*, 141–156. [CrossRef]
45. Bakshi, M.; Oelmüller, R. WRKY transcription factors: Jack of many trades in plants. *Plant Signal. Behav.* **2014**, *9*, e27700. [CrossRef] [PubMed]
46. Bartels, D.; Sunkar, R. Drought and salt tolerance in plants. *Crit. Rev. Plant Sci.* **2005**, *24*, 23–58. [CrossRef]
47. Qiao, Z.; Li, C.L.; Zhang, W. *WRKY1* regulates stomatal movement in drought-stressed *Arabidopsis thaliana*. *Plant Mol. Biol.* **2016**, *91*, 53–65. [CrossRef] [PubMed]
48. Jiang, Y.J.; Yu, D.Q. *WRKY57* regulates *JAZ* genes transcriptionally to compromise *Botrytis cinerea* resistance in *Arabidopsis thaliana*. *Plant Physiol.* **2016**, *171*, 2771–2782. [CrossRef] [PubMed]
49. Jones, J.D.; Dang, J.L. The plant immune system. *Nature* **2006**, *444*, 323–329. [CrossRef] [PubMed]
50. Métraux, J.P.; Nawrath, C.; Genoud, T. Systemic acquired resistance. *Euphytica* **2002**, *124*, 23–243. [CrossRef]
51. Katagiri, F. A global view of defense gene expression regulation—A highly interconnected signaling network. *Curr. Opin. Plant Biol.* **2004**, *7*, 506–511. [CrossRef] [PubMed]
52. Kunkel, B.N.; Brooks, D.M. Cross talk between signaling pathways in pathogen defense. *Curr. Opin. Plant Biol.* **2002**, *5*, 325–331. [CrossRef]
53. Chen, X.T.; Liu, J.; Lin, G.F.; Wang, A.; Wang, Z.G.; Lu, G.D. Overexpression of *AtWRKY28* and *AtWRKY75* in *Arabidopsis* enhances resistance to oxalic acid and *Sclerotinia sclerotiorum*. *Plant Cell Rep.* **2013**, *32*, 1589–1599. [CrossRef] [PubMed]
54. Wang, H.H.; Hao, J.J.; Chen, X.J.; Hao, Z.N.; Wang, X.; Lou, Y.G.; Peng, Y.L.; Guo, Z.J. Overexpression of rice *WRKY89* enhances ultraviolet B tolerance and disease resistance in rice plants. *Plant Mol. Biol.* **2007**, *65*, 799–815. [CrossRef] [PubMed]
55. Qiu, D.Y.; Xiao, J.; Ding, X.H.; Xiong, M.; Cai, M.; Cao, Y.L.; Li, X.H.; Xu, C.G.; Wang, S.P. *OsWRKY13* mediates rice disease resistance by regulating defense-related genes in salicylate- and jasmonate-dependent signaling. *Mol. Plant-Microbe Interact.* **2007**, *20*, 492–499. [CrossRef] [PubMed]
56. Qiu, D.Y.; Xiao, J.; Xie, W.B.; Liu, H.B.; Li, X.H.; Xiong, L.Z.; Wang, S.P. Rice gene network inferred from expression profiling of plants overexpressing *OsWRKY13*, a positive regulator of disease resistance. *Mol. Plant* **2008**, *1*, 538–551. [PubMed]
57. Zheng, Z.Y.; Mosher, S.L.; Fan, B.F.; Klessig, D.F.; Chen, Z.X. Functional analysis of *Arabidopsis* WRKY25 transcription factor in plant defense against *Pseudomonas syringae*. *BMC Plant Biol.* **2007**, *7*, 2. [CrossRef] [PubMed]
58. Jiang, Y.; Deyholos, M.K. Functional characterization of *Arabidopsis* NaCl-inducible WRKY25 and WRKY33 transcription factors in abiotic stresses. *Plant Mol. Biol.* **2009**, *69*, 91–105. [CrossRef] [PubMed]
59. Yokotani, N.; Sato, Y.; Tanabe, S.; Chujo, T.; Shimizu, T.; Okada, K.; Yamane, K.; Shimono, M.; Sugano, S.; Takatsuji, H.; et al. WRKY76 is a rice transcriptional repressor playing opposite roles in blast disease resistance and cold stress tolerance. *J. Exp. Bot.* **2013**, *64*, 5085–5097. [CrossRef] [PubMed]
60. Liu, X.F.; Song, Y.Z.; Xing, F.Y.; Wang, N.; Wen, F.J. *GhWRKY25*, a group I *WRKY* gene from cotton, confers differential tolerance to abiotic and biotic stresses in transgenic *Nicotiana benthamiana*. *Protoplasma* **2015**, *253*, 1–17. [CrossRef] [PubMed]
61. Wang, X.; Yan, Y.; Li, Y.Z.; Chu, X.Q.; Wu, C.G.; Guo, X.Q. *GhWRKY40*, a multiple stress-responsive cotton *WRKY* gene, plays an important role in the wounding response and enhances susceptibility to *Ralstonia solanacearum* infection in transgenic *Nicotiana benthamiana*. *PLoS ONE* **2014**, *9*, e93577. [CrossRef] [PubMed]

62. Mukhtar, M.S.; Deslandes, L.; Auriac, M.; Marco, Y.; Somssich, I.E. The Arabidopsis transcription factor WRKY27 influences wilt disease symptom development caused by *Ralstonia solanacearum*. *Plant J.* **2008**, *56*, 935. [CrossRef] [PubMed]

63. Su, Y.C.; Guo, J.L.; Ling, H.; Chen, S.S.; Wang, S.S.; Xu, L.P.; Allan, A.C.; Que, Y.X. Isolation of a novel peroxisomal catalase gene from sugarcane, which is responsive to biotic and abiotic stresses. *PLoS ONE* **2014**, *9*, e84426. [CrossRef] [PubMed]

64. Li, H.; Gao, Y.; Xu, H.; Dai, Y.; Deng, D.Q.; Chen, J.M. *ZmWRKY33*, a WRKY maize transcription factor conferring enhanced salt stress tolerances in *Arabidopsis*. *Plant Growth Regul.* **2013**, *70*, 207–216. [CrossRef]

65. Scarpeci, T.E.; Zanor, M.I.; Zanor, M.I.; Mueller-Roeber, B.; Valle, E.M. Overexpression of *AtWRKY30* enhances abiotic stress tolerance during early growth stages in *Arabidopsis thaliana*. *Plant Mol. Biol.* **2013**, *83*, 265–277. [CrossRef] [PubMed]

66. Marchlerbauer, A.; Bo, Y.; Han, L.Y.; He, J.; Lanczycki, C.J.; Lu, S.N.; Chitsaz, F.; Derbyshire, M.K.; Geer, R.C.; Gonzales, N.R.; et al. CDD/SPARCLE: Functional classification of proteins via subfamily domain architectures. *Nucleic Acids Res.* **2017**, *45*, 200–203. [CrossRef] [PubMed]

67. Gasteiger, E.; Hoogland, C.; Gattiker, A.; Duvaud, S.; Wilkins, M.R.; Appel, R.D.; Bairoch, A. Protein identification and analysis tools on the ExPASy server. In *The Proteomics Protocols Handbook*; Humana Press: New York, NY, USA, 2005; pp. 571–607.

68. Combet, C.; Blanchet, C.; Geourjon, C.; Deléage, G. NPS@: Network protein sequence analysis. *Trends Biochem. Sci.* **2000**, *25*, 147–150. [CrossRef]

69. Petersen, T.N.; Brunak, S.; Heijne, G.; Nielsen, H. SignalP 4.0: Discriminating signal peptides from transmembrane regions. *Nat. Methods* **2011**, *8*, 785–786. [CrossRef] [PubMed]

70. Nielsen, H. Predicting secretory proteins with SignalP. *Methods Mol. Biol.* **2017**, *1611*, 59–73. [PubMed]

71. Krogh, A.; Larsson, B.; Heijne, G.V.; Sonnhammer, E.L.L. Predicting transmembrane protein topology with a hidden markov model: Application to complete genomes. *J. Mol. Biol.* **2001**, *305*, 567–580. [CrossRef] [PubMed]

72. Kumar, S.; Stecher, G.; Tamura, K. MEGA7: Molecular evolutionary genetics analysis version 7.0 for bigger datasets. *Mol. Biol. Evol.* **2016**, *33*, 1870–1874. [CrossRef] [PubMed]

73. Wang, C.; Deng, P.; Chen, L.; Wang, X.; Ma, H.; Hu, W.; Yao, N.C.; Feng, Y.; Chai, R.H.; Yang, G.X.; et al. A wheat WRKY transcription factor TaWRKY10 confers tolerance to multiple abiotic stresses in transgenic tobacco. *PLoS ONE* **2013**, *8*, e65120. [CrossRef] [PubMed]

74. Curtis, M.D.; Grossniklaus, U. A gateway cloning vector set for high-throughput functional analysis of genes in planta. *Plant Phyosiol.* **2003**, *133*, 462–469. [CrossRef] [PubMed]

75. Hwang, I.S.; Hwang, B.K. Requirement of the cytosolic interaction between pathogenesis-related protein10 and leucine-rich repeat protein1 for cell death and defense signaling in pepper. *Plant Cell* **2012**, *24*, 1675–1690.

76. Fan, Z.Q.; Kuang, J.F.; Fu, C.C.; Shan, W.; Han, Y.C.; Xiao, Y.Y.; Ye, Y.J.; Lu, W.J.; Lakshmanan, P.; Duan, X.W.; et al. The banana transcriptional repressor *MaDEAR1* negatively regulates cell wall-modifying genes involved in fruit ripening. *Front. Plant Sci.* **2016**, *7*, 1021. [CrossRef] [PubMed]

77. Shen, C.J.; Wang, S.K.; Bai, Y.H.; Wu, Y.R.; Zhang, S.N.; Chen, M.; Guilfoyle, T.J.; Wu, P.; Qi, Y.H. Functional analysis of the structural domain of ARF proteins in rice (*Oryza sativa* L.). *J. Exp. Bot.* **2010**, *61*, 3971–3981. [CrossRef] [PubMed]

78. Schütze, K.; Harter, K.; Chaban, C. Bimolecular fluorescence complementation (BiFC) to study protein-protein interactions in living plant cells. *Methods Mol. Biol.* **2009**, *479*, 189–202. [PubMed]

79. Mersereau, M.; Pazour, G.J.; Das, A. Efficient transformation of *Agrobacterium tumefaciens* by electroporation. *Gene* **1990**, *90*, 149–151. [CrossRef]

80. Livak, K.J.; Schmittgen, T.D. Analysis of relative gene expression data using real-time quantitative PCR and the $2^{-\Delta\Delta C_T}$ method. *Methods* **2001**, *25*, 402–408. [CrossRef] [PubMed]

81. Su, Y.C.; Wang, Z.Q.; Li, Z.; Xu, L.P.; Que, Y.X.; Dai, M.J.; Chen, Y.H. Molecular cloning and functional identification of peroxidase gene *ScPOD02* in sugarcane. *Acta Agron. Sin.* **2017**, *43*, 510–521. [CrossRef]

82. Earley, K.W.; Haag, J.R.; Pontes, O.; Opper, K.; Juehne, S.; Song, K.; Pikaard, C.S. Gateway-compatible vectors for plant functional genomics and proteomics. *Plant J.* **2006**, *45*, 616–629. [CrossRef] [PubMed]

83. Peng, Q.; Su, Y.C.; Ling, H.; Ahmad, W.; Gao, S.W.; Guo, J.L.; Que, Y.X.; Xu, L.P. A sugarcane pathogenesis-related protein, *ScPR10*, plays a positive role in defense responses under

Sporisorium scitamineum, SrMV, SA, and MeJA stresses. *Plant Cell Rep.* **2017**, *36*, 1427–1440. [CrossRef] [PubMed]

84. Lai, Y.; Dang, F.F.; Lin, J.; Yu, L.; Shi, Y.L.; Xiao, Y.H.; Huang, M.K.; Lin, J.H.; Chen, C.C.; Qi, A.H. Overexpression of a *Chinese cabbage BrERF11*, transcription factor enhances disease resistance to *Ralstonia solanacearum*, in *tobacco*. *Plant Physiol. Biochem.* **2013**, *62*, 70–78. [CrossRef] [PubMed]

5

Transcriptome Analyses in Different Cucumber Cultivars Provide Novel Insights into Drought Stress Responses

Min Wang [1,2,†], Biao Jiang [1,2,†], Qingwu Peng [1], Wenrui Liu [1], Xiaoming He [1], Zhaojun Liang [1] and Yu'e Lin [1,*]

[1] Vegetable Research Institute, Guangdong Academy of Agricultural Sciences, Guangzhou 510640, China; w.jsun@163.com (M.W.); jiangbiao198354@163.com (B.J.); pengqingwu@gdaas.cn (Q.P.); liuwr10@126.com (W.L.); xiaominghe626@163.com (X.H.); liangzhaojun@gdaas.cn (Z.L.)

[2] Guangdong Key Laboratory for New Technology Research of Vegetables, Guangzhou 510640, China

* Correspondence: cucumber200@163.com

† These authors contributed equally to this work.

Abstract: Drought stress is one of the most serious threats to cucumber quality and yield. To gain a good understanding of the molecular mechanism upon water deficiency, we compared and analyzed the RNA sequencing-based transcriptomic responses of two contrasting cucumber genotypes, L-9 (drought-tolerant) and A-16 (drought-sensitive). In our present study, combining the analysis of phenotype, twelve samples of cucumber were carried out a transcriptomic profile by RNA-Seq under normal and water-deficiency conditions, respectively. A total of 1008 transcripts were differentially expressed under normal conditions (466 up-regulated and 542 down-regulated) and 2265 transcripts under drought stress (979 up-regulated and 1286 down-regulated). The significant positive correlation between RNA sequencing data and a qRT-PCR analysis supported the results found. Differentially expressed genes (DEGs) involved in metabolic pathway and biosynthesis of secondary metabolism were significantly changed after drought stress. Several genes, which were related to sucrose biosynthesis (*Csa3G784370* and *Csa3G149890*) and abscisic acid (ABA) signal transduction (*Csa4M361820* and *Csa6M382950*), were specifically induced after 4 days of drought stress. DEGs between the two contrasting cultivars identified in our study provide a novel insight into isolating helpful candidate genes for drought tolerance in cucumber.

Keywords: *Cucumis sativus* L.; RNA-Seq; DEGs; sucrose; ABA; drought stress

1. Introduction

Drought stress generally occurs when soil water is deficient, leading to a continuous loss of water by transpiration or evaporation [1]. Water deficiency, a key limiting factor in plant growth and development, impacts plant elongation and expansion growth [2,3]. In order to survive under drought stress, plants have to make corresponding adjustments by regulating gene expression of stress-related and signal transduction pathways [4–6], such as reactive oxygen species (ROS)-related genes [7], transcription factors (TFs) [8], and the abscisic acid (ABA) signal transduction pathway [9,10].

Cucumber (*Cucumis sativus* L.), one of the most important vegetable crops in Cucurbitaceae, is originally from the southern Himalayas and shows a preference for warm and moist environment [11]. Previous studies about cucumber resistance on drought have been carried out in different aspects [12–15]. Carbon monoxide (CO) is involved in hydrogen gas (H_2)-induced adventitious root development under stimulated drought stress and alleviates oxidative damage by altering relative physiological index [12]. *CsCER1* is involved in the fruit cuticle synthesis, and overexpressing the gene has been shown to improve the drought tolerance under water-deficiency conditions [13]. Exogenously applied hydrogen peroxide could considerably enhance the cucumber drought resistance by increasing the plant's antioxidative defense system and its capacity for osmotic adjustment [14]. Tobacco PR-2d promoter/*uidA* (GUS) gene is induced in transgenic cucumber and improves the response to biotic and abiotic stimuli [15].

Comparing transcriptome by RNA-seq of various genotypes in different species is one of the most suitable techniques for exploring resistant genes under abiotic stress and elucidating the role of various biological pathways, as well as mechanisms for influencing tolerance to adverse environments [16,17]. When compared with microarray and expressed sequence tag, advantages of RNA-seq showed determination of alternative splicing (AS) events, novel transcripts and digital gene expression at the isoform level [18,19]. In cucumber, the RNA-seq method has been widely employed for performing crucial agricultural functions such as fruit development [20], parthenocarpy [21], flower sex expression [22], and other plant responses to abiotic stresses [23–25].

A transcriptome profiling reveals a mechanism of fruit trichome formation, which plays an important role in plant defense against biotic and abiotic stresses [23]. A total of 121 genes were significantly induced under melatonin treatment, which promoted the cucumber lateral root formation under salt stress [24]. Zhao et al. [25] examined over 23,000 transcripts in cucumber leaves, and found that 364 genes were differentially expressed in response to nitrogen deficiency, providing novel insights into the responses of cucumber to N starvation at the global transcriptome level [25]. However, to the best of our knowledge, no research has been performed on the drought stress in cucumber using compared transcriptome.

In this study, we carried out RNA-sequencing analysis in cucumber to explore the transcriptional variations between a drought-tolerant cultivar L-9 and a drought-sensitive cultivar A-16 under normal and drought conditions. Different drought stress-responsive novel transcript isoforms were identified between L-9 and A-16. Furthermore, we analyzed the differential gene expression patterns in response to drought stresses. Functional categorization of differentially expressed transcripts was carried out to reveal various metabolic pathways involved in drought responses. Overall, this study provides a theoretical basis for further study of the regulatory mechanism of drought tolerance in cucumber.

2. Results

2.1. A-16 Cultivar Is Sensitive to Drought Stress

Ten-day seedlings of L-9 and A-16 (120 plants for three biological replicates, respectively) grown under normal condition (Figure 1A) were treated with water deficiency for 7 days and recovered for 3 days (Figure 1B). Both L-9 and A-16 showed vigorous development before drought; however, A-16 began to exhibit wilting at the top of the growth point after drought stress, and its leaves turned chlorotic and yellow (Figure 1A,B). Approximately 13% of the drought treated A-16 plants survived after the subsequent 3-day recovery, compared with 77% of L-9 plants (Figure 1C). There were no difference of malondialdehyde (MDA) and the enzyme superoxide dismutase (SOD) between L-9 and A-16 before drought, while A-16 presented a prominent increase of MDA and significant decrease of SOD at the 4th day after drought treatment (Figure 1D,E).

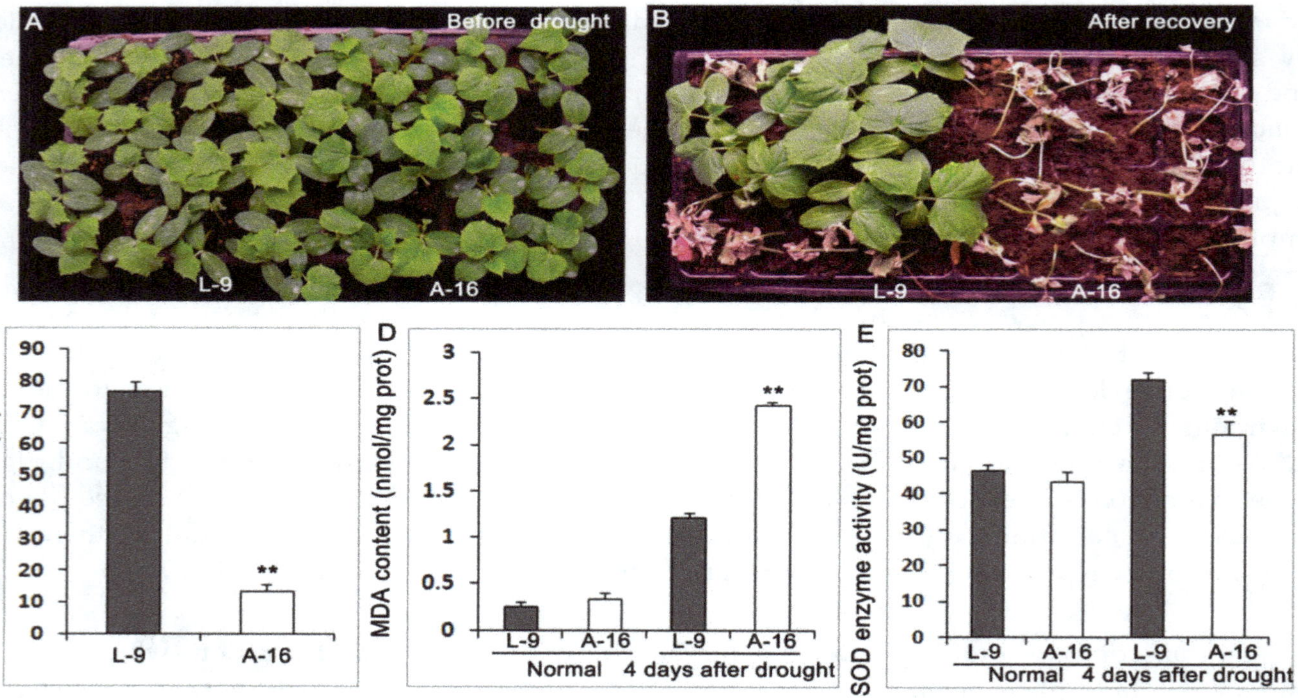

Figure 1. Phenotypes of L-9 and A-16 before drought and after recovery of drought stress. (**A**) L-9 and A-16 plants were grown under normal conditions for 14 days. (**B**) After 7 days drought treatment, seedlings recovered for 3 days. (**C**) Survival rate of plants following the 7-day drought treatment. (**D**,**E**) Measurement of MDA content (**D**) and SOD enzyme activity (**E**) under normal conditions and 4 days after drought. Data is presented as the mean ± standard deviation ($n = 9$). ** $p < 0.01$; Student's t-test.

Before drought, there was no significant difference in chlorophyll content between L-9 and A-16 (Figure 2A). However, the relative content of chlorophyll a decreased to ~34% in L-9 vs. ~52% in A-16, and the chlorophyll b decreased to ~14% and ~33% in L-9 and A-16 after drought treatment, respectively (Figure 2B). These above results indicated that L-9 showed more significant drought tolerance than A-16. In order to compare the ultrastructure of chloroplasts between L-9 and A-16, we used the transmission electron microscopy to observe the leaves at seedling stage. The leaf cells of L-9 contained normal chloroplasts, which showed well-organized lamellar structures with normally stacked grana and thylakoid membranes (Figure 2C–E). However, most cells of A-16 were heteroplastidic, with many more starch grains (Figure 2F–H). These observations implied that the sensitivity to drought stress of A-16 might be related to the abnormal development of chloroplasts in leaves at the early seedling stage.

Additionally, we investigated whether stomatal numbers of A-16 was different from L-9 using scanning electron microscopy (SEM). The result showed that the number of stomas in L-9 (Figure 3A,B) was much less than A-16 (Figure 3C,D) in the same field size, indicating that L-9 lost water more easily when encountering drought stress.

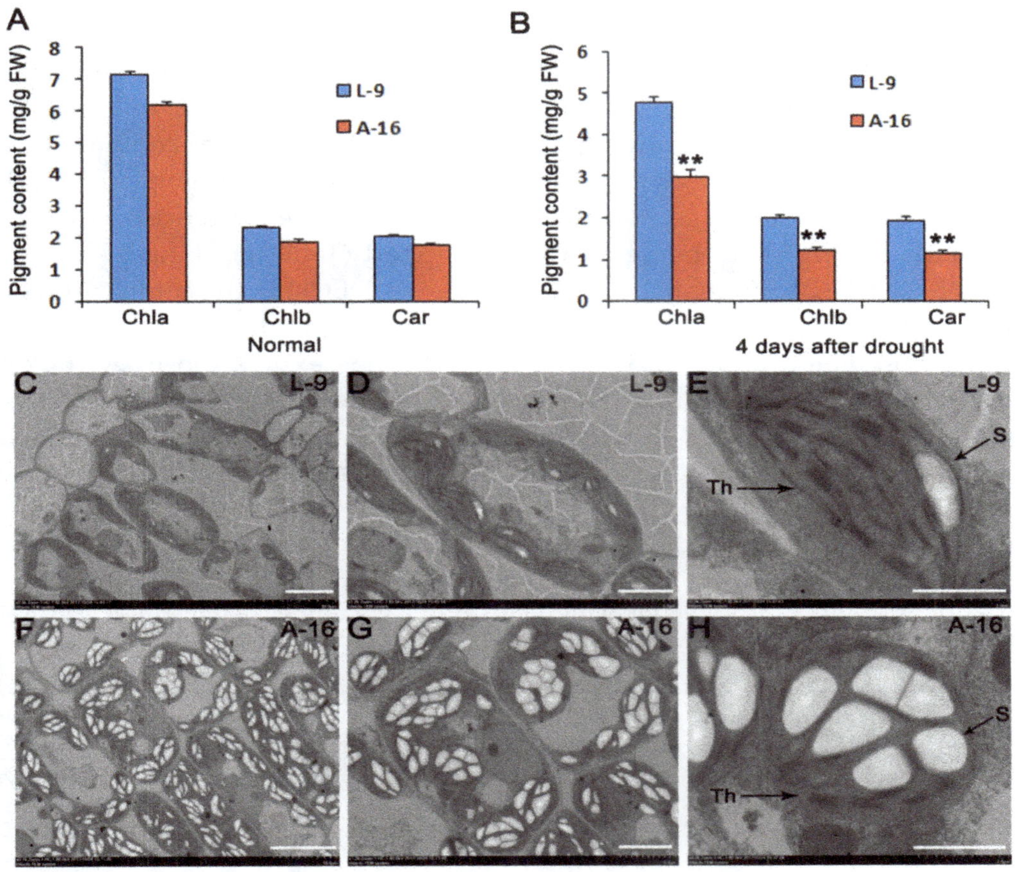

Figure 2. TEM observation of L-9 and A-16 leaves at seedling stage. (**A**) Chlorophyll content of L-9 and A-16 before drought. (**B**) Chlorophyll content of L-9 and A-16 during drought. Data is presented as the mean ± standard deviation ($n = 9$). ** $p < 0.01$; Student's t-test. (**C–H**) Transmission electron microscopic photos of cells from L-9 and A-16. (**C–E**) Mesophyll cells in L-9 plants showed normal, well-ordered chloroplasts. (**F–H**) Cells in A-16 plants displayed some abnormalities and accumulated starch grains. Th: thylakoid, S: starch granule. Bar in (**C,F**): 100 μm. Bar in (**D,G**): 50 μm. Bar in (**E,H**) : 20 μm.

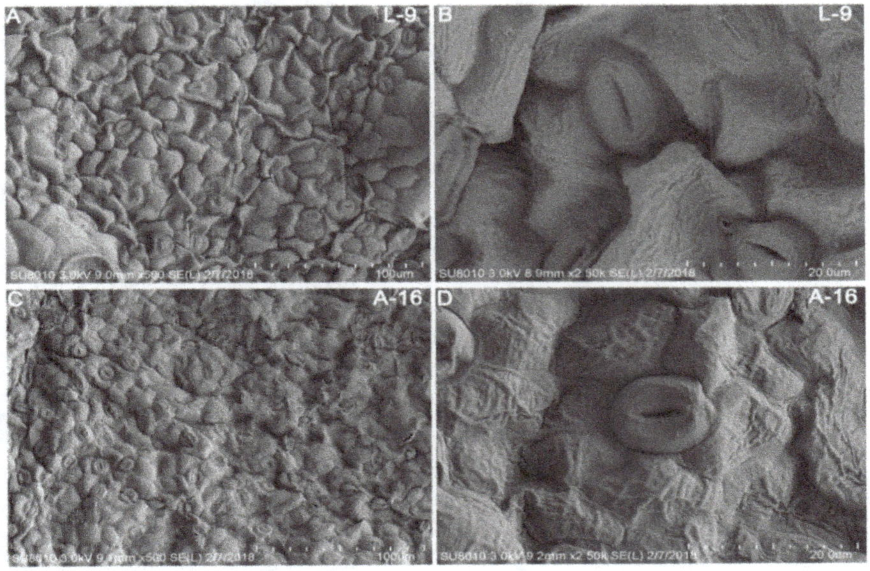

Figure 3. SEM observation of L-9 and A-16 leaves at seedling stage. (**A,B**) Scanning electron microscopy (SEM) images of leaves in L-9. (**C,D**) Scanning electron microscopy (SEM) images of leaves in A-16.

2.2. *Drought Stress Results in Extensive Transcriptomic Reprogramming*

In order to explore the transcriptional variations between L-9 and A-16 under normal and drought conditions, respectively, we carried out RNA-sequencing. A total of about 23 million clean reads were obtained per sample (Table 1) after removing the low-quality and adaptor-containing reads. At least 1.14 Gb clean data were acquired for each sample (Table 1). In total, the expression of 21,019 genes was detected. Approximately 96% of the clean reads were mapped to the reference cucumber genome [26], with more than 68% among them being uniquely mapped (Table 1). Finally, we identified 1008 (Table S1) and 2265 (Table S2) differentially expressed genes (DEGs) in the comparison of L-9 vs. A-16 under normal conditions and drought stress, respectively. Among them, under normal conditions, 466 genes were up-regulated and 542 down-regulated (gene expression in A-16 compared with L-9) (Figure 4A). Additionally, 979 up-regulated and 1286 down-regulated genes were identified during drought stress (Figure 4B). Next, in order to validate the RNA-seq results, we randomly selected 16 DEGs and conducted qRT-PCR analysis. The results showed that there was a strong positive correlation (two tailed, R^2 = 0.973) between the RNA-seq and qRT-PCR result (Figure 5), which indicated the accuracy of the RNA-seq data.

Table 1. Mapping results of RNA sequencing reads of the cucumber between L-9 and A-16 under normal condition (C) and 4 days after drought (D).

Sample	Total Clean Reads	Total Clean Bases (Gb)	Total Mapping Ratio %	Uniquely Mapping Ratio %
A-16_C1	23,001,330	1.15	0.9664	0.8976
A-16_C2	22,799,582	1.14	0.9662	0.8986
A-16_C3	23,275,316	1.16	0.966	0.9013
A-16_D1	23,202,361	1.16	0.963	0.8943
A-16_D2	23,239,914	1.16	0.9625	0.8981
A-16_D3	23,127,940	1.16	0.959	0.8924
L-9_C1	23,343,741	1.17	0.966	0.8971
L-9_C2	23,065,366	1.15	0.9617	0.8901
L-9_C3	23,055,733	1.15	0.9634	0.8923
L-9_D1	22,973,680	1.15	0.9639	0.8927
L-9_D2	23,183,117	1.16	0.963	0.897
L-9_D3	23,037,373	1.15	0.956	0.8932

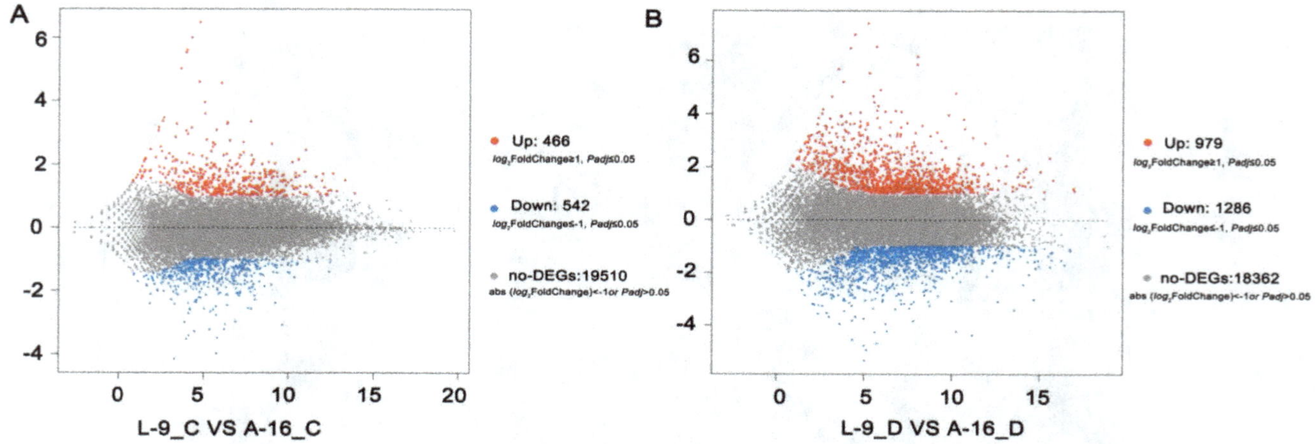

Figure 4. Comparison of different genes expression (DEGs) in leaves between L-9 and A-16 under normal conditions (**A**) and 4 days after drought (**B**). *x*- and *y*-axes represent log2 values of gene expression. Red, brown, and blue correspond to up-regulated, unaltered, and down-regulated gene expression, respectively. If a gene was expressed in just one sample, its expression value in another sample was replaced by the minimum value of all expressed genes in normal and drought samples. The screening threshold is given at the top of the figure.

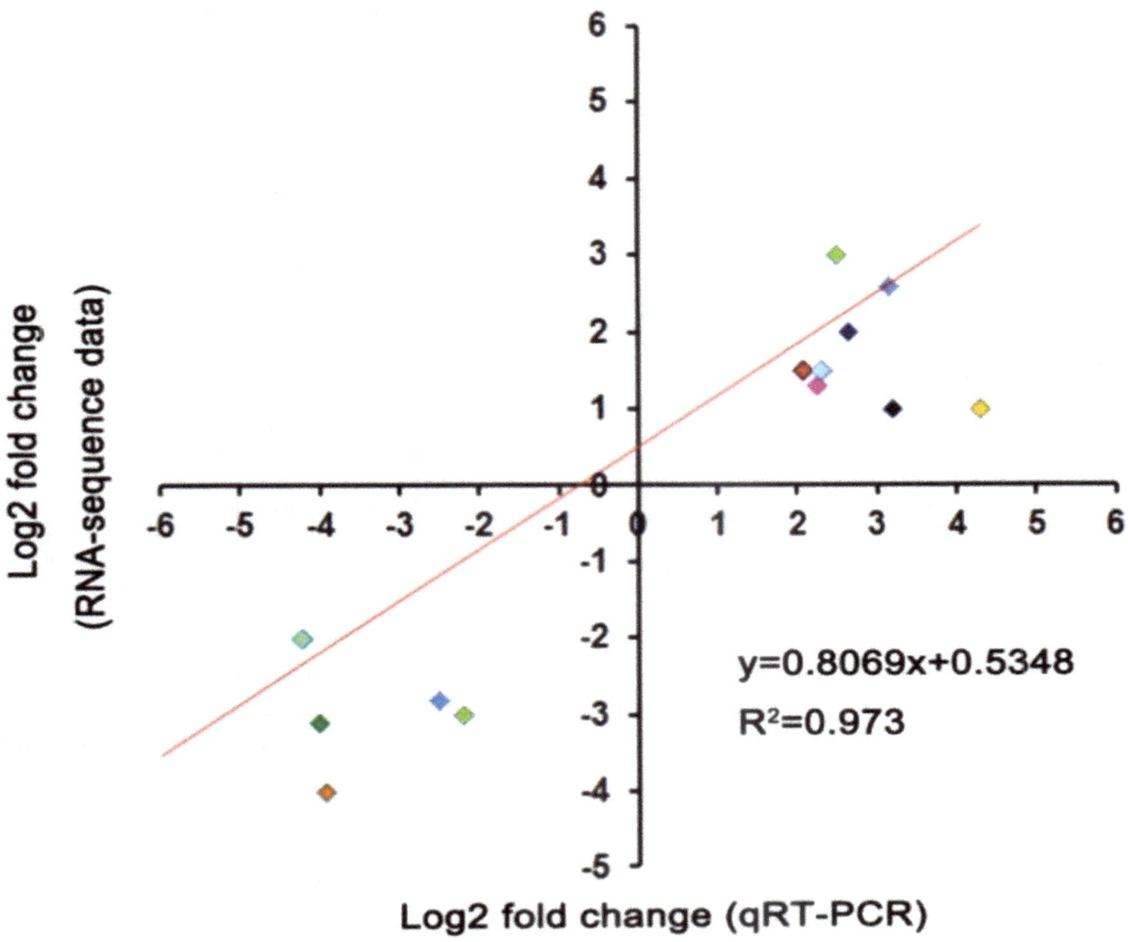

Figure 5. qRT-PCR validation of differentially expressed genes under drought stress. Correlation between the fold change analyzed by RNA-seq (x-axis) and data obtained using qRT-PCR. The different colors represent different genes expression.

2.3. *Functional Classification of Drought-Responsive Genes*

The gene ontology (GO) standardized classification system for gene function was used to analyze DEGs and understand the molecular events involved in drought response. Three categories, including "biological process," "molecular function", and "cellular components", were classified under normal conditions (Figure S1A and Table S3) and drought stress (Figure S1B and Table S4), respectively. The number of the three category genes was prominently increased at 4 days after drought treatment, especially in the metabolic process, membrane, and catalytic activity, followed by subcategories such as cellular process, cell, and binding (Figure S1B).

Next, to examine DEG-associated pathways, they were searched in the KEGG pathway database. The top 20 enriched pathways are shown in Figure 6. The main pathways under normal conditions were "biosynthesis of secondary metabolites", "plant hormone signal transduction", and "MAPK signaling pathway" (Figure 6A and Table S5). When exposed to drought stress, genes related to "metabolic pathways" and "biosynthesis of secondary metabolites" were mostly enriched (Figure 6B and Table S6), indicating that these pathways and processes possibly participated in plant drought resistance. In addition, the category of "starch and sucrose metabolism" was only detected under stress conditions, suggesting these changed genes might contribute to the increased resistance of drought. Under water deficiency, we found that some genes were responsive to water deprivation (Table 2).

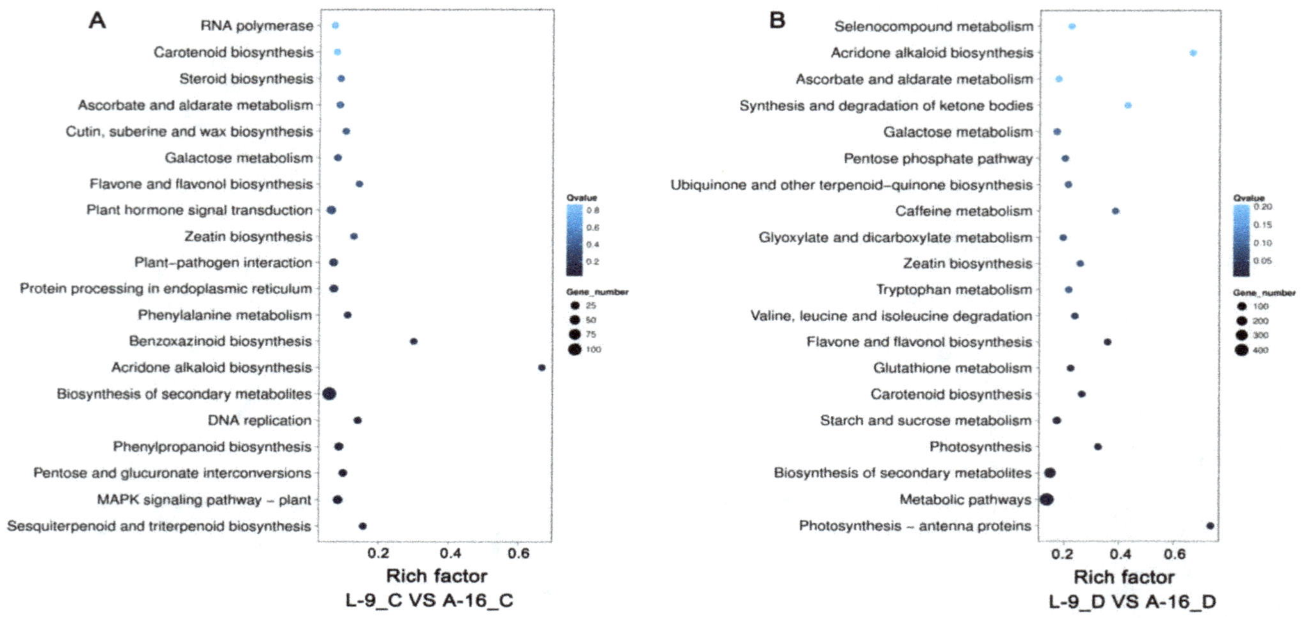

Figure 6. KEGG enrichment of annotated DEGs under three comparisons of normal conditions (**A**) and drought stress (**B**). The *y*-axis indicates the KEGG pathway and the *x*-axis indicates the enrichment factor. A high q-value is represented by light blue, and a low q-value is represented by dark blue.

Table 2. Genes related to sucrose biosynthesis and response to water deprivation.

Gene ID	L-9 Expression	A-16 Expression	Regulation	*p*-Value	Annotation
Csa2G401440	2237.5	1091.6	Down	0.00434039	Sucrose-phosphate synthase
Csa3G784370	3412.9	1102.8	Down	1.20×10^{-6}	Sucrose phosphatase
Csa3G149890	10,202.0	3362.5	Down	3.87×10^{-12}	Glucose-1-phosphate adenylyltransferase
Csa4G001950	2345.5	7726.7	Up	6.84×10^{-6}	Sucrose synthase
Csa4G420150	492.7	239.1	Down	1.17×10^{-5}	4-α-Glucanotransferase
Csa5G568310	4872.5	2423.0	Down	3.78×10^{-6}	Phosphoglucomutase
Csa2G004720	1255.1	2945.9	Up	1.42×10^{-6}	Multiprotein-bridging factor
Csa5G207960	11,815.2	4338.2	Down	1.77×10^{-9}	Omega-3 fatty acid desaturase
Csa3G808370	47.5	102.9	Up	0.00031088	Seed maturation protein LEA 4

2.4. Expression of Genes Involved in Sucrose Biosynthesis and Response to Water Deprivation

Based on the results of GO and KEGG analysis, we chose several DEGs, which were involved in the starch and sucrose synthesis and response to drought stress. A total of 9 transcripts were selected, including 6 genes with sucrose or starch and 3 genes with response to water deprivation (Table 2 and Table S7). The qRT-PCR assay was employed to validate A-16 and L-9 of RNA-seq results under normal and drought stress, respectively. The results showed that no significant changes were detected between these two cultivars before treatment. However, when treated with drought stress for 4 days, six genes were significantly down regulated in A-16, especially genes involved in the sucrose metabolic process, sucrose biosynthetic process, and starch biosynthetic process. The expression of the remaining three genes including genes related to sucrose synthase activity and response to water deprivation, increased significantly in A-16 when compared with L-9 (Figure 7). These results of qRT-PCR were consistent with the RNA-sequencing data.

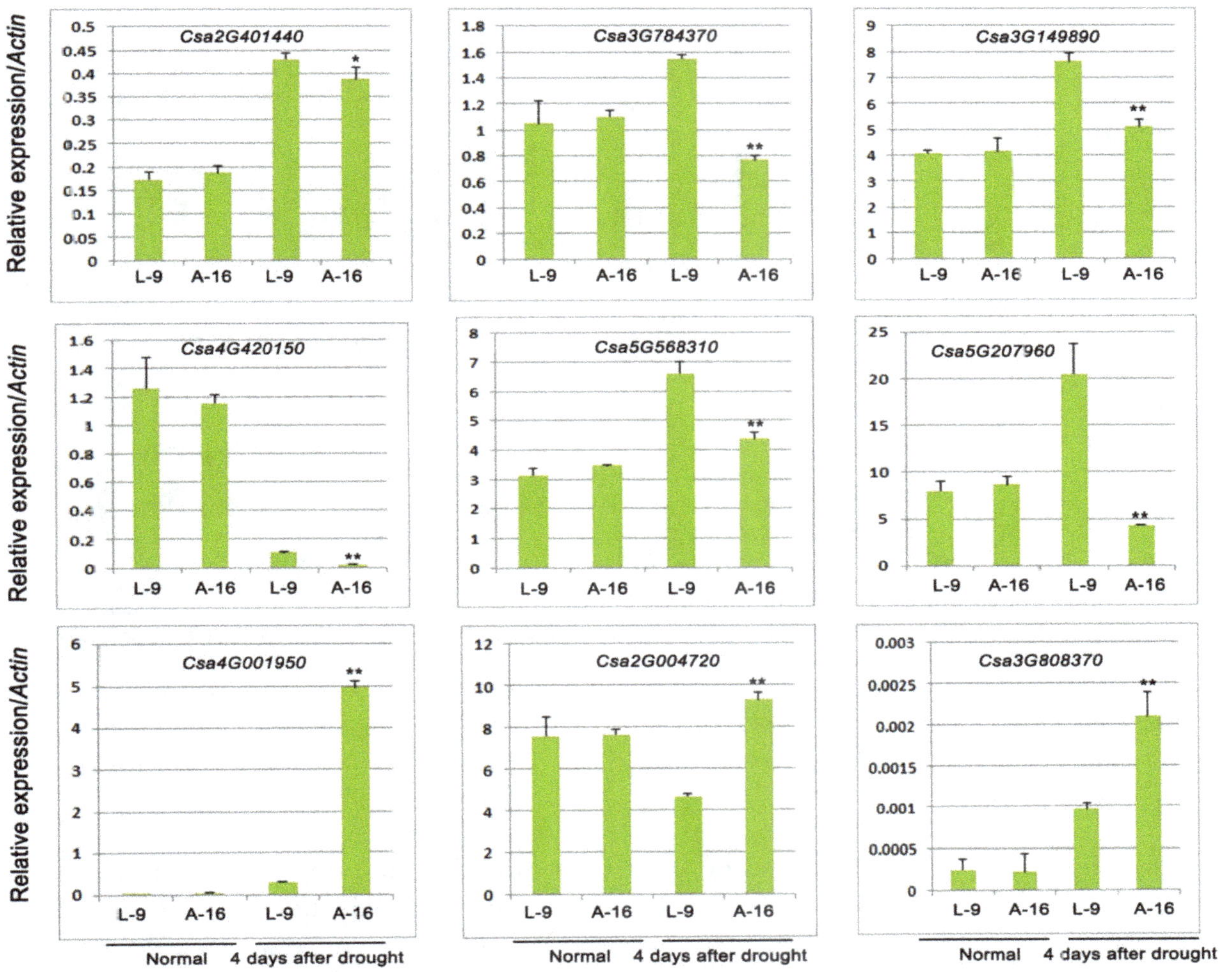

Figure 7. Relative expression of genes related to sucrose biosynthesis and response to water deprivation. Data is presented as the mean $\pm$ standard deviation ($n = 9$). * $0.01 \leq p \leq 0.05$, ** $p \leq 0.01$, Student's t test.

2.5. Analysis of Abscisic Acid (ABA)-Related Genes

Previous studies have reported that plant hormone, especially ABA, plays crucial roles in the regulation of the developmental process and signaling network involved in plant responses to drought stress [27]. Therefore, we selected the ABA-related genes among DEGs of drought stress from RNA-sequencing data (Table 3 and Table S7). In the present study, six genes related to the ABA signaling pathway were verified. The result showed that four genes were up-regulated and two down-regulated prominently (Figure 8), which was consistent with the RNA sequencing results.

Table 3. Genes involved in ABA signaling pathway.

Gene ID	L-9 Expression	A-16 Expression	Regulation	p-Value	Annotation
Csa3G135070	89.6	20.4	Down	3.05×10^{-9}	Calcium-dependent protein kinase
Csa3G133140	463.1	1758.6	Up	1.72×10^{-7}	3-Ketoacyl-CoA thiolase 1
Csa4G361820	1298.2	4273.3	Up	1.04×10^{-17}	NAC domain-containing protein
Csa4G430830	276.8	50.0	Down	1.29×10^{-18}	Calcium-dependent protein kinase-like protein
Csa6G382950	125.3	405.7	Up	4.08×10^{-8}	NAC domain-containing protein
Csa6G408800	47.5	102.9	Up	0.0003109	Circadian clock coupling factor,

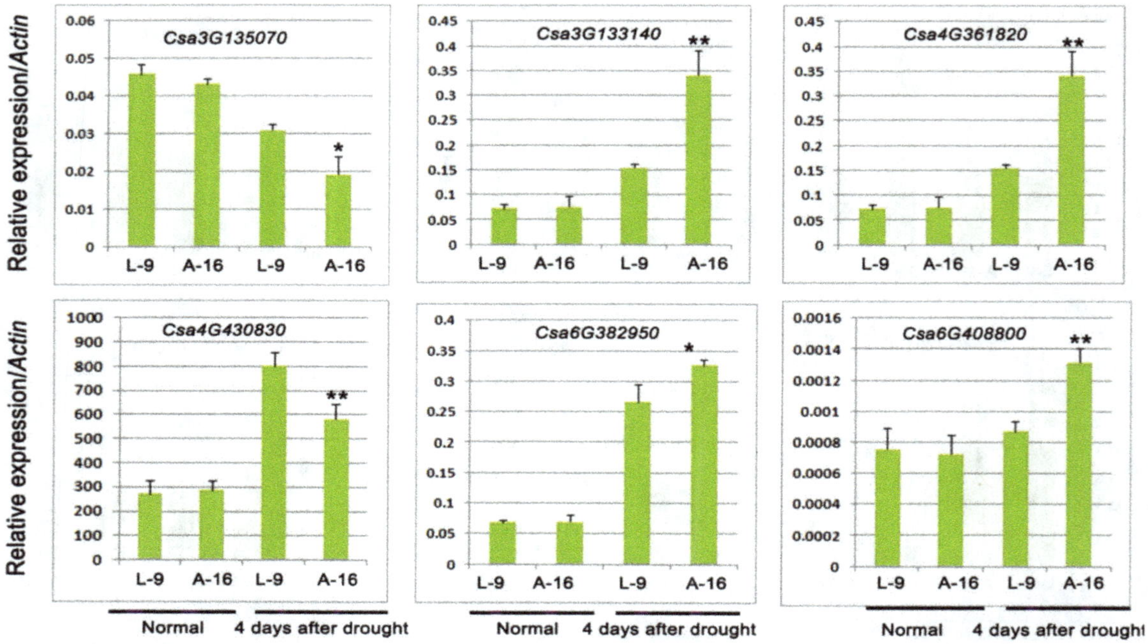

Figure 8. Relative expression of genes involved in ABA signaling pathway. Data is presented as the mean ± standard deviation ($n = 9$). * $0.01 \leq p \leq 0.05$, ** $p \leq 0.01$, Student's t test.

3. Discussion

The analysis and availability of diverse genetic resources could offer important information for understanding the molecular basis of variability in their response to drought stress [16]. In the study, we characterized two cucumber genotypes for their significantly different response to drought (L-9 and A-16) stress. A-16 exerted drought sensibility under water deficiency with increased MDA content and decreased SOD enzyme activity and chlorophyll content. Through the analysis of the transcript level by RNA-seq, we found that the number of DEGs increased significantly at the 4th day after drought treatment. Among them, several DEGs related to the sucrose synthesis and ABA signaling pathway were possibly involved in the drought response tolerance with prominent expression changes between the two cultivars.

3.1. A-16 Has Less Stomata in the Leaf Than L-9

Previous studies have reported that the regulation of stomatal opening and closure is crucial to the normal transpiration and plays an important role in the resistance of drought stress [28]. In rice, *am1* mutant showed drought resistance and highly percentage of completely closed stomata when compared with the wild type [29]. Drought-tolerant variety *dca1* has a lower number of stomata and more completely closed stomata than the control [30]. In our present study, we found that the number of stomata in L-9 was less than in A-16 in the same field size, indicating that L-9 could enhance its tolerance to drought stress by regulating the number of stomata.

3.2. Analysis of Sucrose and Starch Biosynthetic Process in Drought Stress

Sugar metabolism and starch biosynthesis are involved in the plant tolerance under drought stress [31]. Soluble sugar content is identified as a good marker in selecting the durum with drought tolerance [32]. The accumulation of soluble sugars in plant different tissues is reinforced when faced with different environmental stresses [33]. Under water deficiency, the soluble sugar was significantly accumulated in *Arabidopsis* leaves, resulting in its resistance to drought [34]. In our study, we found that most of genes involved in the sucrose and starch biosynthetic process were significantly up-regulated in the drought tolerant cultivar L-9, indicating that more sucrose and starch content might attribute to its resistance on drought stress.

3.3. Analysis of ABA Signal under Drought Stress

ABA plays essential role in the plant drought resistance because it could not only regulate the stomatal closure but also influence genes expression involved in stress-response and metabolic changes [35,36]. NAC transcriptional factors, which respond to ABA, could enhance plant tolerance under water deficiency [37–39]. In rice, both *OsNAC45* and *OsNAC52* were induced by ABA and their overexpressing transgenic plants showed enhanced tolerance to drought and salt treatments [38,40]. Here, we found that the expression of two genes (*Csa4M361820* and *Csa6M382950*), encoding the NAC domain-containing protein, increased prominently in the drought tolerant cultivar L-9, which was consistent with previous studies showing that higher expression of NAC genes could promote plant drought tolerance. Calcium-dependent protein kinase (CDPK), an important group of Ser/Thr protein kinases presents in plants and some protozoans that decode Ca^{2+} signals, are involved in the ABA signal transduction [41,42] and function in the plant response to drought [43,44]. Overexpression of *ZmCK3* (a maize calcium-dependent protein kinase gene) could improve plant survival rates under drought conditions in transgenic *Arabidopsis* [44]. CPK10, interacting with HSP1 (heat shock protein 1), plays important roles in ABA and Ca^{2+} mediated regulation of stomatal movements, leading to different tolerance to water deficiency [43]. *VfCPK1* of *Vicia faba* and *AtCPK11* of *Arabidopsis* are specifically induced by drought and ABA, respectively [45,46]. In this study, the drought-sensitive cultivar A-16 showed significantly decreased expression of *CDPK* genes (*Csa3M135070* and *Csa4M430830*) when compared with L-9, implying that high expression of *CDPK* might contribute to the drought tolerance in L-9.

3.4. Analysis of Cuticular Waxes Biosynthesis under Drought Stress

In addition, we also found that the "Cutin, suberin, and wax biosynthesis" pathway appeared under normal condition. The aerial surfaces of vascular plants are covered with a cuticle layer, including two major types of lipids, cutin and waxes [47]. Cuticular waxes play important roles in ensuring that plants grow and survive under various different biotic and abiotic stresses, which could help plants prevent non-stomatal water loss, and protect them against UV radiation and bacterial and fungal pathogens [48–50]. In the present study, several DEGs were enriched in the cutin pathway involved in lipid mechanism and were significantly down-regulated in A-16 when compared with L-9, suggesting that the decreased expression of related genes in cutin, suberin, and wax biosynthesis might be responsible for A-16's sensitivity to drought stress.

Overall, we firstly carried out RNA-Seq to analyze the regulation mechanism under water deficiency in cucumber. Several crucial genes involved in sucrose biosynthesis and ABA signal transduction were changed during drought stress. Our study not only provided a foundation for the further understanding of the regulation molecular on drought tolerance, but also explored valuable genes involved in drought tolerant, which will contribute to the improvement of drought resistant varieties in cucumber.

4. Materials and Methods

4.1. Plant Materials and Drought Treatment

Two cucumber cultivars, namely L-9 (South China type cucumber variety) and A-16 (North China type cucumber variety), were used in the study. Seeds were germinated overnight on wet filter in a culture dish at 28 °C in a dark environment. After that, the seedlings were grown in a feeding block under 14/10 h with 28/18 °C in day/night, respectively, in a culture room (5500 lux). When plants were grown to the two true leaves stage, they were subjected to lack of water for 7 days. After that, seedlings recovered for 3 days to normal condition. L-9 and A-16 seedlings were 120 for three biological replicates, respectively. Ten normal leaves were sampled from 10 plants before drought treatment, while drought-treated leaves were randomly sampled at the 4th day after drought treatment. Each biological replicate had a total of 10 leaves from 10 plants randomly selected. The samples were immediately frozen in liquid nitrogen and consistently stored at −80 °C until further analysis. In addition, leaf samples of three randomly selected biological replicates were then collected from both L-9 and A-16 plants (twelve samples in total).

4.2. Quantitative Analysis of Chlorophyll Content

Chlorophyll content was measured based on the procedure [51]. In detail, 0.2 g freshly-sampled leaves were homogenized in 5 mL solution with acetone and 0.1 M NH_4OH at a ratio of 9:1 and then centrifuged at 3000× *g* for 20 min. The obtained supernatants were then washed three times using hexane and finally the pigment content was measured by spectrophotometer at the absorption wavelengths of 663 and 645 nm (Beckman Coulter DU-800, Brea, CA, USA). According to the two formulas ($Ca = 13.95 \times D_{665} - 6.88 \times D_{649} \times 6$ (mg/L); $Cb = 24.96 \times D_{649} - 7.32 \times D_{665} \times 5$ (mg/L)), the concentrations of chlorophyll a and chlorophyll b were finally calculated, respectively.

4.3. Analysis of Malondialdehyde (MDA) Content by TBA Method

MDA content was measured according to the following procedures. Briefly, 0.5 g freshly-sampled leaves were dipped into 0.5% trichloroacetic acid (TCA) and ground into powder, then centrifuged at 3000× *g* for 20 min. A total of 2 mL supernatant was added to 2 mL 0.5% thibabituric acid (TBA) 0.5% TCA, after that, the mixture was boiled at 100 °C for 30 min. Then, absorption wavelengths of supernatants on 450 nm, 532 nm, 600 nm were recorded. According to the given formula ($CMDA = 6.45 \times (A_{532} - A_{600}) - 0.56 \times A_{450}$ (μmol/L)), the MDA content was finally calculated.

4.4. Scanning Electron Microscopy (SEM)

Leaves of L-9 and A-16 seedlings under normal conditions were air-dried. The leaf abaxial epidermis was visualized under a HITACHI SU8020 variable pressure scanning electron microscope (SEM) (Hitachi, Tokyo, Japan) and imaged with an H-7500 transmission electron microscope (Hitachi).

4.5. Transmission Electron Microscopy (TEM)

Leaves of L-9 and A-16 seedlings under normal conditions were fixed overnight in 2.5% glutaraldehyde in 0.1 M phosphate buffer (pH 7.4) at 4 °C, then post-fixed in 2% (*v*/*v*) OsO4 in phosphate buffer. A series of 80 nm sections was cut using a Reichert OM2 ultramicrotome (Reichert, Deprew, New York, NY, USA), stained in 2% uranylacetate and 10 mM lead citrate (pH 12), before observation in a HitachiH-7650 (Hitachi) transmission electron microscope.

4.6. BGISEQ-500 Library Construction

A total of twelve samples (three biological replicates each of L-9 and A-16 at normal and drought stress, respectively) were used for RNA extraction with TRIZOL reagent according to the manufacturer's protocol (TaKaRa, Shiga, Japan). Each biological replicate had a total of 10 leaves from 10 plants, selected randomly. After extraction, RNA was then purified (using DNAse) and concentrated using an RNeasyMinElute clean up kit (Qiagen, Duesseldorf, Germany). Then, 2.5 μg RNA of each sample was prepared for constructing BGISEQ-500 library according to the protocol of previous study [52]. Library quality was tested using the Agilent Bioanalyzer (Life Technologies, Carlsbad, CA, USA) 2100 system and the genome reference was the cucumber 9930 genome (http://cucurbitgenomics.org/, Two years).

4.7. Screening and Significant Test for Differentially Expressed Genes (DEGs)

Gene expression level was calculated by quantifying the reads according to the RPKM (reads per kilobase per million reads) method [53]. Then the NOISeq was used to identify DEGs, which existed in the normal and drought stress transcriptome libraries according to the following criteria: fold change $\geq$2 and divergence probability $\geq$0.8. GO enrichment for these DEGs was performed using WEGO software [54]. To further obtain knowledge of DEG biological functions, pathway enrichment analysis was carried out according to the KEGG database [55], the major public pathway-related database.

4.8. Quantitative Real-Time PCR (qRT-PCR) Identification

Quantitative real-time PCR analysis was performed using the total RNA from seedling leaves of both the normal and drought stress treatment. Twenty μL cDNA was obtained using the QuantiTect Reverse Transcription Kit (Qiagen, Duesseldorf, Germany). Quantitative qRT-PCR (20 μL reaction volume) was carried out with 0.5 μL of cDNA, 0.2 μM of primer mix and SYBR Premix Ex Taq Kit (TaKaRa,Shiga, Japan). In an ABI PRISM 7900HT system (Life Technologies, Carlsbad, CA, USA), cucumber α-TUBULIN (*TUA*) gene was used as normal. qRT-PCR was carried out on an ABI 7500 Real-Time PCR System (Applied Biosystems, USA). In addition, all qRT-PCR primers were listed in the Table S8.

4.9. Statistical Analysis

The linux rhel6.7 x64 R-3.4.2 and MEGA6 were used to perform the heat-map and cluster analysis. Significant differences were detected by IBM SPSS Statistics 20 (by Student's *t* test). Relative gene expressions were calculated using the $2^{-\Delta\Delta Ct}$ method [56]. In addition, GraphPad Prism 5 was used for chart preparation.

Author Contributions: M.W. and Y.L. designed the experiment. M.W. and B.J. performed most of the experiments. Q.P., W.L., X.H. and Z.L. performed part of the experiment. M.W. wrote the paper. Y.L. edited the manuscript.

Abbreviations

RNA seq	RNA sequencing
DEGs	Differently expressed genes
ABA	Abscisic acid
MDA	Malondialdehyde
SOD	Enzyme activity of superoxide dismutase
SEM	Scanning electron microscopy
TEM	Transmission electron microscopy
GO	Gene ontology
qRT-PCR	Quantitative Real-Time PCR

References

1. Jaleel, C.A.; Manivannan, P.A.; Wahid, A.; Farooq, M.; Al-Juburi, H.J.; Somasundaram, R.A.; Panneerselvam, R. Drought stress in plants: A review on morphological characteristics and pigments composition. *Int. J. Agric. Biol.* **2009**, *11*, 100–105.
2. Kusaka, M.; Ohta, M.; Fujimura, T. Contribution of inorganic components to osmotic adjustment and leaf folding for drought tolerance in pearl millet. *Phys. Plant.* **2005** *125*, 474–489. [CrossRef]

3. Shao, H.B.; Chu, L.Y.; Shao, M.A.; Jaleel, C.A.; Hong, M.M. Higher plant antioxidants and redox signaling under environmental stresses. *C. R. Biol.* **2008**, *331*, 433–441. [CrossRef] [PubMed]
4. Bohnert, H.J.; Nelson, D.E.; Jensen, R.G. Adaptations to environmental stresses. *Plant Cell* **1995**, *7*, 1099–1111. [CrossRef] [PubMed]
5. Xiong, L.; Zhu, J.K. Molecular and genetic aspects of plant responses to osmotic stress. *Plant Cell Environ.* **2002**, *25*, 131–139. [CrossRef] [PubMed]
6. Shinozaki, K.; Yamaguchi-Shinozaki, K. Gene networks involved in drought stress response and tolerance. *J. Exp. Bot.* **2007**, *58*, 221–227. [CrossRef] [PubMed]
7. Kwak, J.M.; Mori, I.C.; Pei, Z.M.; Leonhardt, N.; Torres, M.A.; Dangl, J.L.; Bloom, R.E.; Bodde, S.; Jone, J.D.G.; Schroeder, J.I. NADPH oxidase AtrbohD and AtrbohF genes function in ROS-dependent ABA signaling in Arabidopsis. *EMBO J.* **2003**, *22*, 2623–2633. [CrossRef] [PubMed]
8. Fujita, Y.; Fujita, M.; Shinozaki, K.; Yamaguchi-Shinozaki, K. ABA-mediated transcriptional regulation in response to osmotic stress in plants. *J. Plant Res.* **2011**, *124*, 509–525. [CrossRef] [PubMed]
9. Ma, Y.; Szostkiewicz, I.; Korte, A.; Moes, D.; Yang, Y.; Christmann, A.; Grill, E. Regulators of PP2C phosphatase activity function as abscisic acid sensors. *Science* **2009**, *324*, 1064–1068. [CrossRef] [PubMed]
10. Park, S.Y.; Fung, P.; Nishimura, N.; Jensen, D.R.; Fujii, H.; Zhao, Y.; Lumba, S.; Santiago, J.; Rodrigues, A.; Tsz-fung, F.C.; et al. Abscisic acid inhibits type 2c protein phosphatases via the pyr/pyl family of start proteins. *Science* **2009**, *324*, 1068–1071. [CrossRef] [PubMed]
11. Malepszy, S. Cucumber (*Cucumis Sativus* L.). In *Crops II*; Bajaj, Y.P.S., Ed.; Springer: Berlin/Heidelberg, Germany, 1988.
12. Chen, Y.; Wang, M.; Hu, L.; Liao, W.; Dawuda, M.M.; Li, C. Carbon monoxide is involved in hydrogen gas-induced adventitious root development in cucumber under simulated drought stress. *Front. Plant Sci.* **2017**, *8*, 128. [CrossRef] [PubMed]
13. Wang, W.; Zhang, Y.; Xu, C.; Ren, J.; Liu, X.; Black, K.; Gai, X.; Wang, Q.; Ren, H. Cucumber *ECERIFERUM1*, (*CsCER1*), which influences the cuticle properties and drought tolerance of cucumber, plays a key role in VLC alkanes biosynthesis. *Plant Mol. Biol.* **2015**, *87*, 219–233. [CrossRef] [PubMed]
14. Sun, Y.; Wang, H.; Liu, S.; Peng, X. Exogenous application of hydrogen peroxide alleviates drought stress in cucumber seedlings. *S. Afr. J. Bot.* **2016**, *106*, 23–28. [CrossRef]
15. Yin, Z.; Hennig, J.; Szwacka, M.; Malepszy, S. Tobacco PR-2d promoter is induced in transgenic cucumber in response to biotic and abiotic stimuli. *J. Plant Physiol.* **2004**, *161*, 621–629. [CrossRef] [PubMed]
16. Garg, R.; Shankar, R.; Thakkar, B.; Kudapa, H.; Krishnamurthy, L.; Mantri, N.; Varshney, R.K.; Bhatia, S.; Jain, M. Transcriptome analyses reveal genotype-and developmental stage-specific molecular responses to drought and salinity stresses in chickpea. *Sci. Rep.* **2016**, *6*, 19228. [CrossRef] [PubMed]
17. Zhou, Y.; Yang, P.; Cui, F.; Zhang, F.; Luo, X.; Xie, J. Transcriptome Analysis of Salt Stress Responsiveness in the Seedlings of Dongxiang Wild Rice (*Oryza rufipogon* Griff.). *PLoS ONE* **2016**, *11*, e0146242. [CrossRef] [PubMed]
18. Cui, J.Y.; Gunewardena, S.S.; Yoo, B.; Liu, J.; Renaud, H.J.; Lu, H.; Zhong, X.B.; Klaassen, C.D. RNA-Seqreveals different mRNA abundance of transporters and their alternative transcript isoforms during liver development. *Toxicol. Sci.* **2012**, *127*, 592–608. [CrossRef] [PubMed]
19. Zhao, S.; Fung-Leung, W.P.; Bittner, A.; Ngo, K.; Liu, X. Comparison of RNA-Seq and microarray in transcriptomeprofiling of activated T cells. *PLoS ONE* **2014**, *9*, e78644.
20. Ando, K.; Grumet, R. Transcriptional profiling of rapidly growing cucumber fruit by 454-pyrosequencing analysis. *J. Am. Soc. Hortic. Sci.* **2010**, *135*, 291–302.
21. Li, J.; Wu, Z.; Cui, L.; Zhang, T.; Guo, Q.; Xu, J.; Jia, L.; Lou, Q.; Huang, S.; Li, Z.; et al. Transcriptome comparison of global distinctive features between pollination and parthenocarpic fruit set reveals transcriptional phytohormone cross-talk in cucumber (*Cucumis sativus* L.). *Plant Cell Physiol.* **2014**, *55*, 1325–1342. [CrossRef] [PubMed]
22. Guo, S.; Zheng, Y.; Joung, J.G.; Liu, S.; Zhang, Z.; Crasta, O.R.; Sobral, B.W.; Xu, Y.; Huang, S.; Fei, Z. Transcriptome sequencing and comparative analysis of cucumber flowers with different sex types. *BMC Genom.* **2010**, *11*, 384. [CrossRef] [PubMed]
23. Chen, C.; Liu, M.; Jiang, L.; Liu, X.; Zhao, J.; Yan, S.; Yang, S.; Ren, H.; Liu, R.; Zhang, X. Transcriptome profiling reveals roles of meristem regulators and polarity genes during fruit trichome development in cucumber (*Cucumis sativus* L.). *J. Exp. Bot.* **2014** *65*, 4943–4958. [CrossRef] [PubMed]

24. Zhang, N.; Zhang, H.J.; Zhao, B.; Sun, Q.Q.; Cao, Y.Y.; Li, R.; Wu, X.X.; Weeda, S.; Li, L.; Ren, S.; et al. The RNA-seq approach to discriminate gene expression profiles in response to melatonin on cucumber lateral root formation. *J. Pineal Res.* **2014**, *56*, 39–50. [CrossRef] [PubMed]
25. Zhao, W.; Yang, X.; Yu, H.; Jiang, W.; Sun, N.; Liu, X.; Liu, X.; Zhang, X.; Wang, Y.; Gu, X. RNA-Seq-based transcriptome profiling of early nitrogen deficiency response in cucumber seedlings provides new insight into the putative nitrogen regulatory network. *Plant Cell Physiol.* **2015**, *56*, 455–467. [CrossRef] [PubMed]
26. Huang, S.; Li, R.; Zhang, Z.; Li, L.; Gu, X.; Fan, W.; Lucas, W.J.; Wang, X.; Xie, B.; Ni, P.; et al. The genome of the cucumber, *Cucumis sativus* L. *Nat. Genet.* **2009**, *41*, 1275–1281. [CrossRef] [PubMed]
27. Bari, R.; Jones, J.D. Role of plant hormones in plant defence responses. *Plant Mol. Biol.* **2009**, *69*, 473–488. [CrossRef] [PubMed]
28. Lim, C.W.; Baek, W.; Jung, J.; Kim, J.H.; Lee, S.C. Function of ABA in Stomatal Defense against Biotic and Drought Stresses. *Int. J. Mol. Sci.* **2015**, *16*, 15251–15270. [CrossRef] [PubMed]
29. Sheng, P.; Tan, J.; Jin, M.; Wu, F.; Zhou, K.; Ma, W.; Heng, Y.; Wang, J.; Guo, X.; Zhang, X.; et al. Albino midrib 1, encoding a putative potassium efflux antiporter, affects chloroplast development and drought tolerance in rice. *Plant Cell Rep.* **2014**, *33*, 1581–1594. [CrossRef] [PubMed]
30. Cui, L.G.; Shan, J.X.; Shi, M.; Gao, J.P.; Lin, H.X. DCA1 acts as a transcriptional co-activator of DST and contributes to drought and salt tolerance in rice. *PLoS Genet.* **2015**, *11*, e1005617. [CrossRef] [PubMed]
31. Mohammadkhani, N.; Heidari, R. Drought-induced Accumulation of Soluble Sugars and Proline in Two Maize Varieties. *World Appl. Sci. J.* **2008**, *3*, 448–453.
32. Al Hakimi, A.; Monneveux, P.; Galiba, G. Soluble sugars, proline and relative water content (RCW) as traits for improving drought tolerance and divergent selection for RCW from triticumpolonicum into triticum durum. *J. Genet. Breed.* **1995**, *49*, 237–244.
33. Prado, F.E.; Boero, C.; Gallardo, M.; González, J.A. Effect of nacl on germination, growth, and soluble sugar content in chenopodium quinoa willd. seeds. *Bot. Bull. Acad. Sin.* **2000**, *41*, 27–34.
34. Sperdouli, I.; Moustakas, M. Interaction of proline, sugars, and anthocyanins during photosynthetic acclimation of Arabidopsis thaliana to drought stress. *J. Plant Physiol.* **2012**, *169*, 577–585. [CrossRef] [PubMed]
35. Seki, M.; Umezawa, T.; Urano, K.; Shinozaki, K. Regulatory metabolic networks in drought stress responses. *Curr. Opin. Plant Biol.* **2007**, *10*, 296–302. [CrossRef] [PubMed]
36. Danquah, A.; de Zelicourt, A.; Colcombet, J.; Hirt, H. The role of ABA and MAPK signaling pathways in plant abiotic stress responses. *Biotechnol. Adv.* **2014**, *32*, 40–52. [CrossRef] [PubMed]
37. Hu, H.; Dai, M.; Yao, J.; Xiao, B.; Li, X.; Zhang, Q.; Xiong, L. Overexpressing a NAM, ATAF, and CUC (NAC) transcription factor enhances drought resistance and salt tolerance in rice. *Proc. Natl. Acad. Sci. USA* **2006**, *103*, 12987–12992. [CrossRef] [PubMed]
38. Zheng, X.; Chen, B.; Lu, G.; Han, B. Overexpression of a NAC transcription factor enhances rice drought and salt tolerance. *Biochem. Biophys. Res. Commun.* **2009**, *379*, 985–989. [CrossRef] [PubMed]
39. Liu, G.; Li, X.; Jin, S.; Liu, X.; Zhu, L.; Nie, Y.; Zhang, X. Overexpression of rice NAC gene SNAC1 improves drought and salt tolerance by enhancing root development and reducing transpiration rate in transgenic cotton. *PLoS ONE* **2014**, *9*, e86895. [CrossRef] [PubMed]
40. Gao, F.; Xiong, A.; Peng, R.; Jin, X.; Xu, J.; Zhu, B.; Chen, J.; Yao, Q. OsNAC52, a rice NAC transcription factor, potentially responds to ABA and confers drought tolerance in transgenic plants. *Plant Cell Tissue Organ Cult.* **2010**, *100*, 255–262. [CrossRef]
41. Zhou, X.; Zhang, H. Roles of calcium-dependent protein kinases in ABA-regulation of stomatal moment in Poplar. *Sci. Technol. Eng.* **2004**, *4*, 80–83. (In Chinese)
42. Zhu, S.Y.; Yu, X.C.; Wang, X.J.; Zhao, R.; Li, Y.; Fan, R.C.; Shang, Y.; Du, S.Y.; Wang, X.F.; Wu, F.Q.; et al. Two Calcium-Dependent Protein Kinases, CPK_4 and CPK_{11}, Regulate Abscisic Acid Signal Transduction in *Arabidopsis*. *Plant Cell* **2007**, *19*, 3019–3036. [CrossRef] [PubMed]
43. Zou, J.J.; Wei, F.J.; Wang, C.; Wu, J.J.; Ratnasekera, D.; Liu, W.X.; Wu, W.H. Arabidopsis calcium-dependent protein kinase CPK10 functions in abscisic acid- and Ca^{2+}-mediated stomatal regulation in response to drought stress. *Plant Physiol.* **2010**, *154*, 1232–1243. [CrossRef] [PubMed]
44. Wang, C.T.; Song, W. *ZmCK$_3$*, a maize calcium-dependent protein kinase gene, endows tolerance to drought and drought stresses in transgenic *Arabidopsis*. *J. Plant Biochem. Biotechnol.* **2014**, *23*, 249–256. [CrossRef]
45. Liu, G.; Chen, J.; Wang, X. *VfCPK1*, a gene encoding calcium-dependent protein kinase from *Vicia faba*, is induced by drought and abscisic acid. *Plant Cell Environ.* **2006** *29*, 2091–2099. [CrossRef] [PubMed]

46. Huang, K.; Peng, L.; Liu, Y.; Yao, R.; Liu, Z.; Li, X.; Yang, Y.; Wang, J. Arabidopsis calcium-dependent protein kinase AtCPK1 plays a positive role in salt/drought-stress response. *Biochem. Biophys. Res. Commun.* **2017**, *49*. [CrossRef] [PubMed]
47. Bernard, A.; Joubès, J. Arabidopsis cuticular waxes: Advances in synthesis, export and regulation. *Prog. Lipid Res.* **2013**, *52*, 110–129. [CrossRef] [PubMed]
48. España, L.; Heredia-Guerrero, J.A.; Reina-Pinto, J.J.; Fernández-Muñoz, R.; Heredia, A.; Domínguez, E. Transient silencing of CHALCONE SYNTHASE during fruit ripening modifies tomato epidermal cells and cuticle properties. *Plant Physiol.* **2014**, *166*, 1371–1386. [CrossRef] [PubMed]
49. Reisige, K.; Gorzelanny, C.; Daniels, U.; Moerschbacher, B.M. The C28 aldehyde octacosanal is a morphogenetically active component involved in host plant recognition and infection structure differentiation in the wheat stem rust fungus. *Physiol. Mol. Plant Pathol.* **2006**, *68*, 33–40. [CrossRef]
50. Shepherd, T.; Wynne, G.D. The effects of stress on plant cuticular waxes. *New Phytol.* **2006**, *171*, 469–499. [CrossRef] [PubMed]
51. Suzuki, Y.; Makino, A. Availability of rubisco small subunit up-regulates the transcript levels of large subunit for stoichiometric assembly of its holoenzyme in rice. *Plant Physiol.* **2012**, *160*, 533–540. [CrossRef] [PubMed]
52. Fehlmann, T.; Reinheimer, S.; Geng, C.; Su, X.; Drmanac, S.; Alexeev, A.; Zhang, C.; Backes, C.; Ludwig, N.; Hart, M.; et al. cPAS-based sequencing on the BGISEQ-500 to explore small non-coding RNAs. *Clin. Epigenet.* **2016**, *8*, 123. [CrossRef] [PubMed]
53. Li, B.; Dewey, C.N. RSEM: Accurate transcript quantification from RNA-Seq data with or without a reference genome. *BMC Bioinform.* **2011**, *12*, 323. [CrossRef] [PubMed]
54. Ye, J.; Fang, L.; Zheng, H.; Zhang, Y.; Chen, J.; Zhang, Z.; Wang, J.; Li, S.; Li, R.; Bolund, L.; et al. WEGO: A web tool for plotting GO annotations. *Nucleic Acids Res.* **2006**, *34*, W293. [CrossRef] [PubMed]
55. Kanehisa, M.; Araki, M.; Goto, S.; Hattori, M.; Hirakawa, M.; Itoh, M.; Katayama, T.; Kawashima, S.; Okuda, S.; Tokimatsu, T.; et al. KEGG for linking genomes to life and the environment. *Nucleic Acids Res.* **2008**, *36*, 480–484. [CrossRef] [PubMed]
56. Vandesompele, J.; de Preter, K.; Pattyn, F.; Poppe, B.; van Roy, N.; de Paepe, A.; Speleman, F. Accuratenormalization of real-time quantitative RT-PCR data by geometric averaging of multiple internal controlgenes. *Genome Biol.* **2002**, *3*, research0034.1. [CrossRef] [PubMed]

6

BvcZR3 and *BvHs1^{pro-1}* Genes Pyramiding Enhanced Beet Cyst Nematode (*Heterodera schachtii* Schm.) Resistance in Oilseed Rape (*Brassica napus* L.)

Xuanbo Zhong [1], Qizheng Zhou [1], Nan Cui [1], Daguang Cai [2] and Guixiang Tang [1,*]

[1] Zhejiang Provincial Key Laboratory of Crop Genetic Resources, Institute of Crop Science, Zhejiang University, Hangzhou 310058, Zhejiang, China; 21716135@zju.edu.cn (X.Z.); 21716136@zju.edu.cn (Q.Z.); 21816113@zju.edu.cn (N.C.)

[2] Department of Molecular Phytopathology, Christian-Albrechts-University of Kiel, Hermann Rodewald Str. 9, D-24118 Kiel, Germany; dcai@phytomed.uni-kiel.de

[*] Correspondence: tanggx@zju.edu.cn

Abstract: Beet cyst nematode (*Heterodera schachtii* Schm.) is one of the most damaging pests in sugar beet growing areas around the world. The *Hs1^{pro-1}* and *cZR3* genes confer resistance to the beet cyst nematode, and both were cloned from sugar beet translocation line (A906001). The translocation line carried the locus from *B. procumbens* chromosome 1 including *Hs1^{pro-1}* gene and resistance gene analogs (RGA), which confer resistance to *Heterodera schachtii*. In this research, *BvHs1^{pro-1}* and *BvcZR3* genes were transferred into oilseed rape to obtain different transgenic lines by *A. tumefaciens* mediated transformation method. The *cZR3Hs1^{pro-1}* gene was pyramided into the same plants by crossing homozygous *cZR3* and *Hs1^{pro-1}* plants to identify the function and interaction of *cZR3* and *Hs1^{pro-1}* genes. In vitro and in vivo cyst nematode resistance tests showed that *cZR3* and *Hs1^{pro-1}* plants could be infested by beet cyst nematode (BCN) juveniles, however a large fraction of penetrated nematode juveniles was not able to develop normally and stagnated in roots of transgenic plants, consequently resulting in a significant reduction in the number of developed nematode females. A higher efficiency in inhibition of nematode females was observed in plants expressing pyramiding genes than in those only expressing a single gene. Molecular analysis demonstrated that *BvHs1^{pro-1}* and *BvcZR3* gene expressions in oilseed rape constitutively activated transcription of plant-defense related genes such as *NPR1* (non-expresser of *PR1*), *SGT1b* (enhanced disease resistance 1) and *RAR1* (suppressor of the *G2* allele of *skp1*). Transcript of *NPR1* gene in transgenic *cZR3* and *Hs1^{pro-1}* plants were slightly up-regulated, while its expression was considerably enhanced in *cZR3Hs1^{pro-1}* gene pyramiding plants. The expression of *EDS1* gene did not change significantly among transgenic *cZR3*, *Hs1^{pro-1}* and *cZR3Hs1^{pro-1}* gene pyramiding plants and wild type. The expression of *SGT1b* gene was slightly up-regulated in transgenic *cZR3* and *Hs1^{pro-1}* plants compared with the wild type, however, its expression was not changed in *cZR3Hs1^{pro-1}* gene pyramiding plant and had no interaction effect. *RAR1* gene expression was significantly up-regulated in transgenic *cZR3* and *cZR3Hs1^{pro-1}* genes pyramiding plants, but almost no expression was found in *Hs1^{pro-1}* transgenic plants. These results show that nematode resistance genes from sugar beet were functional in oilseed rape and conferred BCN resistance by activation of a CC-NBS-LRR R gene mediated resistance response. The gene pyramiding had enhanced resistance, thus offering a novel approach for the BCN control by preventing the propagation of BCN in oilseed rape. The transgenic oilseed rape could be used as a trap crop to offer an alternative method for beet cyst nematode control.

Keywords: *Hs1^{pro-1}*; *cZR3*; gene pyramiding; *Heterodera schachtii*; resistance

1. Introduction

Beet cyst nematode (*Heterodera schachtii* Schm.) is an important pest of sugar beet that can cause significant reductions in yield. Unlike most cyst nematode, the beet cyst nematode (BCN) has a wide host range and can infect more than 218 [1] plant species, including family Brassicaceae and Chenopodiaceae such as sugar beet (*Beta vulgaris* L.), oilseed rape (*Brassica napus* L.), and spinach (*Spinacia oleracear*) [2]. In general, nematodes can be controlled by treating with nematicides, growing resistant cultivars, and crop rotation with non-host or trap crops [3]. However, control via nematicides application is difficult and expensive because nematode larvae and eggs are well protected. In addition, the use of nematicides increases the threat of environmental pollution and some are prohibited worldwide. Growing resistant cultivars with a monogenic resistance can induce the emergence of more virulent nematode pathotypes. Oilseed rape (*Brassica napus* L.) is a good host for BCN and always rotates with sugar beet in the agro-farming system. Therefore, breeding for BCN resistant oilseed rape is of great value and importance [4].

A previous investigation on 111 *Brassica* germplasm lines revealed that all lines are susceptible to *H. schachtii* [5]. BCN resistance is found only in a few spices of *Brassiceae* including oil radish (*Raphanussativus* L. *ssp.oleiferus* DC.) and white mustard [6] (*Sinapisalba* L.). Oil radish shows complete resistance and is often used as a trap crop to mitigate the degree of damage in infested fields [2]. Resistant/trap crop could stimulate the hatching of larvae, which invaded the roots, but prevents larvae from fulfilling their life cycle and thus lowers BCN populations. A dominant nematode gene $Hs1^{Rph}$, which is located in the radish chromosome *d*, confers beet cyst nematode resistance [7]. Efforts have been made to transfer this resistance gene into oilseed rape genome by intergeneric hybridization [4,6,8–10]. However, expression of the resistance gene in a hybrid species remains a great challenge due to the extremely low and unstable inheritance [7]. Cloning of the cyst nematode resistance genes and their subsequent transfer into the rapeseed could be one of the most feasible strategies to induce nematode resistance in oilseed rape.

$Hs1^{pro\text{-}1}$ gene is the first beet cyst nematode resistance gene cloned from sugar beet translocation line (A906001) by a map-based cloning strategy [11]. The translocation line carried the locus from *B. procumbens* chromosome 1 that confers resistance to *H. schachtii* in sugar beet (*Beta vulgaris* L.) [11]. The $Hs1^{pro\text{-}1}$ gene is different from other nematode resistance genes amid the presence of the NBS-LRR structure, such as *Mi* [12], *Gpa2* [13] and *Hero* [14], due to the presence of the NBS-LRR structure. The resistance mechanism of $Hs1^{pro\text{-}1}$ gene is based on the gene-for-gene relationship [15]. The transcript of $Hs1^{pro\text{-}1}$ gene is upregulated about fourfold after one day of nematode infection. However, no considerable change in the transcript accumulation of $Hs1^{pro\text{-}1}$ is recorded in uninfected roots of resistant beet plants [15]. McLean et al. [16] published a complete sequence of the $Hs1^{pro\text{-}1}$ protein, which included an additional 176 amino acid at N-terminal extension conferred resistance to soybean (*Glycine max* L.) cyst nematode (*Heterodera glycines*).

Hunger et al. [17] cloned 47 resistance gene analogs (RGAs) from genomic DNA of sugar beet. Most cloned resistance genes belong to nucleotide-binding site leucine-rich repeat (NBS-LRR) gene family [18,19]. Members of this family have been isolated in both dicot and monocot plants, exhibiting resistance to a variety of plant pathogens, including bacteria, fungi, viruses, and nematodes [20]. The cloning of conserved sequence of NBS domain has been successfully applied to isolate resistance gene candidates or resistance gene analogs (RGAs) from plant genomes, using a degenerate primer-based PCR strategy. For example, from sugar beet translocation line (A906001), *cZR3* was cloned using a degenerated primer-based PCR strategy [21,22]. The gene is similar to a subset of CC-NBS-LRR resistance proteins, including nematode resistance genes *Mi* from tomato and *Gpa2* from potato [12,21]. The phylogenetic analysis shows that this gene originates from the ancestral gene from which *I2C1* [23] (vascular wilt disease resistance), *Xa1* [24] (bacterial blight resistance) and *Cre3* [25] (nematode resistance) were also originated. For CC-NBS domains, RGAs are required for initiation of a necrotic hypersensitive response, as well as for cell death on the initial infection sites [26,27]. RGAs are ubiquitous in plant genomes, and often cluster with close linkage to active resistance genes [28,29]. Because no complete resistance could be observed so far by transgenic sugar beet plants (Cai unpublished data), it is proposed that additional genes are required to confer full resistance [30]. We hypothesized that the *cZR3* gene may interact with the $Hs1^{pro\text{-}1}$ gene to confer resistance against sugar beet nematode in oilseed rape. Therefore, we transferred the $Hs1^{pro\text{-}1}$ and *cZR3* genes into oilseed rape using hypocotyl explants by *A. tumefaciens* mediated transformation method and pyramided the $cZR3Hs1^{pro\text{-}1}$ genes by crossing homozygous transgenic *cZR3* and *Hs1* plants. The beet cyst nematode resistance tests in vitro and in vivo were detected in *cZR3*, $Hs1^{pro\text{-}1}$ and *cZR3* $Hs1^{pro\text{-}1}$ pyramiding plants. Possible resistance mechanisms of $Hs1^{pro\text{-}1}$, *cZR3* and $cZR3Hs1^{pro\text{-}1}$ gene pyramiding were also discussed in the present study.

2. Results

2.1. Generation of Transgenic Plants and Identification

The recombinant plasmid-DNA carrying $Hs1^{pro\text{-}1}$ *and cZR3* genes (Figure 1A,B) were used for transformation to generate transgenic $Hs1^{pro\text{-}1}$ and *cZR3* plants. Figure 1C–F shows the process of transgenic $Hs1^{pro\text{-}1}$ and *cZR3* plants generation. The hypocotyl segments were pre-cultured on the CIM to induce callus formation for two days and then infected with *A. tumefaciens*. After co-cultivation, explants were cultivated on SIM (Figure 1C) supplemented with 500 mg/L Carb to eliminate the extra bacterium and 50 mg/L Kan to select the positive transgenic plants. The adventitious shoots were observed from the cut end of hypocotyl explants 30 days after cultivation in SIM (Figure 1D). The adventitious shoots were cut off and transferred to shoot elongation medium supplemented with 500 mg/L Carb and 50 mg/L Kan. The putative green transgenic shoots were observed 30 after cultivation in SEM (Figure 1E). The green shoots were transferred to rooting medium to develop the roots (Figure 1F). The rooted plantlets were transferred to the soil and the survived putative transgenic plants were assayed by PCR (Figure 1G,H) and Southern blot (Figure 1I). In general, 194 explants were transformed with pAM194-Hs1 vector and four rooting seedlings were obtained, of which three were identified as transgenic seedlings by PCR (Figure 1G) and Southern blot analysis (Figure 1I). A total of 162 explants were transformed with pAM194-cZR3 vector and 12 rooting seedlings were obtained, of which three were identified as transgenic seedlings by PCR (Figure 1H) and Southern blot analysis (Figure 1I).

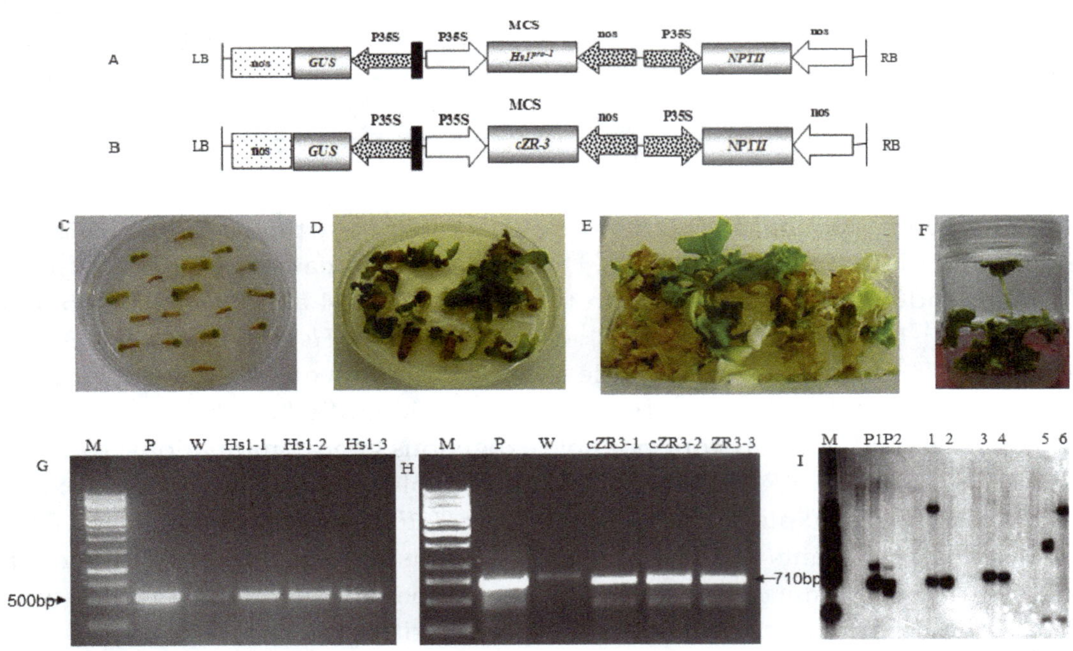

Figure 1. Generation and identification of transgenic $Hs1^{pro\text{-}1}$ and *cZR3* plants from hypocotyle explants of B. *napus* cv. *Zheshuang 758* by *A. tumefaciens* mediated transformation. (**A**,**B**) Diagram of

T-DNA region of the binary vector pAM194-*Hs1^{pro-1}* (**A**) and pAM194-*cZR3* (**B**). LB, T-DNA left border; RB, right border; P35S, Cauliflower mosaic virus (CaMV) 35S promoter; NOS, nopaline synthase terminator; MCS, multi clone site including XhoI restriction site; *Hs1^{pro-1}*, *Beta vulgaris Hs1^{pro-1}* gene open read fragment sequence; *cZR3*, *Beta vulgaris* resistance gene sequence; *GUS*, β-glucuronidase report gene; *NPTII*, neomycin phosphotranferase gene for kanamycin resistance. (**C**) Callus induction from hypocotyl explants after co-cultivation with *A. tumefaciens*. (**D**) Shoots regenerated on SIM. (**E**) Shoots elongation on SEM. (**F**) Shoots rooted on RM. (**G**,**H**) PCR assay for transgenic *Hs1^{pro-1}* plants (**G**) and *cZR3* plants (**H**) M means 1 kb DNA ladder; P means *Hs1* or *cZR3* plasmid DNA, W means non-transgenic plant DNA; Hs1-1, Hs1-2 and Hs1-3 mean T0 independent transgenic *Hs1^{pro-1}* plants; cZR3-1, cZR3-2 and cZR3-3 mean T0 independent transgenic *cZR3* plants. (**I**) Southern blot assay for T0 independent transgenic *cZR3* and *Hs1^{pro-1}* lines, P1P2 means *Hs1* and *cZR3* plasmid DNA, respectively; 1, 3, and 5 mean T0 independent transgenic *Hs1* plants; 2, 4, and 6 mean T0 independent transgenic *cZR3* plants.

2.2. Generation of cZR3Hs1^{pro-1} Gene Pyramiding Plants

T1 seeds were obtained by self-pollination of the six T0 independent transgenic *Hs1^{pro-1}* (*Hs1^{pro-1}*-1, -2, -3) and *cZR3* (*cZR3*-1, -2, -3) oilseed rape plants. The segregation ratio of T1 progeny was determined by PCR analysis using *Hs1^{pro-1}* and *cZR3* genes specific primers. The exogenous gene began to segregate in T1 progeny and could be inherited to the next generation (Table 1). The results show the segregation ratio of *Hs1^{pro-1}*-1, -3 and *cZR3*-1, -2, -3 were nearly 3:1 and the χ^{c2} test was positive.

Table 1. Segregation rates of T1 progeny of three independent transgenic *Hs1^{pro-1}* and *cZR3* lines.

Lines	Number of Plants			Expected Ratio	*p*-Value
	T1	*Hs1^{pro-1}* or *cZR3*+	*Hs1^{pro-1}* or *cZR3*−		
Hs1^{pro-1}-1	21	19	2	3:1	$p > 0.05$
Hs1^{pro-1}-2	6	1	5	3:1	$p < 0.05$
Hs1^{pro-1}-3	18	16	2	3:1	$p > 0.05$
cZR-3-1	8	6	2	3:1	$p > 0.05$
cZR-3-2	10	8	2	3:1	$p > 0.05$
cZR-3-3	37	27	10	3:1	$p > 0.05$

Hs1^{pro-1} *cZR3* genes pyramiding plants were generated by artificial hybridization using *cZR3* plant as female and *Hs1^{pro-1}* plants as male. Five cross combinations carried out between *cZR3* and *Hs1^{pro-1}* independent transgenic lines. In total, 20, 15, 27, 14 and 35 seeds were harvested for F1 *cZR3*-1 × *Hs1^{pro-1}*-1-2, *cZR3*-2.1 × *Hs1^{pro-1}*-1-3, *cZR3*-2.2 × *Hs1^{pro-1}*-1-3, *cZR3*-3 × *Hs1^{pro-1}*-1-2 and *cZR3*-3 × *Hs1^{pro-1}*-3, respectively (Table 2). Segregation of F2 crossed progeny occurred and the heterozygous *cZR3Hs1^{pro-1}* gene pyramiding was identified by PCR analysis using specific gene primers (Figure 2). There were four possible exogenous gene combinations: e.g., Type I: plants, containing both *Hs1^{pro-1}* and *cZR3* genes, Type II and III: plants including a single exogenous gene *Hs1^{pro-1}* or *cZR-3* and Type IV: plants including neither *Hs1^{pro-1}* nor *cZR3*. Of all 111 F2 individuals, we identified 32 with *cZR3*-plants and 19 with *Hs1^{pro-1}*- plants, 47 with *cZR3Hs1^{pro-1}* genes pyramiding plants, and 13 neither *Hs1^{pro-1}* nor *cZR3* genes. The segregation ratio of F2 cross progeny was 47:32:19:13 (9:3:3:1) which were consistent with Mendel's laws of inheritance.

Table 2. Segregation ratios in the F2 generation derived from a cross between homozygous transgenic *Hs1*$^{pro-1}$ and *cZR3* plants.

Cross Combination	Number of Plants					Expected Ratio	*p*-Value
	F2	*cZR3Hs1*$^{pro-1}$+	*cZR3* +	*Hs1*$^{pro-1}$+	*cZR3Hs1*$^{pro-1}$ –		
cZR3-1 ×*Hs1*$^{pro-1}$-2	20	10	3	4	3	9:3:3:1	$p > 0.05$
cZR3-2.1 ×*Hs1*$^{pro-1}$-3	15	1	12	1	1	9:3:3:1	$p < 0.05$
cZR3-2.2 ×*Hs1*$^{pro-1}$-3	27	16	4	4	3	9:3:3:1	$p > 0.05$
cZR3-3 ×*Hs1*$^{pro-1}$-2	14	6	2	2	4	9:3:3:1	$p < 0.05$
cZR3-3 ×*Hs1*$^{pro-1}$-3	35	15	8	9	3	9:3:3:1	$p > 0.05$

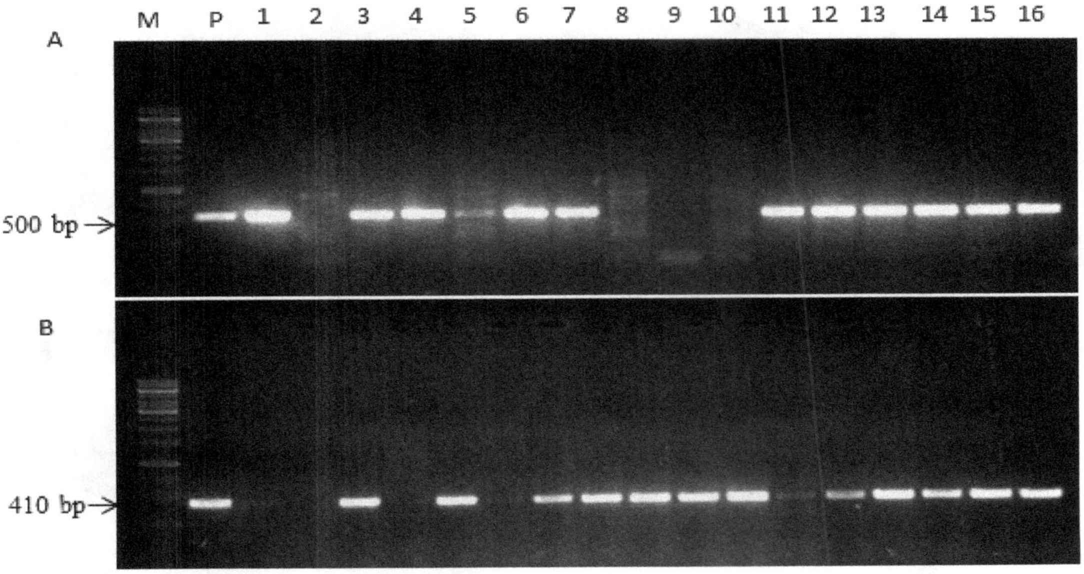

Figure 2. Identification of *cZR3Hs1*$^{pro-1}$ pyramiding homozygous plants in F2 crossing progeny by PCR assay using specific *cZR3* and *Hs1*$^{pro-1}$ gene primers. The amplification fragment length of *Hs1*$^{pro-1}$ and *cZR3* gene was 500 bp and 410 bp, respectively. (**A**) Identification of *Hs1*$^{pro-1}$ gene in F2 crossing progeny. (**B**) Identification of *cZR3* gene in F2 crossing progeny. M represents a molecular marker (1 kb); P represents plasmid DNA for *Hs1*$^{pro-1}$ or *cZR3* binary vector; 1–16 represent different F2 rapeseed plants.

2.3. *Beet Cyst Nematode Test In Vitro and In Vivo*

To determine the resistance to BCN of the *cZR3* and *Hs1* genes and *cZR3Hs1* gene pyramiding in oilseed rape, BCN resistance tests in vitro and in vivo were performed in T2 *cZR3*, *Hs1*$^{pro-1}$ and F3 *cZR3Hs1*$^{pro-1}$ pyramiding seeds. For this, transgenic and gene pyramiding seeds were geminated on agar plates containing 150 mg/L kanamycin for selection of transgenic plants (Figure 3). The surviving seedlings were transferred to six-well plates for nematode infection experiments (Figure 4A), where each plant was inoculated with 200 infective nematode juveniles and was repeated three times for individual plant. The J2 penetration rate was determined one week after nematode inoculation. On average 15–22% of inoculated BCN J2 juveniles penetrated the plants containing exogenous genes as well as the wild type plants. However, a significant difference was observed between the transgenic and the control plants by counting developed females six weeks after nematode inoculation. Most juveniles in transgenic plants were not fully developed and became smaller and translucent, whereas the well-developed nematode females were easily recognizable on wild type plants (Figure 4B). These results demonstrate that *Hs1*$^{pro-1}$ and *cZR3* could confer a certain level of resistance to BCN in oilseed rape. There was a significant difference in the reduction of the number of developed females among transgenic generations and *cZR3Hs1*$^{pro-1}$ gene pyramiding plants. On average, 10.8 ± 0.9 developed females were found in each wild type plant, while 5.3 ± 1.4 developed females in *cZR3*, 7.1 ± 2.4 in *Hs1*$^{pro-1}$, and 4.1 ± 1.3 in *cZR3:Hs1*$^{pro-1}$ pyramiding plants were counted (Figure 5). Thus, $cZR3Hs1^{pro\text{-}1}$ gene pyramiding could enhance the BCN resistance by decreasing the number of developed females per plant.

Figure 3. Selection of transgenic rapeseed plants on MS germination containing 150 mg/L kanamycin medium (A) and histochemical GUS staining (B). The healthy and green plants were used for nematode infestation in vitro and in vivo.

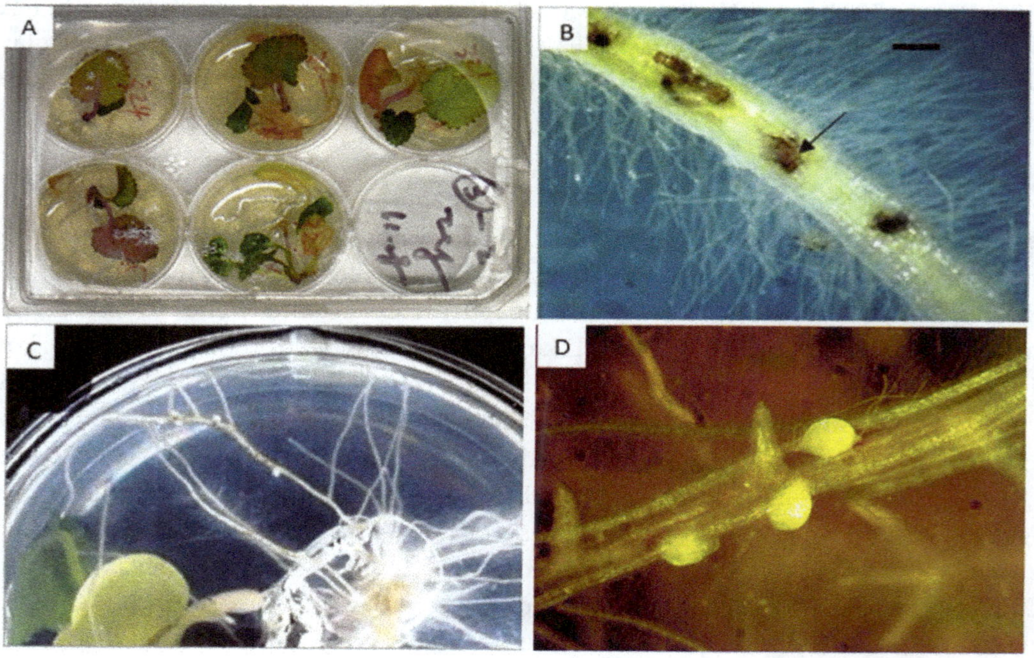

Figure 4. Beet cyst nematode resistance analysis in vitro. (A) The kanamycin resistance transgenic plants were planted on the six-well plate with KNOP medium. (B) Second stage juveniles (J2s) were inoculated near the plant root and larval penetration could be seen in the oilseed rape roots (dark arrow). The black bar equals 500 μm. (C) Female cyst nematode developed on the root surface after 21 days of inoculation with J2. (D) The stereo microscope figures of developed female on oilseed rape root.

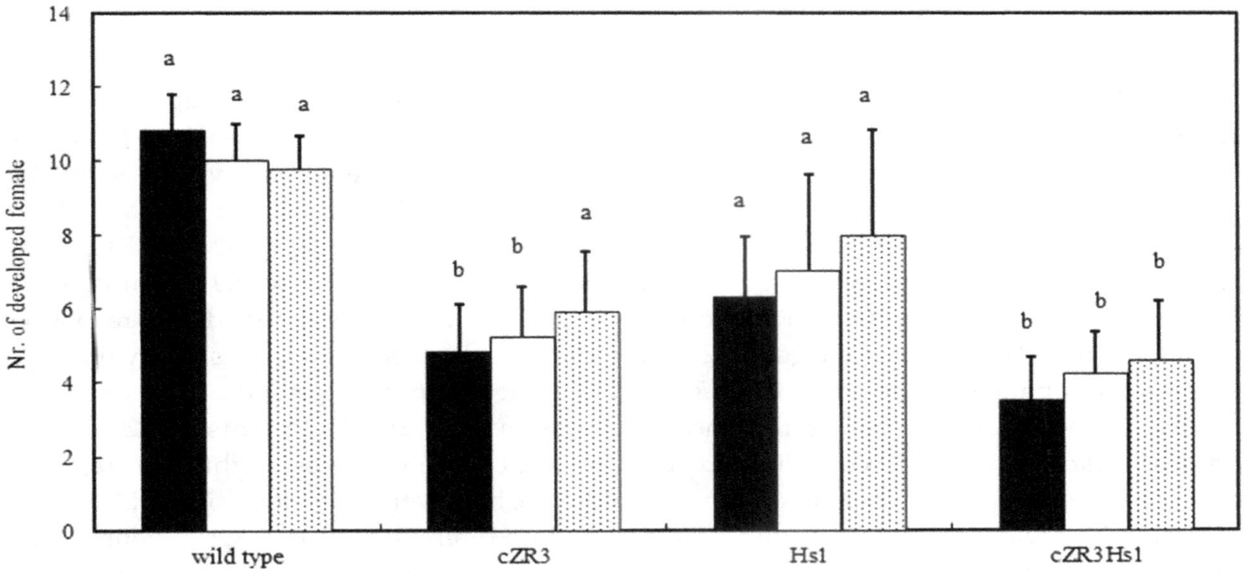

Figure 5. Numbers of developed female per plant on transgenic and wild type plants according to cyst nematode test in vitro. Wild type means non-transgenic oilseed rape (*B. napus* L.); transgenic plants consist of cZR3 and Hs1; and cZR3Hs1 means *cZR3*, *Hs1*$^{pro\text{-}1}$ and *cZR3Hs1*$^{pro\text{-}1}$ populations. The black, white, dotted rectangles represent independent wildtype plants and transgenic lines. The averages ± standard errors from three separate replicates are shown. The values with different letters are significantly different at $p \leq 0.05$ as determined by Duncan's test (a, b).

BCN resistance tests in vivo were performed under the greenhouse condition. Six weeks after BCN inoculation, the number of developed females per plant was counted. The number of developed females in wild type was 50 ± 4.7, whereas the number of developed females among *cZR3*, *Hs1*$^{pro\text{-}1}$ and *cZR3Hs1*$^{pro\text{-}1}$ was 38 ± 7.4, 33 ± 3.7 and 24 ± 5.2, respectively (Figure 6). There was a significant difference in the number of developed females among wild type, *cZR3* and *Hs1*$^{pro\text{-}1}$ transgenic and *cZR3Hs1*$^{pro\text{-}1}$ pyramiding plants.

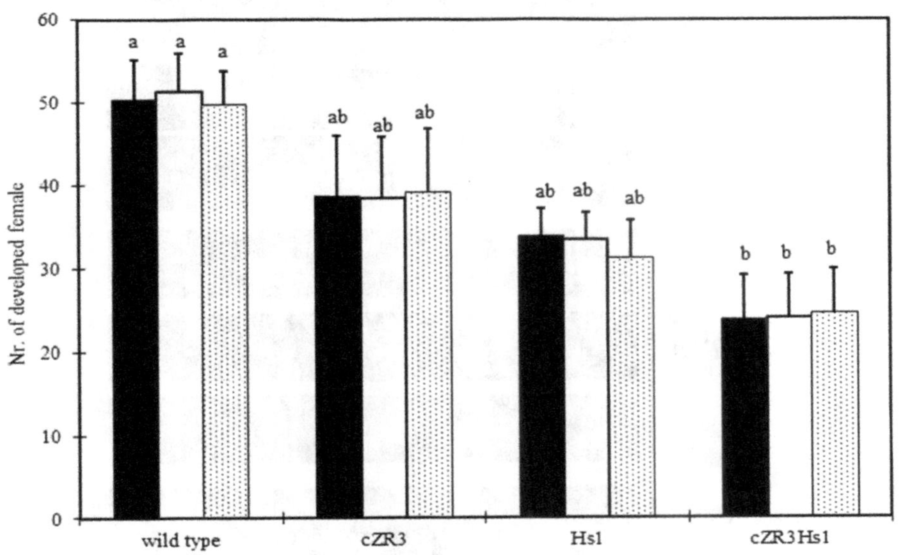

Figure 6. Numbers of developed female per plant on transgenic and wild type plants according to cyst nematode test in vivo. Wild type means non-transgenic oilseed rape (*B. napus* L.); transgenic plants consist of cZR3 and Hs1; and cZR3Hs1 means *cZR3*, *Hs1*$^{pro\text{-}1}$ and *cZR3Hs1*$^{pro\text{-}1}$ plants. The black, white, dotted rectangles represent independent wildtype plants and transgenic lines. The averages ± standard errors from three separate replicates are shown. Values with different letters are significantly different at $p \leq 0.05$ as determined by Duncan's test (a, b).

2.4. Determination of RGA-Mediated Signaling Pathways

To clarify whether expression of *cZR3* and *Hs1$^{pro\text{-}1}$* activates a specific signaling pathway, the transcript levels of four key genes involve in distinct defense pathways were analyzed in transgenic *cZR3, Hs1* and *cZR3Hs1* gene pyramiding plants compared with those in wild type by semi-RT-PCR (Figure 7). The gene list included *NPR1* (non-expresser of *PR1*) and *EDS1* (enhanced disease resistance 1), which are key regulators of resistance responses triggered by TIR-NBS-LRR-R-proteins mediated response and *SGT1* (suppressor of the *G2* allele of *skp1*) as well as *RAR1* (required for *Mla12* resistance), which are both involved in the non-TIR-(CC) NBS-LRR-R proteins mediated responses [31]. The expression of *NPR1* gene in transgenic *cZR3* and *Hs1$^{pro\text{-}1}$* plants seemed to be slightly up-regulated and there was enhanced expression in *cZR3Hs1$^{pro\text{-}1}$* gene pyramiding plants. The expression of *EDS1* gene did not change significantly among transgenic *cZR3* and *Hs1* plants, *cZR3Hs1$^{pro\text{-}1}$* gene pyramiding plants and wild type. The expression of *SGT1b* gene was slightly up-regulated in transgenic and *Hs1$^{pro\text{-}1}$* plants compared with the wild type, however, its response in *cZR3Hs1$^{pro\text{-}1}$* gene pyramiding plant was similar to the wild type and had no interaction effect. *RAR1* gene expression was significantly up-regulated in *cZR3* transgenic plants, but almost no change in expression was found in *Hs1$^{pro\text{-}1}$* transgenic plants compared with wild type plants. However, *RAR1* gene expression was also significantly enhanced in *cZR3Hs1$^{pro\text{-}1}$* gene pyramiding plants. Furthermore, the relative expression of *SGT1b* and *RAR1* genes was also analyzed. Compared with wild type, expression level of *SGT1b* in *cZR3* transgenic plants and *cZR3Hs1$^{pro\text{-}1}$* pyramiding plants was decreased slightly, while it was increased in *Hs1* transgenic plants (Figure 8A). The expression of *RAR1* gene by qRT-PCR analysis showed that *RAR1* was slightly expressed in wild type and *Hs1$^{pro\text{-}1}$* transgenic plants. The expression of *RAR1* in *cZR3* transgenic plant was about 50 times higher than the wild type, while the expression of *RAR1* was enhanced by *cZR3Hs1$^{pro\text{-}1}$* gene pyramiding, which was about 150 times higher than the wild type (Figure 8B).

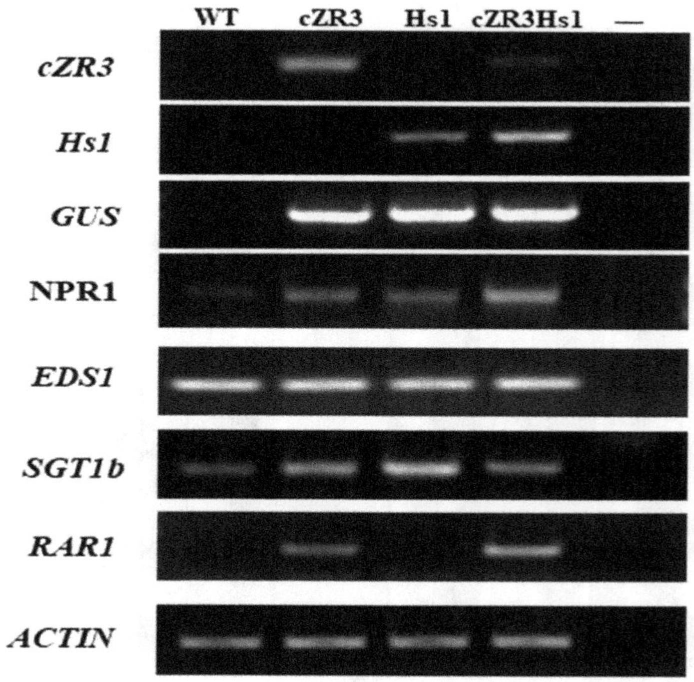

Figure 7. Expression of the key defense related genes in wild-type, transgenic *cZR3* and *Hs1$^{pro\text{-}1}$*, and *cZR3Hs1$^{pro\text{-}1}$* gene pyramiding plants. The expression levels of *cZR3, Hs1$^{pro\text{-}1}$, GUS, NPR1, EDS1, SGT1* and *RAR1* were determined by semi-quantitative RT-PCR with independent transgenic and pyramiding lines, while oilseed rape wild type *zheshuang 758* served as control. "–" represents the negative control.

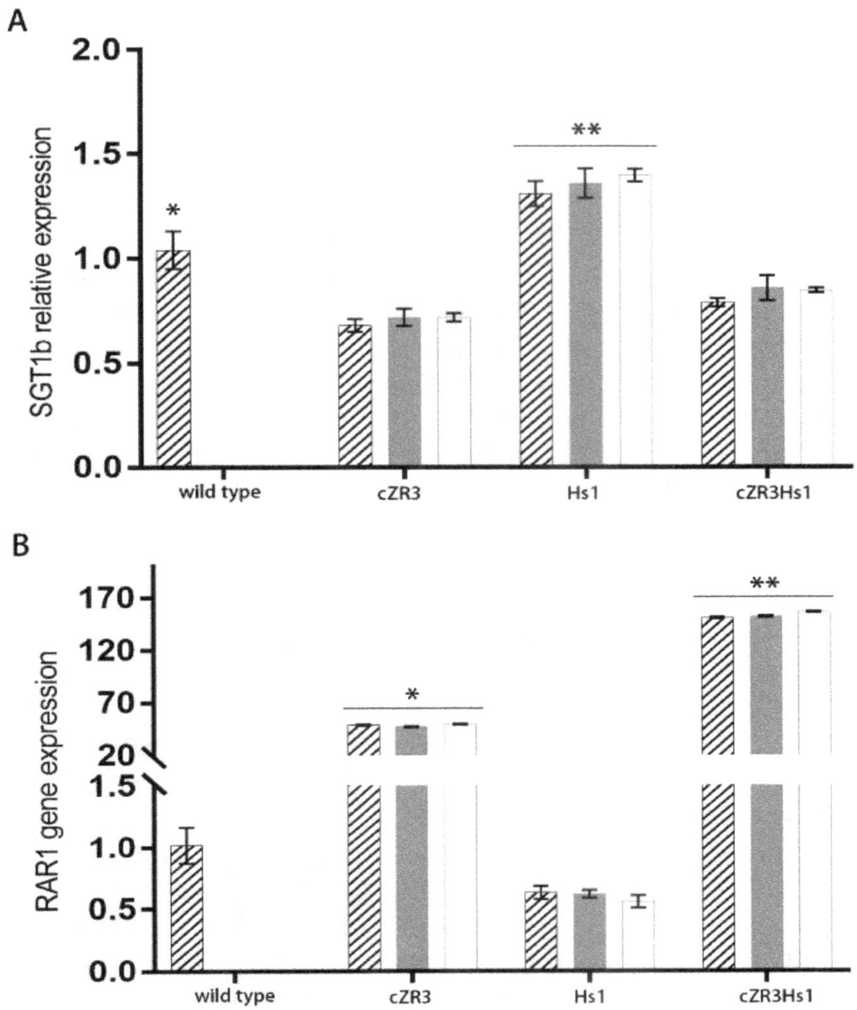

Figure 8. The relative expression of *SGT1b* (**A**) and *RAR1* (**B**) gene among wild type, *cZR3* and $Hs1^{pro\text{-}1}$, and $cZR3Hs1^{pro\text{-}1}$ pyramiding plants by qRT-PCR. The diagonal stripes, grey, white rectangles represent independent wildtype plants and transgenic lines. The averages ± standard errors from three separate replicates are shown. Significant difference is indicated by different symbols (* and **), as determined by Duncan's test.

3. Discussion

Crops are attacked by nematodes causing considerable economic losses worldwide. The estimated worldwide losses due to plant parasitic nematodes are about $125 billion annually [32]. An integrated strategy, e.g., including trap crop planting, nematicide application and cultivation of resistant sugar beet varieties, could control beet cyst nematode infestation in sugar beets [33]. Effective nematode control could be achieved by cultivating a nematode resistant variety that can reduce nematode populations up to 70% in field conditions [33]. Breeding of resistant cultivars is the most desired and promising alternative because regeneration of whole transgenic sugar beet plants with *A. tumefaciens* mediated transformation is laborious and timing consuming. In the present study, we transferred $BvHs1^{pro\text{-}1}$ and *BvcZR3* resistance genes, into oilseed rape to develop cyst nematode resistance crop. In vitro and in vivo nematode tests showed the developed female reduced 50.93%, 34.26%, 24.0%, and 34.0%, respectively, in *BvcZR3* and $BvHs1^{pro\text{-}1}$ transgenic plants. Our results are consistent with previous studies, where *A. rhizogenes*-mediated transformation was performed to generate transgenic $Hs1^{pro\text{-}1}$ and *cZR3* hairy roots in sugar beet [11,21].

Different R genes often confer resistance to different isolates, races or biotypes of beet cyst nematode. The simultaneous expression of different R genes could broaden the resistance spectrum, as it may provide resistant against various races or isolates [34]. The multiple genes pyramiding strategy has been applied in genetic engineering to achieve durable resistance against phytopathogen [35,36] and nematode resistance [37,38]. Bharathia et al. [36] reported *Allium sativum* (*asal*) and *Galanthus nivalis* (*gna*) lectin genes pyramided into rice lines through sexual crosses between two stable transgenic rice lines, which endowed the pyramided rice lines with enhanced resistance to major sap sucking insects. Urwin et al. [38] transferred cowpea trypsin inhibitor (*CpTI*) and a cystatin gene (*Oc-ID86*) into *Arabidopsis* and the *CpTI Oc-ID86* transgenic plants displayed enhanced resistance against *H. schachtii*. Chan et al. [37] demonstrated via dual gene overexpression system that utilizing a plant cysteine proteinase inhibitor (*CeCPI*) and a fungal chitinase (*PjCHI-1*) in tomato (*Solanum lycopersicum*) can enhance resistance against root-knot nematode (*Meloidogyne incognita*). The $Hs1^{pro\text{-}1}$ gene encodes a plasma membrane protein with an extensive leucine-rich region, which contains a transmembrane spanning domain and a short hydrophobic C-terminal domain. It could be speculated that $Hs1^{pro\text{-}1}$ resides in the plasma membrane as a receptor with its N-terminus toward the extracellular space [11]. Similar to recently cloned nematode R-genes, including *Mi* [12] and *Gpa2* [21], *cZR3* belongs to the CC-NBS-LRR class of R-protein lacking a signal sequence [21]. Previous studies show that no complete resistance is observed in $Hs1^{pro\text{-}1}$ transgenic sugar beet plants and a second gene is proposed to be essential for the resistance [30]. In this study, $cZR3Hs1^{pro\text{-}1}$ genes pyramiding plants were generated through sexual crosses made between two transgenic $Hs1^{pro\text{-}1}$ and *cZR3* plants of oilseed rape. Cyst nematode resistance test in vitro and in vivo showed that the number of females that developed per plant on transgenic $cZR3Hs1^{pro\text{-}1}$ genes pyramiding plants were reduced 52.0% and 62.0%, respectively, and the reduction percentage was significantly different with transgenic *cZR3* and $Hs1^{pro\text{-}1}$ and wild type plants. The possible functional model for $cZR3Hs1^{pro\text{-}1}$ mediated response was based on *cZR3*, which may function as a co-receptor with $Hs1^{pro\text{-}1}$ together recognizing the *Avr* products released by nematodes into the cytoplasm initiating signal transduction that finally leads to resistance response.

Thus far, little is known about the complex regulatory role of cloned nematode R-genes. It is generally believed that these genes recognize nematode effectors and trigger specific signaling pathways that lead to resistance responses [39]. The disease resistance mechanisms in model plant *Arabidopsis thaliana* were extensively studied, and the gene-for-gene hypothesis has been proposed for a long time. R genes interact with the corresponding pathogenic toxic genes, thus causing local reactive oxygen species accumulation, programmed cell death and local allergic reactions [40]. On the one hand, these reactions limit the growth and extend of pathogenic bacteria in infected sites. On the other hand, these reactions release signaling molecules to surrounding cells, further inducing expression of defense genes to boost whole-plant resistance. Some crucial genes mediated by R gene have been found in recent years, such as *RAR1*, *NPR1*, *EDS1* and *SGT1*. Studies have shown that the CC-NBS-LRR R gene mediates disease resistance reaction associated with *RAR1* and *SGT1*, while *NPR1*, *EDS1* and *PAD4* are related with TIR-NBS-LRR R genes [31]. This study proved that cZR-3 and $Hs1^{pro\text{-}1}$ could independently activate a *RAR1/SGT1* dependent signaling pathway in plants, which is essential for a CC-NBS-LRR R gene mediated resistance response. Hence, the present study provides a new approach to develop BCN resistance in oilseed rape plants based on stacking of $cZR3Hs1^{pro\text{-}1}$ genes that confer a high level of BCN resistance in transgenic plants.

4. Materials and Methods

4.1. Plant Transformation

Direct shoots regeneration from hypocotyl explants of semi-winter-type oilseed rape cultivar *zheshuang 758* were transformed by *A. tumefaciens*, as described by Tang et al. [41]. The PAM194-$Hs1^{pro\text{-}1}$ and PAM194-*cZR3* vector contains the complete $Hs1^{pro\text{-}1}$ ORF (Gene accession number U79733) and *cZR3* (Gene accession number DQ907613), respectively, driven by cauliflower 35S promoter and nopaline synthase terminator, the neomycin phosphotranferase marker gene (*nptII*) and β-glucuronidase (GUS) reporter gene (Figure 1A,B) driven by the cauliflower 35S promoter. Both vectors were presented by Prof. Daguang Cai from Institute of Molecular Phytopathology, Kiel University, Germany. The binary vectors were then transformed into the competent cells of *A. tumefaciens* strain GV3101 by the freeze–thaw method.

In brief, we used about 10 mm hypocotyl segments was obtained by growing seeds of oilseed rape seeds on germination medium [42] (2.22 g/L 1/2MS basal medium from Phytotechnology Laboratories®, Shawnee Mission, Lenexa, KS, USA, 20 g/L sucrose, 8 g/L Agar, pH 5.8) at 25 °C and 16 h light/8 h dark periods for about 4–5 days as explants. The isolated explants were pre-cultured on the callus induction medium (CIM) (MS 4.43 g/L, 1 mg/L 2,4-D, 30 g/L sucrose, 8 g/L Agar, pH 5.6) at 25 °C and 16 h light/8 h dark periods for 3 days. The pre-cultured explants were infected by *A. tumefaciens* suspension (OD600 0.3–0.4) for 10 min, then subsequently blotted on sterile filter paper and placed on CIM at 22 °C under dim light co-cultivation for 3 days. After co-cultivation, the explants were placed on SIM media for shoot generation (SIM) (MS 4.43 g/L, 4 mg/L N-6-benzylaminopurine (BAP), 0.1 mg/L 1-Naphthaleneacetic acid (NAA), 5 mg/L silver nitrate ($AgNO_3$), 500 mg/L carbenicillin (Carb), 50 mg/L kanamycin (Kan), 30 g/L sucrose, 8 g/L Agar, pH 5.8) at 25 °C and 16 h light/8 h dark periods for 2 weeks. The media were replaced every two weeks until multiple shoots were generated. The multiple induced shoots were isolated and transferred to shoot elongation media [43] (SEM) (B5 basal medium from Phytotechnology laboratories® 3.21 g/L, 10 g/L sucrose, 50 mg/L Kan, 9 g/L Agar, pH 5.8). The elongated shoots were transferred to SEM to develop the roots. The rooting plants were first grown in 10 cm diameter pots containing nutrient soil (Pindstrup, Ryomgaard, Denmark). The surviving plants were transferred to greenhouse to vernalize at 4 °C 16 h light/8 h dark periods for 40 days. The vernalized plants were transferred to the greenhouse to grow normally to harvest seeds at 25 °C and 16 h light/8 h dark periods. When the plants began to flower, the inflorescence of each independent putative transformed line was covered with transparent plastic bags and the plants were allowed self-pollination to acquire T1 seeds.

4.2. PCR Analysis

Forty-day-old rapeseed leaves were used to extract genomic DNA from transgenic and wild-type plants using CTAB method. PCR analyses were performed using specific gene primers for $Hs1^{pro\text{-}1}$ (accession number: U79733.1) (F: 5′-GGCACCATCCAAACTCGG-3′, R: 5′-CGAATAAGTGAGAGGATC-3′), *cZR3* (accession number: DQ907613) (F: 5′-GGCAAAACTGCTCTTGCC-3′ and R: 5′-AGCCCTATCAATAACTCC-3′) and *cZR3* (F: 5′-AGTTATTGATAGGGCTATGG-3′ and R: 5′-ATACTTGAAGCAGTCAGG-3′). And the size of amplifying products are 500-bp [11], 710-bp and 410-bp [22] respectively. The PCR reaction was performed at 94 °C for 3 min, followed by 35 cycles of 94 °C for 1 min, 55 °C ($Hs1^{pro\text{-}1}$ and *cZR3*) for 1 min and 72 °C for 1 min 20 s with a final extension at 72 °C for 10 min. PCR-amplified products were analyzed on 1% agarose gel, stained with ethidium bromide and fluoresce under ultraviolet light.

4.3. Southern Blot Analysis

Genomic DNA of each independent transgenic *cZR3* and $Hs1^{pro\text{-}1}$ lines was restricted by Hind III (Takara Bio Inc., Dalian, China) at 37 °C for 5 h, and was separated on 0.7% agarose gel overnight and transferred onto a Hybond N^+ nylon membrane (GE Healthcare, RPN 303B, Piscataway, NJ, USA) using the alkaline transfer buffer as recommended by the manufacturer. Southern blots were hybridized using 32P-labeled *cZR-3* and $Hs1^{pro\text{-}1}$ DNA fragment as probe at 62 °C overnight. The blots were washed twice with 0.5× saline sodium citrate (SSC), 0.2% *w/v* SDS for 30 min and together with the film exposed at −70 °C for 48 h [44].

4.4. Histochemical GUS Assays

The histochemical *GUS* assay was carried out according to Jefferson et al.'s [45] method. Ten milliliters of X-Gluc solution including 50 mM Na_3PO_4 buffer (pH 7.0), 0.2 mg/mL X-Gluc (5-bromo-4-chloro-3-indolyl β-D-glucuronide) and two drops of Triton-100 were added in the Petri-dish, which contained the putative transgenic leaves and plants. The leaves and plants were covered with X-Gluc solution and incubated at 37 °C for 16 h. Afterwards, the samples were washed using 70% ethanol to remove the chlorophyll. The GUS staining signals were evaluated under a stereomicroscope (Stemi SV 11, Zeiss, Jena, Germany).

4.5. Genes Pyramiding and Progeny Analysis

Hs1$^{pro\text{-}1}$ and *cZR3* genes pyramiding were generated by artificial hybridization using T0 putative transgenic *cZR3* line as female and T0 putative transgenic *Hs1*$^{pro\text{-}1}$ line as male. F1 seeds obtained from the crossed inflorescence and *cZR3Hs1*$^{pro\text{-}1}$ positive F2 plant were screened for the presence of *cZR3Hs1*$^{pro\text{-}1}$ genes by PCR.

T1 seeds obtained from self-pollination of 3 T0 independent *Hs1*$^{pro\text{-}1}$ and *cZR3* plants were sown in 9 cm Petri-dishes containing germination medium with 150 mg/L kanamycin. The surviving plants were transferred to pots containing peat moss (Pindstrup, Balozi, Lativa) and maintained in green house. All T1 were screened for the presence of *Hs1*$^{pro\text{-}1}$ or *cZR3* genes by PCR. Subsequent generations were obtained by self-pollination of transgenic plants and confirmed by PCR.

4.6. Propagation of H. Schactii

In vitro cultured mustard (*Sinapis albacv.* Albatros) was used as host plants for beet cyst nematode (*H. schactii*) stock propagation. The mustard seeds were surface sterilized in 70% ethanol and in 5% NaClO solution, containing 2–3 drops of Tween 20 for 10 and min, then rinsed at least 3 times with sterilized water. The sterilized seeds were sown on half-strength MS media with 0.8% agar on 9 cm Petri-dishes at 25 °C in the dark. After 7-days of germination, the seedlings were transferred to 15 cm Petri-dishes containing 0.2× KNOP medium with 2% (*w*/*v*) sucrose and 0.8% (*w*/*v*) Daishin agar (Duchefa, Haarlem, The Netherlands) and placed in dark for 4 weeks at 25 °C [46]. The cysts were picked from the fields and J2 larvae hatched on 50 μm gauze stimulated by incubating cysts in 3 mM $ZnCl_2$ for 7–12 days. The J2 larvae were collected with 10 μm gauze, surface-sterilized with 0.1% $HgCl_2$, washed four times in sterile water and re-suspended in 0.2% (*w*/*v*) Gelrite (Duchefa). The sterile J2 larvae were directly inoculated to the mustard roots and the cysts were propagated in the mustard roots in vitro. These plates were used as a stock for beet cyst nematode.

4.7. Nematode Resistance Assay In Vitro and In Vivo

T2 transgenic *Hs1*$^{pro\text{-}1}$, *cZR3* and F3 *Hs1*$^{pro\text{-}1}$*cZR3* gene pyramiding oilseed rape plants were used for nematode infestation analysis, under both in vitro and in vivo. Nematode resistance analysis in vitro was performed according to Sijmons et al. [47]. First, the T1 transgenic *Hs1*$^{pro\text{-}1}$ and *cZR3* and F3 *Hs1*$^{pro\text{-}1}$*cZR3* gene pyramiding seeds were surface sterilized, and then germinated on 15 mm Petri dishes containing half-length MS germination medium supplemented with 150 mg/L kanamycin to select the positive transgenic plants. Ten days after germination, the surviving seedlings were transferred to a six-well plate containing KNOP medium and cultured for further 7 days at 23 ± 1 °C with a fluorescent light illumination regime of 16 h/8 h (day/night, 100 μ·mol m^{-2} s^{-1}) and the relative humidity was 75%. Two hundred sterile infective J2 larvae were inoculated to each plant by a veterinary syringe. The number of females that developed per plant was counted 6 weeks after inoculation under a stereomicroscope (Stemi SV11, Zeiss, Jena, Germany). At least three 6-well plates were performed for each transgenic line and wild type as biological replicates. For nematode resistance test in vivo, the T2 transgenic *Hs1*$^{pro\text{-}1}$, *cZR3* and F3 *Hs1*$^{pro\text{-}1}$*cZR3* genes pyramiding seeds were germinated on the germination medium with 150 mg/L kanamycin for 10 days to select the positive transgenic ones. The positive transgenic single plants were transplanted in plastic tubes (3 cm × 4 cm × 20 cm) filled with silver sand and moistened with Steiner I nutrient solution [48] and cultured in a greenhouse at 25 °C 16 h/8 h (light/dark) periods for further 2 weeks. Each plant was then inoculated with 2 mL suspension containing approximately 2000 pre-hatched J2 BCN larvae by a veterinary syringe. Six weeks after BCN inoculation, the root system was washed free from sand by high-speed tap water and the number of developed BCN cysts was counted. Each line and wild type were replicated 15 plants at least.

4.8. Semi Real Time PCR Analysis

For real time PCR analysis, the total RNA from leaves and roots was extracted with Trizol (Gibco, BRL life technologies, Grand Island, NY, USA) according to the manufacturer's protocol. First strand cDNA was synthesized with 1 μg of purified total RNA using PrimeScriptTMRT reagent Kit with gDNA Eraser (TaKaRa). Real-time PCR reaction was carried out according to SYBR Premix Ex Taq II (TaKaRa). Specific primers of defensing genes were designed according to sequences of *Arabidopsis* defensing genes in NCBI and summarized in Table 3. The semi-quantitative PCR was performed in 50 μL reactions consisting of 2.5 μL 10 ng/μL cDNA, 5 μL 10× buffer, 0.5 μL 10 mM dNTPs, 5 μL each of 10 pmol/μL primer, 2.5 units of Taq polymerase and 31.5 μL H_2O under the PCR program: 94 °C for 50 s, 51 °C for 1 min and 72 °C for 1 min for 25 cycles, followed by 10 min at 72 °C. Amplicons were separated on a 1% (w/v) agarose gel and visualized under UV-light. The levels of gene expression were calculated by comparison of the densities of the PCR products, in which house-keeping *ubiquitin* gene served as an internal control and the mRNA levels for each cDNA probe were normalized to the ubiquitin message RNA level (5′-ACTCTCACCGGAAAGACAATC-3′ and 5′-TGACGTTGTCGATGGTGTCAG-3′). Quantitative RT-PCR was carried out using a SYBR Premix Ex22Taq (perfect real time) kit (TaKaRa Biomedicals) on a LightCycler23480 machine (Roche Diagnostics, Rotkreuz, Switzerland) according to the manufacturer's instructions (Roche Diagnostics). The qRT-PCR amplification was performed at 94 °C for 10 s, 58 °C for 10 s and 72 °C for 10 s. All reactions were repeated three times. The relative level of gene expression was calculated using the formula $2^{-\triangle\triangle CP}$ according to Livak and Schmittgen [49].

Table 3. Primers were used in the described gene expression of defense pathway.

Target Gene	Accession Number	Primer Sequence (5′ → 3′)	References
NPR1	AT1G64280.1	TGAATTGAAGATGACGCTGCT AGGCCTTCTTTAGTGTCTCTTGTA	Wu et al. 2012 [50]
PAD4	AT3G52430	GGTCGACGCTGCCATACTCAAACT AGAGAGATTGGTTTCCGAGCAGAG	Youssef et al. 2013 [51]
RAR1	AT5G51700	CGGCTCCTACTTCATCTCCAG AACATCGCAACATTTCCACCCTCT	Tornero et al. 2002 [52]
SGT1b	AT4G11260	CCCAAACCCAATGTCTCATCAG TCCACTTTCTTAGTCCCAACTTCT	Tör et al. 2002 [53]

4.9. Data Analysis

The segregation rates of the T2 progenies of transgenic oilseed plants, as well as the segregation ratios in the F2 generation from crosses between *cZR3* and *Hs1* homozygous transgenic plants were analyzed using the chi-square test to confirm the expected Mendelian segregation pattern of 3:1 (transgenic: non-transgenic plants) and 9:3:3:1. Data were analyzed using IBMSPSS 23.0 statistical system for windows (SPSS Inc., Armonk, NY, USA). Duncan multiple range tests were performed at the 0.05 level of probability.

Author Contributions: Conceptualization, G.T.; Formal analysis, X.Z. and Q.Z.; Investigation, X.Z.; Resources, G.T. and D.C.; Validation, Q.Z.; Visualization, N.C.; Writing—original draft, G.T.; and Writing—review and editing, X.Z.

Abbreviations

BCN	beet cyst nematode
RGAs	resistance gene analogs
CIM	callus induction medium
SIM	shoot induced media
SEM	shoot elongation media
BAP	N-6-benzylaminopurine
NAA	Naphthaleneacetic acid
$AgNO_3$	silver nitrate
Carb	carbenicillin
Kan	kanamycin

References

1. Lilley, C.J.; Atkinson, H.J.; Urwin, P.E. Molecular aspects of cysts nematodes. *Mol. Plant Pathol.* **2005**, *6*, 577–588. [CrossRef] [PubMed]
2. Hemayati, S.S.; Akbar, J.E.; Ghaemi, A.R.; Fasahat, P. Efficiency of white mustard and oilseed radish trap plants against sugar beet cyst nematode. *Appl. Soil Ecol.* **2017**, *119*, 192–196. [CrossRef]
3. Jung, C.; Wyss, U. New approaches to control plant parasitic nematodes. *Appl. Microbiol. Biotechnol.* **1999**, *51*, 439–446. [CrossRef]
4. Lelivelt, C.L.C.; Lange, W. Intergeneric crosses for the transfer of resistance to the beet cyst nematode from *Raphanussativus* to *Brassica napus*. *Euphytica* **1993**, *68*, 111–120. [CrossRef]
5. Nielsen, E.L.; Baltenspencer, D.P.; Kerr, E.D.; Rife, C.L. Host suitability of rapeseed for *Heterodera schachtii*. *J. Nematol.* **2003**, *35*, 35–38. [PubMed]
6. Peterka, H.; Budahn, H.; Schrader, O.; Ahne, R.; Schütze, W. Transfer of resistance against the beet cyst nematode from radish (*Raphanussativus*) to rape (*Brassica napus*) by monosomic chromosome addition. *Appl. Genet.* **2004**, *109*, 30–41. [CrossRef] [PubMed]
7. Peterka, H.; Budahn, H.; Zhang, S.S.; Li, J.B. Nematode resistance of rape-radish chromosome addition lines. *Nematology* **2010**, *12*, 269–275.
8. Lelivelt, C.L.C.; Leunissen, E.H.M.; Frederiks, H.J.; Helsper, J.P.F.G.; Krens, F.A. Transfer of resistance to the beet cyst nematode (*Heterodera schachtii* Schm.) from *Sinapis alba* L. (white mustard) to the *Brassica napus* L. gene pool by means of sexual and somatic hybridization. *Appl. Genet.* **1993**, *85*, 688–696. [CrossRef]
9. Brown, J.; Brown, A.P.; Davis, J.B.; Erickson, D. Intergeneric hybridization between *Sinapis alba* and *Brassica napus*. *Euphytica* **1997**, *93*, 163–168. [CrossRef]
10. Zhang, S.S.; Peterka, H.; Budahn, H.; Schrader, O.; Li, C.Y. Chromosomal localization of resistance gene in radish against beet cyst nematode and the stability of additional oil radish chromosomes in rape-radish addition lines. *Sci. Agric. Sin.* **2008**, *4*, 93–101.
11. Cai, D.; Kleine, M.; Kifle, S.; Harloff, H.J.; Sandal, N.N.; Marcker, K.A.; Klein-Lankhorst, R.M.; Salentijn, E.M.; Lange, W.; Stiekema, W.J.; et al. Positional cloning of a gene for nematode resistance in sugar beet. *Science* **1997**, *275*, 832–834. [CrossRef]
12. Milligan, S.B.; Bodeau, J.; Yaghoobi, J.; Kaloshian, I.; Zabel, P.; Williamson, V.M. The root knot nematode resistance gene *Mi* from tomato is a member of the leucine zipper, nucleotide binding, leucine-rich family of plant genes. *Plant Cell* **1998**, *10*, 1307–1319. [CrossRef]
13. Van Der Vossen, E.A.; Van Der Voort, J.N.; Kanyuka, K.; Bendahmane, A.; Sandbrink, H.; Baulcombe, D.C.; Bakker, J.; Stiekema, W.J.; Klein-Lankhorst, R.M. Homologues of a single resistance-gene cluster in potato confer resistance to distinct pathogens: A virus and a nematode. *Plant J.* **2000**, *23*, 567–576. [CrossRef]
14. Ernst, K.; Kumar, A.; Kriseleit, D.; Kloos, D.U.; Phillips, M.S.; Ganal, M.W. The broad-spectrum potato cyst nematode resistance gene (*Hero*) from tomato is the only member of a large gene family of NBS-LRR genes with an annual amino acid repeat in the LRR region. *Plant J.* **2002**, *31*, 127–136. [CrossRef]
15. Thurau, T.; Kifle, S.; Jung, C.; Cai, D. The promoter of the nematode resistance gene $Hs1^{pro\text{-}1}$ activates a nematode-responsive and feeding site specific gene expression in sugar beet (*Beta vulgaris* L.) and *Arabidopsis thaliana*. *Plant Mol. Biol.* **2003**, *52*, 643–660. [CrossRef]

16. McLean, M.D.; Hoover, G.J.; Bancroft, B.; Makhmoudova, A.; Clark, S.M.; Welacky, T.; Simmonds, D.H.; Shelp, B.J. Identification of the full-length *Hs1^{pro-1}* coding sequence and preliminary evaluation of soybean cyst nematode resistance in soybean transformed with *Hs1^{pro-1}* cDNA. *Can. J. Bot.* **2007**, *85*, 437–441. [CrossRef]
17. Hunger, S.; Gaspero, G.D.; Mohring, S.; Bellin, D.; Schafer-Pregl, R.; Borchardt, D.C.; Durel, C.E.; Werber, M.; Weisshaar, B.; Salamini, F.; et al. Isolation and linkage analysis of expressed disease-resistance gene analogues of sugar beet (*Beta vulgaris* L.). *Genome* **2003**, *46*, 70–82. [CrossRef]
18. Leister, D.; Kurth, J.; Laurie, D.A.; Yano, M.; Sasaki, T.; Devos, K.; Graner, A.; Schulze-Lefert, P. Rapid reorganization of resistance gene homologues in cereal genomes. *Proc. Natl. Acad. Sci. USA* **1998**, *95*, 370–375. [CrossRef]
19. Timmerman-Vaughan, G.M.; Frew, T.J.; Weeden, N. Characterization and linkage mapping of R-gene analogous DNA sequences in pea (*Pisum sativum* L.). *Appl. Genet.* **2000**, *101*, 241–247. [CrossRef]
20. Dangl, J.L.; Jones, J.D. Plant pathogens and integrated defence responses to infection. *Nature* **2001**, *411*, 826–833. [CrossRef]
21. Tian, Y.Y.; Fan, L.J.; Thurau, T.; Jung, C.; Cai, D. The absence of TIR-type resistance gene analogues in the sugar beet (*Beta vulgaris* L.) genome. *Mol. Evol.* **2004**, *58*, 40–53. [CrossRef]
22. Lein, J.C.; Asbach, K.; Tian, Y.; Schulte, D.; Li, C.; Koch, G.; Jung, C.; Cai, D.G. Resistance gene analogues are clustered on chromosome 3 of sugar beet and cosegregate with QTL for rhizomania resistance. *Genome* **2007**, *50*, 61–71. [CrossRef]
23. Ori, N.; Eshed, Y.; Paran, I.; Presting, G.; Aviv, D.; Tanksley, S.; Zamir, D.; Fluhr, R. The I2C family from the wilt disease resistance locus *I2* belongs to the nucleotide binding, leucine-rich repeat superfamily of plant resistance genes. *Plant Cell* **1997**, *9*, 521–532. [CrossRef]
24. Yoshimura, S.; Yamanouchi, U.; Katayose, Y.; Toki, S.; Wang, Z.X.; Kono, I.; Kurata, N.; Yano, M.; Iwata, N.; Sasaki, T. Expression of Xa1, a bacterial blight-resistance gene in rice, is induced by bacterial inoculation. *Proc. Natl. Acad. Sci. USA* **1998**, *95*, 1663–1668. [CrossRef]
25. Lagudah, E.S.; Moullet, O.; Appels, R. Map-based cloning of a gene sequence encoding a nucleotide binding domain and a leucine-rich region at the *Cre3* nematode resistance locus of wheat. *Genome* **1997**, *40*, 659–665. [CrossRef]
26. Whitham, S.; Dinesh-Kumar, S.P.; Choi, D.; Hehl, R.; Corr, C.; Baker, B. The product of the tobacco mosaic virus resistance gene N: Similarity to toll and the interleukin-1 receptor. *Cell* **1994**, *78*, 1101–1115. [CrossRef]
27. Dinesh-Kumar, S.P.; Baker, B.J. Alternatively spliced N resistance gene transcripts: Their possible role in tobacco mosaic virus resistance. *Proc. Natl. Acad. Sci. USA* **2000**, *97*, 1908–1913. [CrossRef]
28. Paal, J.; Henselewski, H.; Muth, J.; Meksem, K.; Menéndez, C.M.; Salamini, F.; Ballvora, A.; Gebhardt, C. Molecular cloning of the potato *Gro1-4* gene conferring resistance to pathotype Ro1 of the root cyst nematode *Globodera rostochiensis*, based on a candidate gene approach. *Plant. J.* **2004**, *38*, 285–297. [CrossRef]
29. Calenge, F.; Van der Linden, C.G.; Van de Weg, E.; Schouten, H.J.; Van Arkel, G.; Denance, C.; Durel, C.E. Resistance gene analogues identified through the NBS-profiling method map close to major genes and QTL for disease resistance in apple. *Appl. Genet.* **2005**, *110*, 660–668. [CrossRef]
30. Schulte, D.; Cai, D.; Kleine, M.; Fan, L.; Wang, S.; Jung, C. A complete physical map of wild beet (*Beta procumbens*) translocation in sugar beet. *Mol. Gen. Genom.* **2006**, *275*, 504–511. [CrossRef]
31. Hammond-Kosack, K.E.; Parker, J.E. Deciphering plant-pathogen communication: Fresh perspectives for molecular resistance breeding. *Curr. Opin. Biotechnol.* **2003**, *14*, 177–193. [CrossRef]
32. Chitwood, D.J. Research on plant-parasitic nematode biology conducted by the United States Department of Agriculture–Agricultural Research Service. *Pest. Manag. Sci.* **2003**, *59*, 748–753. [CrossRef]
33. Hauer, M.; Koch, H.J.; Krüssel, S.; Mittler, S.; Märländer, B. Integrated control of *Heterodera schachtii* Schmidt in Central Europe by trap crop cultivation, sugar beet variety choice and nematicide application. *Appl. Soil Ecol.* **2016**, *99*, 62–69. [CrossRef]
34. Feechan, A.; Kocsis, M.; Riaz, S.; Zhang, W.; Gadoury, D.M.; Walker, M.A.; Dry, I.B.; Reisch, B.; Cadle-Davidson, L. Strategies for RUN1 deployment using RUN2 and REN2 to manage grapevine powdery mildew informed by studies of race specificity. *Phytopathology* **2015**, *105*, 1104–1113. [CrossRef]
35. Abdeen, A.; Virgòs, A.; Olivella, E.; Villanueva, J.; Avilé, X.; Gabarra, R.; Prat, S. Multiple insect resistance in transgenic tomato plants over-expressing two families of plant proteinase inhibitors. *Plant Mol. Biol.* **2005**, *57*, 189–202. [CrossRef]

36. Bharathia, Y.; Kumara, S.V.; Pasalub, I.C.; Balachandranb, S.M.; Reddya, V.D.; Raoa, K.V. Pyramided rice lines harbouring *Allium sativum* (*asal*) and *Galanthus nivalis* (*gna*) lectin genes impart enhanced resistance against major sap-sucking pests. *J. Biotechnol.* **2011**, *152*, 63–71. [CrossRef]
37. Chan, Y.L.; He, Y.; Hsiao, T.T.; Wang, C.J.; Tian, Z.H.; Yeh, K.W. Pyramiding taro cystatin and fungal chitinase genes driven by a synthetic promoter enhances resistance in tomato to root-knot nematode Meloidogyne incognita. *Plant Sci.* **2015**, *231*, 74–81. [CrossRef]
38. Urwin, P.E.; McPherson, M.J.; Atkinson, H.J. Enhanced transgenic plant resistance to nematodes by dual proteinase inhibitor constructs. *Planta* **1998**, *204*, 472–479. [CrossRef]
39. Williamson, M.V.; Kumar, A. Nematode resistance in plants: The battle underground. *Trends Genet.* **2006**, *22*, 396–403. [CrossRef]
40. Glazebrook, J. Contrasting mechanisms of defense against biotrophic and necrotrophic pathogens. *Annu. Rev. Phytopathol.* **2005**, *43*, 205–227. [CrossRef]
41. Tang, G.X.; Kneck, K.; Yang, X.F.; Qin, Y.B.; Zhou, W.J.; Cai, D. A two-step protocol for shoot regeneration from hypocotyl explants of oilseed rape and its application for *Agrobacterium*-mediated transformation. *Biol. Plant.* **2011**, *55*, 21–26. [CrossRef]
42. Murashige, T.; Skoog, F. A revised medium for rapid growth and bioassays in tobacco tissue culture. *Physiol. Plant.* **1962**, *15*, 474–493. [CrossRef]
43. Gamborg, O.L.; Miller, R.A.; Ojima, K. Nutrient requirement of suspensions cultures of soybean root cells. *Exp. Cell Res.* **1968**, *50*, 151. [CrossRef]
44. Rogers, S.O.; Bendich, A.J. Extraction of DNA from milligram amounts of fresh, herbarium and mummified plant tissues. *Plant Mol. Biol.* **1985**, *5*, 69–76. [CrossRef]
45. Jefferson, R.A. Assaying chimeric gene in plants: The GUS gene fusion system. *Plant. Mol. Rep.* **1987**, *5*, 387–405. [CrossRef]
46. Reski, R.; Abel, W.O. Induction of budding on chloronemata and caulonemata of the moss, *Physcomitrella patens*, using isopentenyladenine. *Planta* **1985**, *165*, 354–358. [CrossRef]
47. Steiner, A.A. A universal method for preparing nutrient solutions of a certain desired composition. *Plant Soil* **1961**, *15*, 134–154. [CrossRef]
48. Sijmons, P.C.; Grundler, F.M.W.; Mende, N.; Burrows, P.R. *Arabidopsis thaliana* as a new model host for plant parasitic nematodes. *Plant J.* **2010**, *1*, 245–254. [CrossRef]
49. Livak, K.J.; Schmittgen, T.D. Analysis of relative gene expression data using real-time quantitative PCR and $2^{-\triangle(\triangle CP)}$ method. *Methods* **2001**, *25*, 402–408. [CrossRef]
50. Wu, Y.; Zhang, D.; Chu, J. The Arabidopsis NPR1 protein is a receptor for the plant defense hormone salicylic acid. *Cell Rep.* **2012**, *1*, 639–647. [CrossRef]
51. Youssef, R.M.; Macdonald, M.H.; Brewer, E.P. Ectopic expression of AtPAD4 broadens resistance of soybean to soybean cyst and root-knot nematodes. *BMC Plant Biol.* **2013**, *13*, 67. [CrossRef]
52. Tornero, P. RAR1 and NDR1 contribute quantitatively to disease resistance in *Arabidopsis*, and their relative contributions are dependent on the R gene assayed. *Plant Cell* **2002**, *14*, 1005–1015. [CrossRef]
53. Tör, M.; Gordon, P.; Cuzick, A. *Arabidopsis* SGT1b is required for defense signaling conferred by several downy mildew (*Peronospora parasitica*) resistance genes. *Plant Cell* **2002**, *14*, 993–1003. [CrossRef]

7

Molecular Characterization and Overexpression of *SmJMT* Increases the Production of Phenolic Acids in *Salvia miltiorrhiza*

Bin Wang [1,2,†], Junfeng Niu [1,†], Bin Li [1], Yaya Huang [1], Limin Han [1], Yuanchu Liu [1], Wen Zhou [1], Suying Hu [1], Lin Li [1], Donghao Wang [1], Shiqiang Wang [1], Xiaoyan Cao [1,*] and Zhezhi Wang [1,*]

1 National Engineering Laboratory for Resource Development of Endangered Crude Drugs in Northwest China, Key Laboratory of the Ministry of Education for Medicinal Resources and Natural Pharmaceutical Chemistry, College of Life Sciences, Shaanxi Normal University, Xi'an 710119, China; happywangbin2003@163.com (B.W.); niujunfeng6829@126.com (J.N.); libin1989@snnu.edu.cn (B.L.); h_yoyo@snnu.edu.cn (Y.H.); hdd_1981_@163.com (L.H.); shidayuanchu@snnu.edu.cn (Y.L.); wenzhou0229@snnu.edu.cn (W.Z.); husuying0315@163.com (S.H.); shidalilin@snnu.edu.cn (L.L.); wangdonghao@snnu.edu.cn (D.W.); wsq@snnu.edu.cn (S.W.)

2 College of Chemistry, Biology and Materials Science, East China University of Technology, NanChang 330013, China

* Correspondence: caoxiaoyan@snnu.edu.cn (X.C.); zzwang@snnu.edu.cn (Z.W.)

† These authors contributed equally to this work.

Abstract: Jasmonic acid (JA) carboxyl methyltransferase (JMT), a key enzyme in jasmonate-regulated plant responses, may be involved in plant defense and development by methylating JA to MeJA, thus influencing the concentrations of MeJA in plant. In this study, we isolated the *JMT* gene from *Salvia miltiorrhiza*, an important medicinal plant widely used to treat cardiovascular disease. We present a genetic manipulation strategy to enhance the production of phenolic acids by overexpresion *SmJMT* in *S. miltiorrhiza*. Global transcriptomic analysis using RNA sequencing showed that the expression levels of genes involved in the biosynthesis pathway of phenolic acids and MeJA were upregulated in the overexpression lines. In addition, the levels of endogenous MeJA, and the accumulation of rosmarinic acid (RA) and salvianolic acid (Sal B), as well as the concentrations of total phenolics and total flavonoids in transgenic lines, were significantly elevated compared with the untransformed control. Our results demonstrate that overexpression of *SmJMT* promotes the production of phenolic acids through simultaneously activating genes encoding key enzymes involved in the biosynthesis pathway of phenolic acids and enhancing the endogenous MeJA levels in *S. miltiorrhiza*.

Keywords: *SmJMT*; transgenic; *Salvia miltiorrhiza*; overexpression; transcriptome; phenolic acids

---

1. Introduction

Salvia miltiorrhiza Bunge (Lamiaceae) is a well-known traditional Chinese herb with significant medicinal and economic value. Its dry roots or rhizomes (called "danshen" in Chinese) are used to treat various cerebrovascular and cardiovascular diseases in Asian countries, and are widely accepted as a health supplement in western countries [1,2]. Due to its remarkable and reliable therapeutic actions, *S. miltiorrhiza* is being developed as a potential model for research into traditional Chinese medicine [3]. Furthermore, by virtue of high-throughput technologies, the genomic sequence of *S. miltiorrhiza* was published [4].

S. miltiorrhiza contains two major medicinal components that are largely responsible for the observed pharmacological activities; one is a group of lipid-soluble (non-polar, lipophilic) diterpenoids, known as tanshinones, and the other is a water-soluble (polar, hydrophilic) group of phenolic acids, such as rosmarinic acid (RA) and salvianolic acid B (Sal B) [5]. *S. miltiorrhiza* is traditionally processed through extraction with water. Sal B becomes the predominant active ingredient among the phenolic acids, which is designated as a marker component of *S. miltiorrhiza* in the official Chinese Pharmacopoeia. In additional, it was reported that Sal B can provide protection against cardiovascular, neural, and hepatic diseases, as well as certain cancers [6]. Although beneficial to human health, the concentration of Sal B is low in commercial cultivars of *S. miltiorrhiza*, which currently limits its widespread use and medicinal efficiency. Furthermore, Sal B is difficult to purify from complex mixtures, resulting in inefficient chemical synthesis [7].

In *S. miltiorrhiza*, the biosynthetic pathway leading to Sal B and RA is thought to entail both the phenylpropanoid and tyrosine-derived pathways [7]. Enzymes involved in the phenylpropanoid pathway include phenylalanine ammonia-lyase (PAL), cinnamate 4-hydroxylase (C4H), and 4-coumarate/coenzyme A ligase (4CL). Tyrosine aminotransferase (TAT) and hydroxyphenylpyruvate reductase (HPPR) are active in the tyrosine-derived pathway. An additional enzyme, rosmarinic acid synthase (RAS), couples products from the two pathways. This product is then hydroxylated by cytochrome P450 monooxygenase C3′H (CYP98A) to form RA [8]. Many methods were identified for increasing the production rates of such health-promoting phenolic acids, such as genetic engineering [7], hormone induction [9], and biological fermentation [10]. Jasmonate treatment is the most commonly used method of hormone elicitation.

Jasmonates which include jasmonic acid (JA), methyl jasmonate (MeJA), and its cyclopentane derivatives, are a class of plant hormone that regulate many aspects of plant development such as root growth, production of viable pollen, fruit ripening, and senescence [11,12]. They are also involved in plant responses to biotic and abiotic stresses including insect attack, wounding, water deficiency, ultraviolet (UV) light, pathogen infection, and ozone [13–15]. Jasmonates are synthesized in plants via the octadecanoid pathway [14,16] from α-linolenic acid through a series of enzymes, beginning with an oxygenation catalyzed by lipoxygenase (LOX) [17]. The product is then converted to 12-oxo-phytodienoic acid (12-OPDA) by allene oxide synthase (AOS) and alleneoxide cyclase (AOC) [18,19]. Afterward, JA is synthesized from 12-oxo-phytodienoic acid (12-OPDA) through reduction by 12-oxo-phytodienoic acid reductase (OPR) and three steps of β-oxidation, and formation of MeJA is catalyzed by jasmonic acid carboxyl methyltransferase (JMT).

Jasmonic acid carboxyl methyltransferase (JMT) is an *S*-adenosyl-L-methionine-dependent methyltransferase of the SABATH gene family, which could specifically methylate carboxyl groups of small molecules such as jasmonic acid, salicylic acid, and benzoic acid, and named based on the first three identified genes belonging to this family, *SAMT*, *BAMT*, and that coding for theobromine synthase [20]. The *JMT* gene was first identified in *Arabidopsis* [21], and then was successively found in *Capsicum annum* [22], *Populus trichocarpa* [23], strawberry [17], and rice [24]. *JMT* acts as a cellular regulator of the level of physiologically active JA [17,24], which functions in response to several different external environmental stimuli [21] and mediates diverse developmental processes in plants [24,25]. Transgenic *Arabidopsis* lines overexpressing *JMT* had an elevated level of endogenous MeJA, and the transgenic plants showed enhanced levels of resistance against the virulent fungus [21]. Meanwhile, overexpressing *Arabidopsis JMT* in potato increased tuber yield and size [26].

In plants, MeJA, a signal molecule that acts as a second messenger, is proposed to play a role in the elicitation process [27,28], which could lead to the accumulation of secondary metabolites [29]. Furthermore, MeJA could induce plant tissues to provide a responsive system to identify and profile the transcripts and regulation factors involved in secondary metabolite accumulation [28]. When *S. miltiorrhiza* was treated with MeJA, an extensive transcriptional reprogramming of metabolism was triggered, and the biosynthesis of active ingredients was dramatically increased [30]. In addition, it was also shown that the biosynthesis of phenolic acids is stimulated by MeJA treatment [5,9,31,32], and the expression levels of most of the genes involved in the biosynthesis of bioactive compounds are induced by MeJA at different levels [8,9,28,32,33].

In addition, when a plant is subjected to biotic and abiotic stresses, it produces secondary metabolites that function as direct defenses [25], and these metabolites may have important medicinal value. Overexpression of tomato *prosystemin* (*LePS*) i n *S. miltiorrhiza* enhanced resistance the pest, while the production of secondary metabolites and the level of endogenous MeJA was increased [34]. Moreover, as described above, overexpressing *JMT* in plants elevated the level of endogenous MeJA, and transgenic plants exhibited constitutive expression of jasmonate-responsive genes [21], while most genes involved in the biosynthesis of Sal B and RA could be induced by jasmonates in *S. miltiorrhiza* [8,9,28,32,33].

Thus, we inferred that, if the level of endogenous MeJA were elevated, a series of defense responses mediated by MeJA may be elicited, a group of jasmonate-responsive genes would be activated, and the accumulation of secondary metabolites could be enhanced.

Thus, we developed a novel strategy to enhance the production of phenolic acids in *S. miltiorrhiza* simultaneously using genetic manipulation and MeJA induction. On the one hand, we overexpressed *SmJMT* in *S. miltiorrhiza* aiming to activate the expression of genes responsible for biosynthesis of phenolic acids and MeJA. On the other hand, by enhancing the level of endogenous MeJA, we intended to elicit a series of biomechanisms to elevate the accumulation of secondary metabolites. After obtaining the overexpressing *SmJMT* plants, transcriptome analysis was carried out on the above transgenic plants and the untransformed control plants.

The differentially expressed genes (DEGs) involved in phenolic acids biosynthesis and the α-linolenic acid metabolism pathway were identified. Finally, the contents of RA, Sal B, total phenols, total flavonoids, and endogenous MeJA in transgenic and control lines were analyzed using different biological techniques.

2. Results

2.1. Isolation and Sequence Analysis of SmJMT

Using PCR amplification, the full-length complementary DNA (cDNA) of *SmJMT* was obtained and submitted to GenBank with the accession number MH136806. The cDNA fragment contained a 1167-bp open reading frame (ORF), encoding a predicted 389-amino-acid polypeptide with an isoelectric point of 5.98 and a molecular mass of 43.5 kDa. The amino-acid sequence contains all the characteristic elements of the *S*-adenosyl-L-methionine-dependent methyltransferases [35]; two conserved binding sites of motifs I and III of *S*-adenosyl-L-methionine (SAM) [36,37], the signature of SABATH gene family members [20,21], were found using multiple-sequence alignment (Figure 1).

To determine the evolutionary relationship of the *SmJMT* with the members of JMTs from other species, an unrooted phylogenetic tree was constructed using the amino-acid sequences of *S. miltiorrhiza* JMT and 27 JMTs from other species (Figure 2). Phylogenetic analysis showed that *SmJMT* was most closely related to *SiJMT* (*S. indicum* jasmonic acid carboxyl methyltransferase) and *EgJMT* (*Erythranthe guttata* jasmonic acid carboxyl methyltransferase), both of which belong to the Lamiales order (Figure 2). Furthermore, the species belonging to same family were classed into the same clade (Figure 2), suggesting that the cluster relationship of JMT proteins from different species is consistent with the traditional taxonomy.

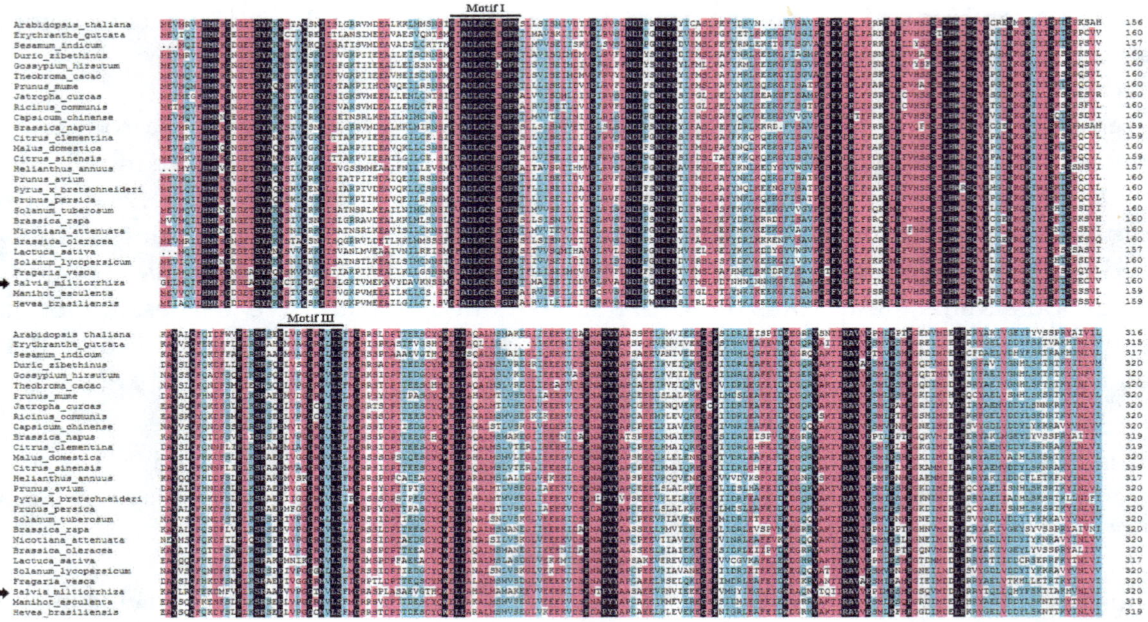

Figure 1. Multiple-sequence alignment of the jasmonic acid (JA) carboxyl methyltransferase (JMT) conserved amino-acid sequences from *Salvia miltiorrhiza* and selected known JMTs from other species, including the binding sites (motifs I and III are indicated) of *S*-adenosyl-L-methionine. The *S. miltiorrhiza* sequences are marked with a black arrow.

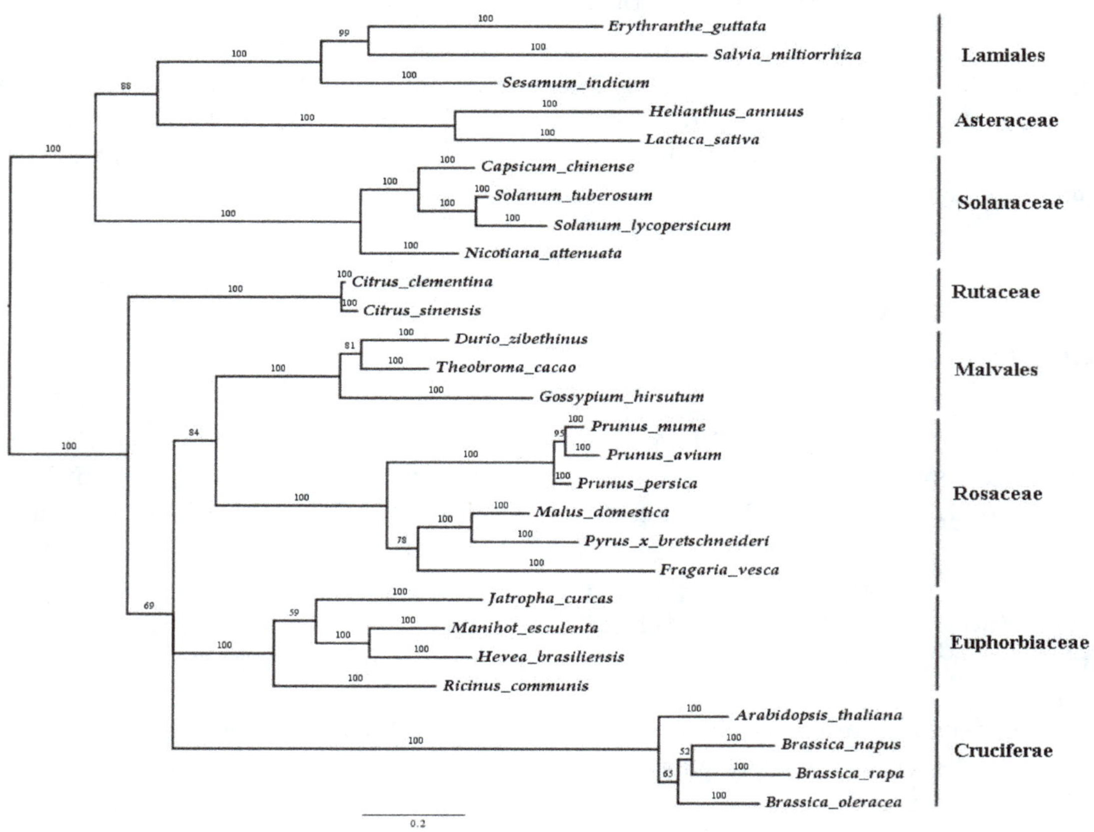

Figure 2. Phylogenetic tree based on JMT from *S. miltiorrhiza* and other species. The tree was constructed using Bayesian inference implemented in MrBayes, based on the amino-acid sequences of *SmJMT* and other species of JMTs under the model of JTT + I + G. The species taxonomy is indicated on the tree.

2.2. Generation of Transgenic *S. miltiorrhiza* Plants

Transgenic *S. miltiorrhiza* plants overexpressing *SmJMT* were obtained in our laboratory by *Agrobacterium*-mediated transformation. After selective culturing on a glufosinate/ammonium medium, resistant plants were confirmed through PCR amplification to contain an expected 929-bp fragment of the CaMV 35S promoter (Figure S1A). Real-time PCR analyses demonstrated that *SmJMT* was obviously overexpressed at the transcriptional level in lines OEJ-2, OEJ-5, OEJ-7, OEJ-8, OEJ-9, and OEJ-10 (OEJ stands for overexpressed *SmJMT*) (Figure S1B). Since expression was significantly higher in OEJ-7 and OEJ-10 than in the other lines and the non-transformed control, we chose them for further analysis. Due to OEJ-10 being the most highly expressed line, we chose it to do the transcriptome sequencing. However, there was no phenotypic change between transgenic lines and control lines.

2.3. Overexpression of *SmJMT* Enhances Production of Salvianolic and Rosmarinic Acids inTransgenic S. miltiorrhiza

To further characterize how the production of phenolic acids was modified in our *SmJMT* overexpression transgenic plants, we extracted the phenolic acids and separated them via LC/MS (Figure S2). The results show that the concentrations of both RA and Sal B, which are the two major hydrophilic active pharmaceutical ingredients in *S. miltiorrhiza,* were increased significantly compared with levels in control samples. We detected the concentrations of RA and Sal B in all the transgenic plants with overexpressed *SmJMT*. The highest concentrations were found in transgenic OEJ-10. Compared with the control, OEJ-10 showed a 3.61-fold increase in RA and a 2.00-fold increase in Sal B (Figure 3A). In OEJ-7, concentrations of RA and Sal B were approximately 1.80- and 1.15-fold higher than those of the control (Figure 3A).

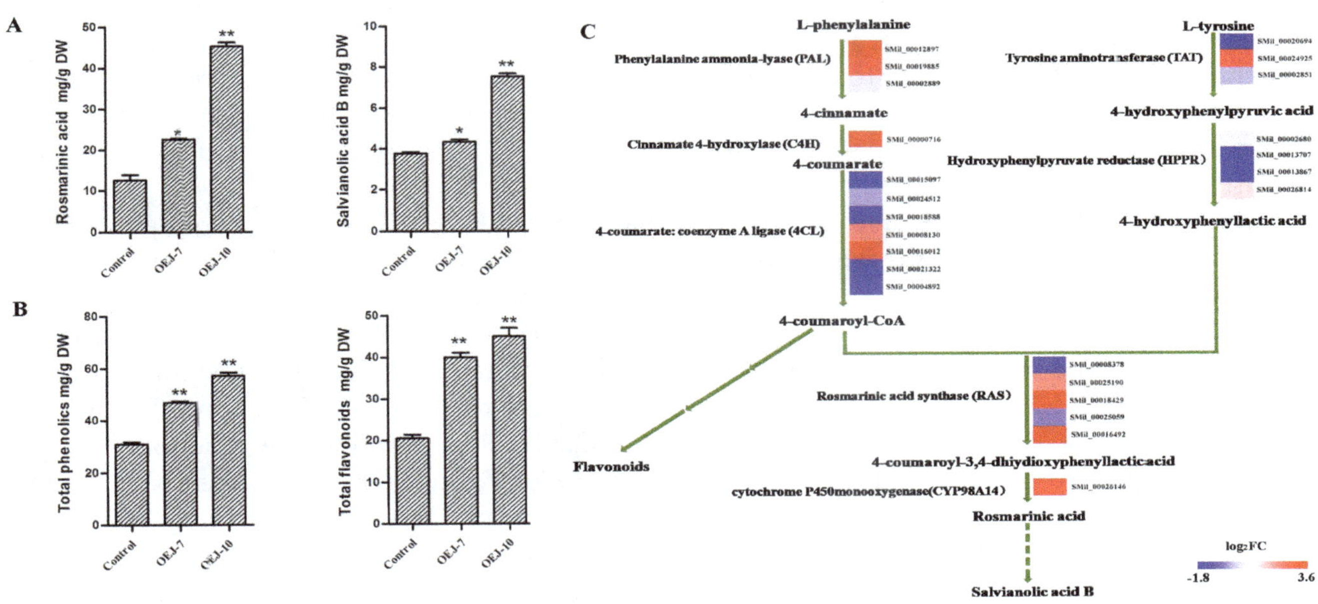

Figure 3. (**A**) Concentrations of rosmarinic acid (RA) and salvianolic acid (Sal B) in root extracts from control and transgenic OEJ-7 and OEJ-10 plants. All data are means of three replicates, with error bars indicating SD. ** Values are significantly different from the control at $p < 0.01$. (**B**) Concentrations of total phenolics and total flavonoids in root extracts from control and transgenic OEJ-7 and OEJ-10 plants. All data are means of three replicates, with error bars indicating SD. * and ** Values are significantly different from the control at $p < 0.05$ and $p < 0.01$, respectively. (**C**) Differentially expressed genes (DEGs) involved in the pathway for salvianolic acid biosynthesis between OEJ-10 and the control. For each gene, relative expression (OEJ-10 versus control) is represented as $\log_2$FC. The color scale is shown at the bottom. Higher expression levels are shown in red.

2.4. Transgenic Plants Show Higher Levels of Total Phenolics and Total Flavonoids

The results above showed that overexpression of *SmJMT* modified the accumulation of two non-flavonoid phenolic acids, RA and Sal B, while the phenolics and flavonoids share an upstream core phenylpropanoid metabolism with Sal B [38]. To determine whether the upregulation of RA and Sal B could cause activation of the phenylpropanoid pathway and provide substrates for the biosynthesis of other types of end product, global assays for phenolics and flavonoids of transgenic plants and control plants were performed. The results showed that total phenolics and total flavonoids accumulated at higher levels in the roots of the overexpression line than in the control sample. The highest concentrations of total phenolics and total flavonoids were also found in transgenic OEJ-10 (Figure 3B). Differences were significant and corresponded to a 1.85-fold increase in the total phenolics content of OEJ-10 roots and a 2.20-fold increase in the total flavonoid content. In OEJ-7, concentrations of total phenolics and total flavonoids were approximately 1.56- and 1.94-fold higher than those of the control (Figure 3B).

2.5. Transcriptomic Analysis of S. miltiorrhiza SmJMT Overexpression and Control Plants

In order to detect the differentially expressed genes between *SmJMT* overexpression transgenic and control plants, and the genes regulated by *SmJMT*, RNA sequencing (RNA-seq) experiments were performed, and the global expression profiles of OEJ-10 and the control were compared. Through Illumina deep sequencing, approximately 26.03 and 26.88 million high-quality clean reads were obtained from OEJ-10 and the control, respectively. The average length of each read was 296 bp. The Q30 values (percentage of sequences with a sequencing error rate <0.1%) for OEJ-10 and the control were 93.45% and 93.38%, respectively. Principal component analysis (PCA) showed that the two groups of samples were distributed in different regions of the spaces, and were distinguished clearly. It indicated that there are differences between the two groups of samples (Figure S3). The control samples were relatively concentrated, indicating that the biological homogeneity of the control samples was better (Figure S3). According to the criteria of differential gene expression screening, there were 2052 genes showing significant differences in expression between the OEJ-10 and control plants, with 998 genes being upregulated and 1054 genes downregulated in OEJ-10 when compared with expression in control (Figure 4 and Table S1).

The Gene Ontology (GO) analysis showed that a total of 14,375 unigenes were annotated in this manner, including 986 DEGs (Figure 5 and Table S2). The GO terms of three categories, namely biological process, cellular component, and molecular function, were assigned to categorize the function of the predicted unique sequences of *S. miltiorrhiza*. In many cases, multiple terms were assigned to the same transcript [39]. This categorization resulted in 1628 DEGs assigned to cellular component, 2573 DEGs to biological process, and 1163 DEGs to molecular function. The GO terms of "binding" (GO: 0005488) and "catalytic activity" (GO: 0003824) of molecular function; "cell part" (GO: 0044464) and "cell" (GO: 0005623) of cellular component; and "cellular process" (GO: 0009987) and "metabolic process" (GO: 0008152) of biological process were predominantly represented (Figure 5 and Table S2). Furthermore, the enriched GO terms "L-phenylalanine metabolic process" (GO: 0009694), "jasmonic acid metabolic process" (GO: 0006558), and "response to extracellular stimulus" (GO: 0009991) correlate well with the biosynthetic pathways for phenolic acids, α-linolenic acid metabolism, and plant defense.

In addition, the Kyoto Encyclopedia of Genes and Genomes (KEGG) analysis showed that a total of 379 DEGs could be to assigned to KEGG pathways, with enrichment in pathways including phenylalanine, tyrosine, and tryptophan biosynthesis (ko00400) and phenylpropanoid biosynthesis (ko00940), which are connected with the biosynthetic pathways for phenolic acids, phenylalanine, and tyrosine (Figure 6 and Table S3). Both GO terms and KEGG pathways of transcriptome analysis were correlated with the phenylalanine metabolic process, indicating that *SmJMT* could be correlated with the biosynthetic pathways for phenolic acids in *S. miltiorrhiza*.

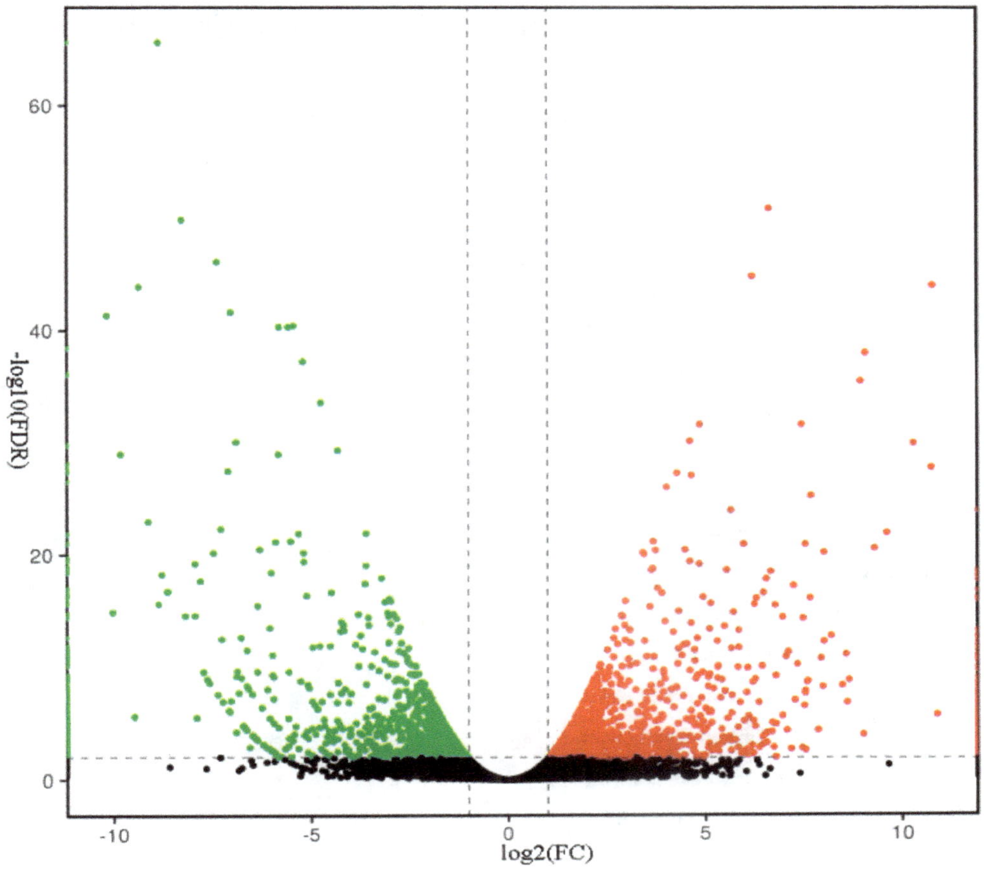

Figure 4. Volcano plot of DEGs between OEJ-10 and the control. Red points represent the DEGs that were upregulated. Black points represent the DEGs without statistically significant differences. Green points represent the DEGs that were downregulated.

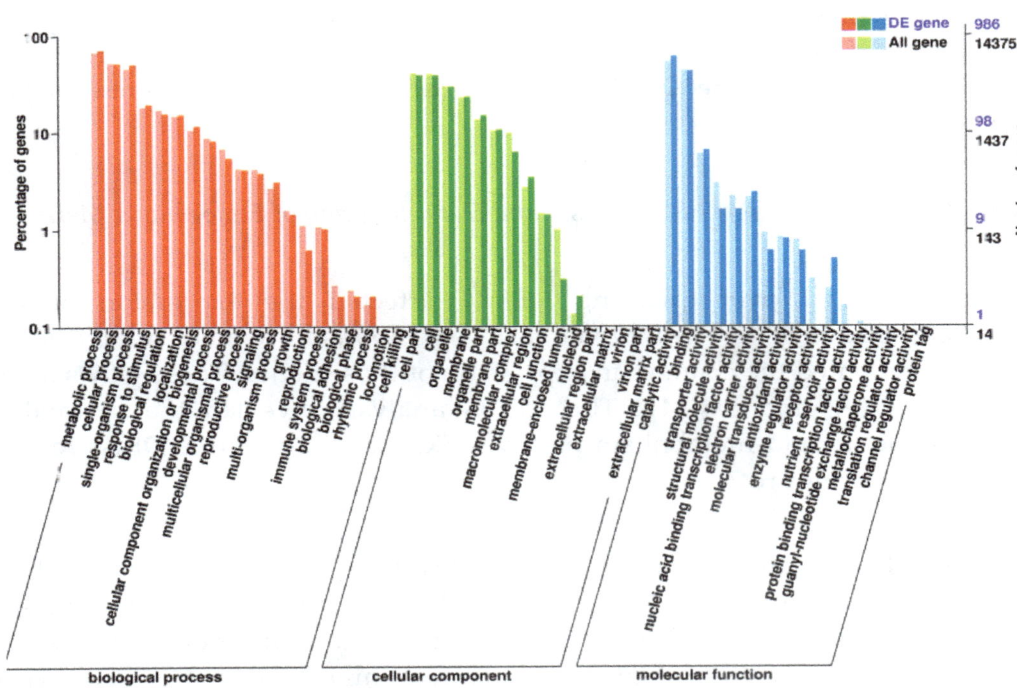

Figure 5. The second Gene Ontology (GO) classification annotation of DEGs between OEJ-10 and the control. The *X*-axis represents GO classification; red represents a biological process, green represents a cellular compcnent, and blue represents a molecular function. The left *Y*-axis represents the percentage of DEGs with respect to all genes. The right *Y*-axis represents the number of genes.

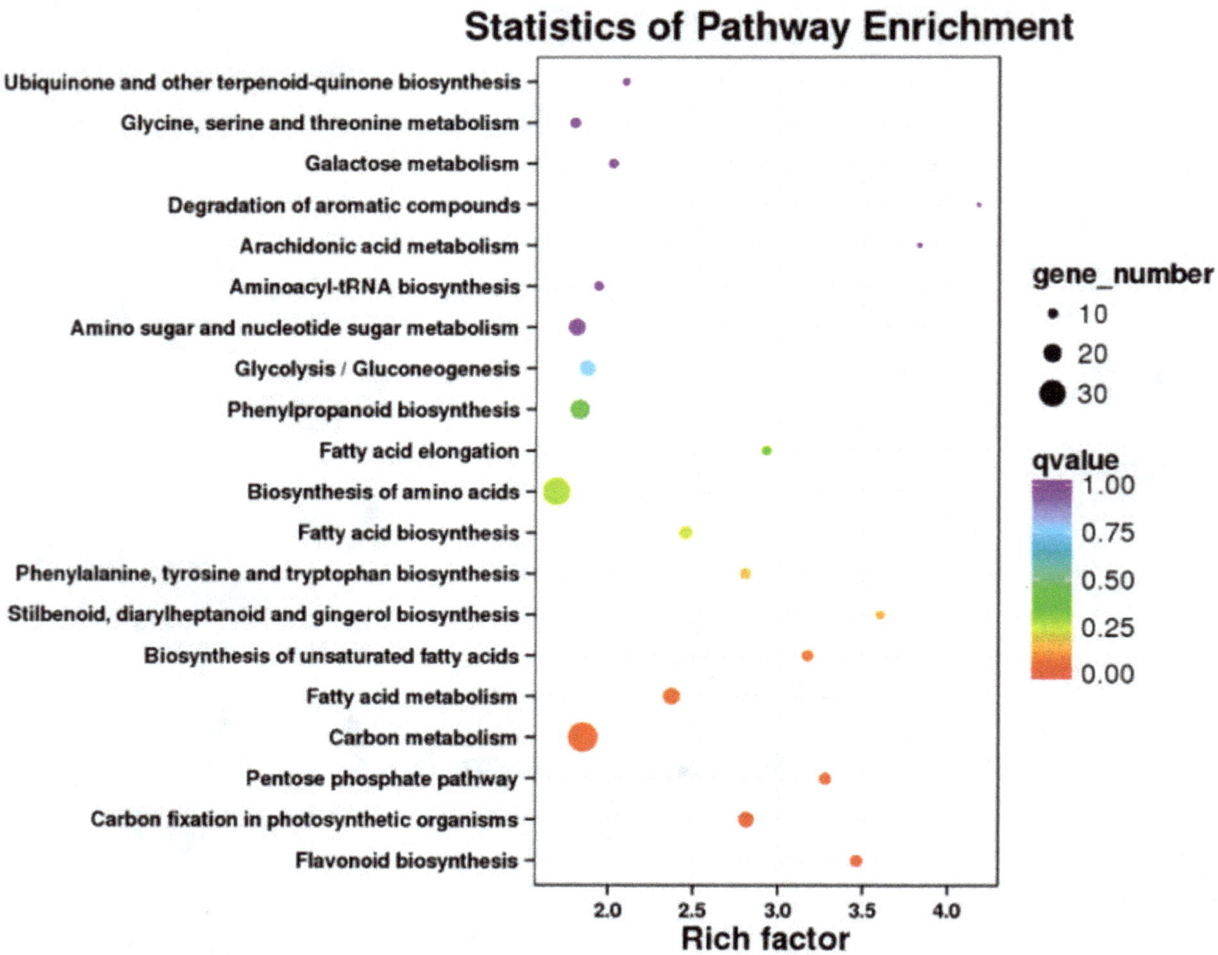

Figure 6. Kyoto Encyclopedia of Genes and Genomes (KEGG) pathway enrichment analysis of DEGs between OEJ-10 and the control. The X-axis represents enrichment factor; the Y-axis represents the pathway. The circles represent the KEGG pathways. The color of the circle represents the q-value. Lower q-values are shown in red.

2.6. DEGs Involved in α-Linolenic Acid Metabolism and Determination of Endogenous MeJA Levels

To determine whether overexpressing *SmJMT* affected the expression of genes closely associated with the pathway of α-linolenic acid metabolism, which finally leads to MeJA biosynthesis, we investigated the changes in expression of those genes. These genes include *SmLOX*, *SmAOS*, *SmAOC*, *SmOPR*, and *SmJMT*. The KEGG analysis revealed that a total of 14 genes assigned to the α-linolenic acid metabolism pathway (ko00592) and three DEGs relevant to MeJA biosynthesis were found (Figure 7A).

Among the DEGs encoding putative *SmAOS*, *SmOPR*, and *SmJMT*, all of them have one unigene transcriptionally activated and being upregulated (Figure 7A and Table S4). For further detecting the endogenous MeJA levels of transgenic plants, the concentrations of endogenous MeJA in fresh leaves from OEJ-10 and control plants were determined using ELISA. According to the optical density (OD) values of samples, the concentrations of MeJA were 3.57 ± 0.08 pmol·g^{-1} for the control and 5.36 ± 0.30 pmol·g^{-1} for OEJ-10, respectively. MeJA concentrations were significantly higher for the transgenic OEJ lines (Figure 7B).

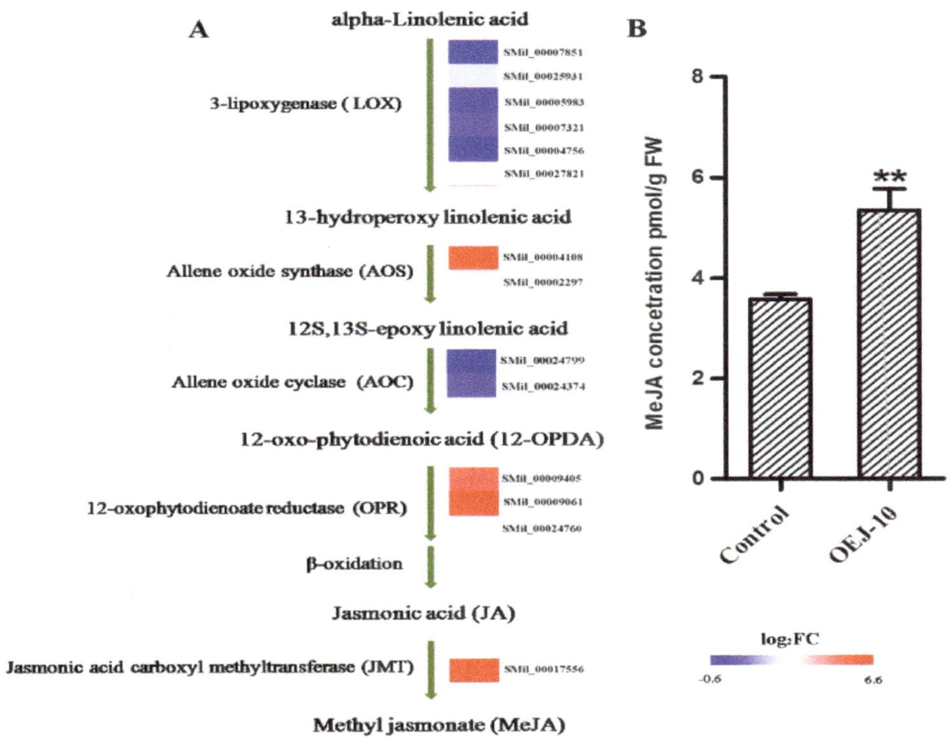

Figure 7. (**A**) DEGs involved in the pathway for MeJA biosynthesis between OEJ-10 and the control. For each gene, relative expression (OEJ-10 versus control) is represented as $\log_2$FC. The color scale is shown at the bottom. Higher expression levels are shown in red. (**B**) Concentrations of MeJA in leaf extracts from transgenic line OEJ-10 and the control. All data are means of three replicates, with error bars indicating SD. ** Values are significantly different from the control at $p < 0.01$.

2.7. DEGs Involved in the Pathway for Salvianolic Acid Biosynthesis

To evaluate whether upregulated expression of *SmJMT* could modify the activation of the key enzymes in the pathway for salvianolic acid biosynthesis, all of the putative enzyme genes in this pathway were examined through transcriptome analysis. A total of 25 unique sequences that encode seven enzymes involved in the biosynthetic pathway of salvianolic acid were present in the RNA-seq dataset, including 12 DEGs (Table S5). These enzymes were SmPAL, SmC4H, Sm4CL, SmTAT, SmHPPR, SmRAS, and SmCYP98A14 (Figure 3C). According to the transcriptome data, among the 12 DEGs in the salvianolic acid biosynthesis pathway, nine were upregulated in the OEJ-10 plants, while three were downregulated (Figure 3C and Table S5). Following comprehensive analysis of the values of counts and the annotation of the RNA-seq dataset, the expression of *SmPAL*$_1$ (*SMil_00019885* Accession No: KF462460), *SmC4H* (*SMil_00000716* Accession No: DQ355979), *Sm4CL*$_3$ (*SMil_00016012* Accession No: KF220556), *SmTAT* (*SMil_00024925* Accession No: DQ334606), *SmRAS* (*SMil_00025190* Accession No: FJ906696), and *SmCYP98A14* (*SMil_00026146* Accession No: HQ316179) demonstrated significant upregulation in OEJ-10, while the expression of *SmHPPR* (*SMil_00002680* Accession No: DQ09974) was not significantly upregulated.

2.8. Confirmation of RNA-Seq Data by qRT-PCR Analysis

To validate the RNA-seq data for differential gene expression between the control and transgenic OEJ lines, the expression levels of genes encoding nine key enzymes involved in the salvianolic acid biosynthesis and α-linolenic acid metabolism were analyzed by qRT-PCR (Figure 8). According to the statistical analysis of qRT-PCR, *SmPAL*$_1$, *SmC4H*, *Sm4CL*$_3$, *SmTAT*, *SmHPPR*, *SmRAS*, *SmCYP98A*, *SmAOS*, and *SmJMT* were upregulated in transgenic lines compared with the control (Figure 8). On the other hand, based on the screening conditions of DEGs, the expression of *SMil_00019885*, *SMil_00000716*, *SMil_00016012*, *SMil_00024925*, *SMil_00025190*, *SMil_00026146*, *SMil_00002680*,

SMil_00004108, and *SMil_00017556* was upregulated in transgenic lines compared with the control (Tables S4 and S5). Thus, through statistical analysis, the relative expression levels of these genes were shown to be consistent with those of the RNA-seq data, which indicated the accuracy of the results of the latter (Figure 8).

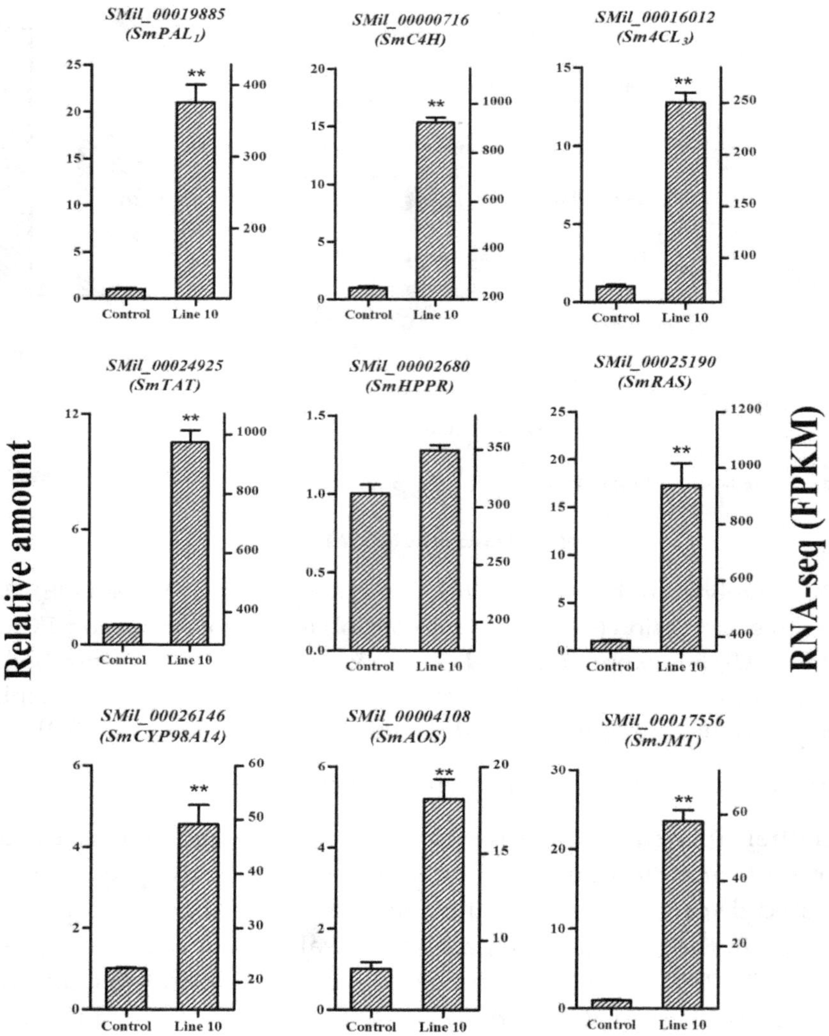

Figure 8. Validation by qRT-PCR of nine genes involved in the salvianolic acid biosynthesis and MeJA biosynthesis pathways in the control and transgenic line OEJ-10. All data are means of three replicates, with error bars indicating SD. ** Values are significantly different from the control at $p < 0.01$.

3. Discussion

Among medicinal plants, *S. miltiorrhiza* is an ideal and representative model for studying transcriptional regulation, and phenolic acid biosynthesis became a new research focus [7]. Fortunately, the *S. miltiorrhiza* genomic sequence was published [4], and this provides a good opportunity for studying the function of many valuable genes in *S. miltiorrhiza*. In this article, we report on the isolation, bioinformatics analysis, molecular characterization, and preliminary function analysis of the *JMT* gene from *S. miltiorrhiza* encoding jasmonic acid carboxyl methyltransferse. Phylogenetic analysis of the SABATH gene family of *S. miltiorrhiza* and *Arabidopsis* suggests that *SmJMT* and *AtJMT* are apparent orthologs and may have the same function [20]. Overexpressing JMT in plants can elevate the level of endogenous MeJA and improve resistance against external environmental stimuli [21,24,26]. Thus, we sought to detect the effect of JMT on the biosynthesis of MeJA in *S. miltiorrhiza* and study the function of JMT in regulating the expression of genes involved in the phenolic acid biosynthesis pathway, as well as the impact on accumulation of secondary metabolites in *S. miltiorrhiza*.

After obtaining the full-length cDNA of *SmJMT*, we constructed a *SmJMT* overexpression vector and transferred it into *S. miltiorrhiza*. Then, after obtaining *SmJMT* overexpression in transgenic *S. miltiorrhiza* lines, we compared the MeJA levels between control and transgenic lines. The results showed that overexpression of *SmJMT* significantly changed the level of endogenous MeJA in *S. miltiorrhiza* (Figure 7B). In addition, transcriptome analysis showed that the three DEGs (*AOS, OPR*, and *JMT*) in α-linolenic acid metabolism were upregulated in transgenic lines, while no genes were downregulated, suggesting that the α-linolenic acid metabolism pathway was obviously activated in transgenic line OEJ-10 (Figure 7A). Furthermore, AOS is the major control point of MeJA biosynthesis and the first specific enzyme [17,40,41]. The transcription level of *SmAOS* was significantly upregulated in OEJ-10 according to the transcriptome data. This strong induction of *SmAOS* may contribute to the biosynthesis of MeJA, which was consistent with the results of the ELISA experiment (Figure 7).

Jasmonic acid carboxyl methyltrans-ferase (JMT) is a key enzyme for jasmonate-regulated plant responses and defense response [17,21,24,42]. Based on previous studies, the expression of primary enzymes involved in the phenylpropanoid and tyrosine-derived pathways are upregulated by jasmonates, which could enhance the biosynthesis of phenolic acids [8,9,28,32,33]. Furthermore, the elicitation of defense responses could lead to the accumulation of secondary metabolites [25,34]. Therefore, we designed a strategy to elevate the content of phenolic acids through overexpression of *JMT* in *S. miltiorrhiza*.

According to the transcriptome data, genes involved in the phenylpropanoid pathway (*SmPAL, SmC4H*, and *Sm4CL*) and in the tyrosine-derived pathway (*SmTAT* and *SmHPPR*), as well as *SmRAS* and *SmCYP98A*, were upregulated in transgenic lines compared with the control (Figure 3C). Meanwhile, the contents of RA, Sal B, total phenolics, and total flavonoids in transgenic lines were significantly elevated over those of the control (Figure 3A,B). The results indicated that overexpression of *SmJMT* significantly increased the contents of phenolic acids by activating the phenylpropanoid and tyrosine-derived pathways. *SmJMT* may play an important role in the regular expression of functional genes contributing to the accumulation of phenolic acids. Our results demonstrate that production of salvianolic acid could be improved by overexpression of *SmJMT* in *S. miltiorrhiza* using genetic engineering techniques. Therefore, our research on *SmJMT* overexpression in *S. miltiorrhiza* provides a good foundation for further study of the function of JMT in plants.

MeJA treatment is the most commonly used method of eliciting herbivore resistance in many different plant species, and a series of JA-mediated defense responses are quickly elicited when plants are exposed to volatile MeJA [43–45]. In brief, when treated with exogenously applied MeJA, plants could elicit most defense-resistant responses by JA [25]. However, following overexpression of the plastidic flax *AOS* cDNA in transgenic potato plants, even though the plants exhibited six- to 12-fold increased levels of JA, this increase did not activate jasmonate-responsive genes [46].

The reason for this may be that the free acid JA could not move across the cellular membrane without a carrier [21,47], while MeJA, which diffuses through the membranes, may act as an intracellular regulator and a diffusible intercellular signal transducer mediating intra- and interplant communications [21]. MeJA also plays its own role in developmental processes and defense responses. Therefore, the effects of treatment with exogenously applied MeJA may be different from those of enhanced endogenous MeJA levels in plants. Improvement of the defense ability of pharmaceutical crops through spraying of jasmonates is not feasible, while strengthening the stress resistance of medicinal herbs by elevating endogenous MeJA levels via genetic engineering would represent a new strategy. Furthermore, the plant defense is related to plant secondary metabolism. The improvement of plant defense could promote the accumulation of secondary metabolites. The defense mechanisms, resistance effects, and plant responses need to be further studied.

4. Materials and Methods

4.1. Isolation of the *SmJMT* Gene

Plant materials of leaves from *S. miltiorrhiza* were collected following a method described previously [20]. Total RNA was extracted from the *S. miltiorrhiza* leaf tissue with a Plant RNA Kit (OMEGA, Houston, TX, USA). RNA quantity was determined using a NanoDrop 2000C Spectrophotometer (Thermo Scientific, Wilmington, DE, USA). First-strand cDNA was synthesized using a Prime-Script RT Master Mix (TaKaRa, Beijing, China) according to the manufacturer's protocol. The full-length cDNA coding sequence for *SmJMT* was amplified from leaf cDNA with gene-specific primers *SmJMT*-F and *SmJMT*-R (Table S6), both of which were designed according to the phylogenetic analysis of the *SmSABATH* gene family [20] and the unigene sequence (*SMil_00017556*) available in the *S. miltiorrhiza* genomic database [4]. Amplification was achieved using PrimeSTAR® HS DNA Polymerase (TaKaRa, Beijing, China), and the PCR reaction was performed on a FlexCycler thermocycler (Analytikjena, Jena, Germany) in a 50-μL final volume comprising 2.5 U/μL PrimeSTAR® HS DNA Polymerase, 100 ng of first-strand cDNA, 500 nM each primer, 10 μL of 5× PrimeSTAR Buffer, and 2.5 mM deoxynucleoside triphosphate (dNTP) mixture, under the following conditions: cDNA was denatured at 94 °C for 3 min followed by 35 cycles of amplification (94 °C for 30 s, 51 °C for 30 s and 72 °C for 72 s), and then extension at 72 °C for 10 min. The PCR fragments were purified using a DNA Gel Extraction Kit (Tiangen Beijing, China), inserted into pMD19T-vector (Takara, Beijing, China), and then sequenced by Beijing Genomics Institute (Shenzhen, Guangdong, China).

4.2. Multiple-Sequence Alignment and Phylogenetic Analysis

Multiple-sequence alignment of the JMT conserved amino-acid sequences from *S. miltiorrhiza* and selected known JMTs from other species (Table S7) was performed with the DNAMAN program (Lynnon Corporation, San Ramon, CA, USA). Conserved blocks were obtained with the online program Gblocks 0.91b (http://www.phylogeny.fr/one_task.cgi?task_type=gblocks) [48]. Phylogenetic trees were constructed using Bayesian inference implemented in MrBayes [49,50] with the amino-acid sequences of the *SmJMT* and homolog from other species under the model of JTT + I + G. The model was chosen using the program ProtTest [51]. The phylogenetic tree was represented with the help of Treeview1.61 software [52].

4.3. Vector Construction and Transformation

In order to construct *SmJMT* overexpression vectors, the pMD19T-*JMT* plasmid was used as a template to amplify *SmJMT* with the primers pDONR207-*SmJMT*-F/pDONR207-*SmJMT*-R (Table S6), which contained *att*B1/*att*B2 sites. PCR amplification followed the description above. The PCR products were purified and cloned into entry vector pDONR207, using the BP recombination reaction, and then transferred into the destination vector pEarleyGate202 with the LR reaction according to the protocol from the Gateway Technology manufacturer (Invitrogen, Carlsbad, CA, United States) (Figure S4). The pDONR207-*SmJMT* and pEarleyGate202-*SmJMT* plasmid were sequenced by Beijing Genomics Institute (Shenzhen, Guangdong, China). Finally, the pEarleyGate202-*SmJMT* plasmid was transferred into *Agrobacterium tumefaciens* EHA105 using the freeze–thaw method [53].

An *Agrobacterium*-mediated gene transfer method was performed to generate transgenic plants with leaf explants from sterile *S. miltiorrhiza* cultured on Murashige and Skoog (MS) basal medium under the conditions described previously [54,55]. After transformation, MS with 1 $mg \cdot L^{-1}$ naphthalene acetic acid, 10 $mg \cdot L^{-1}$ 6-benzyl-aminopurine, 10 $mg \cdot L^{-1}$ glufosinate/ammonium, and 200 $mg \cdot L^{-1}$ cefotaxime as the selection medium was used to culture the transgenic explants. They were transferred to fresh selection medium at one-week intervals. Developing shoots were excised and placed on $\frac{1}{2}$ MS selection medium supplemented with 10 $mg \cdot L^{-1}$ glufosinate/ammonium, and 200 $mg \cdot L^{-1}$ cefotaxime for root induction [34]. After two weeks, the rooted transgenic plants were propagated through several generations to expand the culture on the MS basic medium.

4.4. PCR Detection and qRT-PCR Analysis

Genomic DNA was obtained from leaves of one-month-old transgenic and control plants using the CTAB Plant Genomic DNA Rapid Extraction Kit (Aidlab, Beijing, China) according to the manufacturer's protocol. A pair of gene-specific primers (pEarleyGate202-35S-F/R) (Table S6) was used to amplify a 929-bp fragment from genomic DNA of transgenic and control plants. PCR reactions were also performed on a FlexCycler thermocycler (Analytikjena, Jena, Germany) with a 20-μL final volume comprising 10 μL of 2× Taq PCR master Mix (Tiangen, Beijing, China), 500 nM each primer, and 100 ng of template DNA. All PCR reactions were carried out as follows: preheating at 94 °C, then 35 cycles of amplification at 94 °C for 30 s, 58 °C for 30 s, and 72 °C for 1 min, followed by a final elongation of 10 min at 72 °C. The positive control was the pEarleyGate202-*SmJMT*vector, while genomic DNA from wild-type plants served as the negative control. Amplified products were electrophoresed on a 1.0% agarose gel.

Total RNA extraction, RNA quantity determination, and first-strand cDNA synthesis from transgenic and control plants followed the method described above. Quantitative PCR was carried out on a Light Cycler 96 Instrument (Roche, Basel, Switzerland) in a 20-μL final volume comprising 10 μL of SYBR Premix Ex Taq II (Takara, Beijing, China), 20 ng of first-strand cDNA, and 500 nM each primer. The reactions were performed in triplicate under the following conditions: initial thermal cycling at 95 °C for 30 s, followed by 45 cycles of 95 °C for 10 s and 60 °C for 30 s. *Smβ-actin* (DQ243702) was selected as a reference gene [20,56]. Relative expression was calculated by the $2^{-\Delta\Delta Ct}$ method [57], and the relative expression levels were analyzed as means ± standard deviation (SD) of the biological triplicates. The lengths of the amplicons were between 100 and 250 bp. Quantitative primers are listed in Table S6.

4.5. Extraction of MeJA and Determination of Its Concentration

Fresh leaves from transgenic plants lines and control plants were used to investigate the MeJA concentration levels. Tissues (100 mg) were ground in liquid nitrogen, and 9 mL of phosphate-buffered saline (PBS; pH 7.4) was added. The extracts were then centrifuged at 8000× *g* at 4 °C for 30 min, and the upper layer was collected. Based on the method of detecting the endogenous jasmonic acid (JA) in *S. miltiorrhiza* [38], the endogenous MeJA of both transgenic and control plants was measured with a Plant MeJA ELISA Kit (mlbio, Shanghai, China) according to the manufacturer's protocol. A standard curve of optical density (OD) versus MeJA concentration at 0, 125, 250, 500, 1000, and 2000 pmol·L^{-1} was produced by testing a set of calibration standards. By comparing OD values with the standard curve, the content of MeJA in each sample was determined. The intensity of the final reaction color was measured spectrophotometrically at 450 nm to calculate the final MeJA concentration [38].

4.6. Extraction of Total Phenolics and Total Flavonoids and Determination of Their Concentrations

The roots from two-month-old transgenic and control *S. miltiorrhiza* plantlets were air-dried at 20 ± 2 °C. The dried root sample (20 mg) was ground into a powder, and mixed with 1 mL of methanol and acetone (7:3, *v*/*v*) in an ultrasonic bath for 1 h. The mixture was centrifuged at 6000× *g* for 3 min and the supernatant was collected.

The total phenolics content was measured using a modified colorimetric Folin–Ciocalteu method [58] with gallic acid as the standard. The extract solution (100 μL) was mixed in a centrifuge tube with 500 μL of the Folin–Ciocalteu reagent in darkness for 8 min, then incubated with 400 μL of sodium carbonate solution (7.5% *w*/*v*) at 40 °C for 30 min. Absorbance was measured at 765 nm against a reagent blank without the extract, and measurements were carried out in triplicate. The calibration equation for gallic acid was $y = 0.0073x + 0.1023$ ($R^2 = 0.9995$).

Total flavonoid content was determined following the procedure of Dewanto et al. [59] with epicatechin as the standard. The test sample (100 μL) was placed in a centrifuge tube, and 800 μL of 60% ethanol was added followed by 20 μL of 5% $NaNO_2$. After 6 min, 20 μL of 10% $AlCl_3$ was added. After another 6 min, 60 μL of 4% NaOH was added, and then the solution was mixed and the absorbance was measured at 510 nm against a reagent blank without the extract; measurements were carried out in triplicate. The calibration equation for epicatechin was $y = 0.0022x + 0.0737$ ($R^2 = 0.9900$).

4.7. LC/MS Analysis of Phenolic Compounds

Roots harvested from transgenic *S. miltiorrhiza* transplanted for two months and control plantlets were air-dried at 20 ± 2°C. Dried roots were then ground to a fine powder in a mechanical grinder with a 2-mm-diameter mesh. Samples (30 mg) were extracted with 500 μL of 75% methanol in an ultrasonic bath for 20 min, and then centrifuged at 12000× *g* for 6 min. The residual pellet was re-extracted twice, and all supernatants were combined. Finally, the extracted samples were filtered through a 0.2-μm Millipore filter and analyzed by LC/MS.

For LC/MS, extracts were applied to an Agilent 1260 HPLC system coupled to an Agilent 6460 QQQ LC–MS system (Agilent Technologies, Palo Alto, CA, USA), an HPLC system equipped with a pump (Agilent G1312B), an auto-sampler (Agilent G7127A), and a column temperature controller (Agilent G1316A). Chromatography separation was performed with a Welch Ultimate XB-C_{18} column (150 × 2.1 mm, 3 μm particle size) at a flow rate of 0.4 mL·min^{-1} (temperature 30 °C) and 5 μL of sample was injected. The mobile phase comprised Solvent A (acetonitrile) and Solvent B (0.1% formic acid in deionized water), and followed a solvent gradient profile: 0–6 min, A 20–60% and B 80–50%; 6–7 min, A 60–20% and B 50–80%; 7–10 min, A 20% and B 80%. The retention times were 3.8 min for RA and 4.1 min for Sal B.

Mass spectra were acquired by a QQQ-MS instrument with an Agilent Jet Stream (AJS) electrospray ionization (ESI) source. Multiple reaction monitoring (MRM) mode was used for the quantification. For phenolic acids, the ionization mode was negative and the selected transitions of m/z were 359.1→161.1 for RA and 717.2→519.2 for Sal B. The fragmentor voltage was 130 V and the collision energy was 18 eV. The drying gas flow was 10 L·min^{-1}, and the nebulizer pressure was set to 45 psi at a capillary temperature and voltage of 350 °C and 3500 V, respectively. The sheath gas flow was 11 L·min^{-1} at a temperature of 350 °C.

4.8. Transcriptome Analysis and Identification of Differentially Expressed Genes (DEGs)

Two-month-old transgenic and control *S. miltiorrhiza* plants were collected from three biological replicates. Total RNA extraction and RNA quantity determination followed the method described above. The cDNA library preparation and construction were performed as described previously [60]. The libraries were sequenced by Biomarker Technologies Co., Ltd. (Beijing, China) with an Illumina HiSeq4000 platform. The raw transcriptome data were submitted to the National Center for Biotechnology Information (NCBI) with the accession number of SRP155681.

After sequencing of the transcriptome, clean paired-end reads were mapped to the reference genome of *S. miltiorrhiza* [4] using TopHat v2.0.12 [61]. The *S. miltiorrhiza* genome and gene model annotation files were downloaded from genome websites directly (http://www.ndctcm.org/shujukujieshao/2015-04-23/27.html) [4]. Principal component analysis (PCA) was employed to investigate the correlation of biological duplication [62]. Analysis of differential gene expression was performed by Cufflinks [63], and fragments per kilobase of transcript per million mapped reads (FPKM) was used to normalize gene expression levels [64]. The DESeq R package was used to analyze the differential expression of the transgenic and control plants [65]. The false discovery rate (FDR) was controlled by p-values, which were corrected by the Benjamini–Hochberg procedure. Genes with $|\log_2\text{fold-change}| \geq 2$ and an adjusted p-value < 0.01, as found by DESeq were considered differentially expressed. Gene Ontology (GO) enrichment analysis of DEGs was performed using the topGO R package [66]. GO terms with corrected p-values < 0.01 were considered significantly enriched in the DEGs. To identify significantly over-represented metabolic pathways or signal transduction pathways, all DEGs were mapped to terms in the KEGG (Kyoto Encyclopedia of Genes and Genomes) database [67], and the pathway enrichment analysis was conducted using KOBAS [68].

4.9. Verification of RNA-Seq Data by qRT-PCR

To validate the RNA-seq data, seven unigenes involved in salvianolic acid biosynthesis (*SMil_00019885*, *SMil_00000716*, *SMil_00016012*, *SMil_00024925*, *SMil_00002680*, *SMil_00025190*, and *SMil_00026146*) and two unigenes involved in α-linolenic acid metabolism (*SMil_00004108* and *SMil_00017556*) were selected for expression analysis through qRT-PCR experiments. Transcriptome data showed that mostly those genes had significant changes in expression between control and transgenic OEJ lines. The qRT-PCR experiment implementation and statistical analysis followed the description above. Quantitative primers are listed in Table S6.

Supplementary Materials
Table S1: Overview of up- and down-regulated genes; Table S2: Results of topGO enrichment analysis of DEGs; Table S3: Overview of the significant enrichment of the KEGG pathways; Table S4: DEGs involved in the α-Linolenic acid metabolism; Table S5: DEGs involved in the salvianolic acid biosynthesis; Table S6: Primer pairs used in the paper; Table S7: List of the JMT from different species; Figure S1: Expression of *SmJMT* in control and overexpression transgenic lines; Figure S2: Principal Component Analysis (PCA) about the correlation of biological duplication of the transcriptome sequencing samples; Figure S3: Mass chromatograms of standards (RA and Sal B) and samples (Control, OEJ-7 and OEJ-10); Figure S4: The *SmJMT*-overexpression vectors construction process with Gateway technology.

Author Contributions: The experiments were conceived and organized by Z.W., B.W., J.N., and X.C. B.W., Y.H., and L.H. performed the experiments. Y.L., W.Z., L.L., S.H., S.W., and D.W. contributed to the data analysis. The paper was written by B.W. and B.L. All authors discussed and approved the final manuscript.

Acknowledgments: This study was supported by the National Natural Science Foundation of China (31670299 and 31800259), the Funds of Jiangxi provincial education department (GJJ170434), the major Project of Shaanxi Province, China (2017ZDXM-SF-005), the Shaanxi province traditional Chinese medicine standardized planting base construction project (2013K14-03-01), the Natural Science Basic Research Plan in Shaanxi Province of China (2018JQ3033), and the Fundamental Research Funds for the Central Universities (GK201806006, GK201302044 and 1301031470).

References

1. Zhou, L.; Zuo, Z.; Chow, M.S. Danshen: An overview of its chemistry, pharmacology, pharmacokinetics, and clinical use. *J. Clin. Pharmacol.* **2005**, *45*, 1345–1359. [CrossRef] [PubMed]
2. Geng, Z.H.; Huang, L.; Song, M.B.; Song, Y.M. Cardiovascular effects in vitro of a polysaccharide from Salvia miltiorrhiza. *Carbohydr. Polym.* **2015**, *121*, 241–247. [CrossRef] [PubMed]
3. Wang, Q.H.; Chen, A.H. Salviae Miltiorrhiza: A Model Organism for Chinese Traditional Medicine Genomic Studies. *Acta Chin. Med. Pharmacol.* **2009**, *37*, 1–4.
4. Xu, H.; Song, J.; Luo, H.; Zhang, Y.; Li, Q.; Zhu, Y.; Xu, J.; Li, Y.; Song, C.; Wang, B. Analysis of the genome sequence of the medicinal plant Salvia miltiorrhiza. *Mol. Plant* **2016**, *9*, 949–952. [CrossRef] [PubMed]
5. Ma, Y.; Yuan, L.; Wu, B.; Li, X.; Chen, S.; Lu, S. Genome-wide identification and characterization of novel genes involved in terpenoid biosynthesis in Salvia miltiorrhiza. *J. Exp. Bot.* **2012**, *63*, 2809–2823. [CrossRef] [PubMed]
6. Ho, J.H.; Hong, C.Y. Salvianolic acids: Small compounds with multiple mechanisms for cardiovascular protection. *J. Biomed. Sci.* **2011**, *18*, 30. [CrossRef] [PubMed]
7. Zhang, Y.; Yan, Y.-P.; Wu, Y.-C.; Hua, W.-P.; Chen, C.; Ge, Q.; Wang, Z.-Z. Pathway engineering for phenolic acid accumulations in Salvia miltiorrhiza by combinational genetic manipulation. *Metab. Eng.* **2014**, *21*, 71–80. [CrossRef] [PubMed]
8. Wang, B.; Sun, W.; Li, Q.; Li, Y.; Luo, H.; Song, J.; Sun, C.; Qian, J.; Zhu, Y.; Hayward, A.; et al. Genome-wide identification of phenolic acid biosynthetic genes in Salvia miltiorrhiza. *Planta* **2015**, *241*, 711–725. [CrossRef] [PubMed]
9. Xiao, Y.; Gao, S.; Di, P.; Chen, J.; Chen, W.; Zhang, L. Methyl jasmonate dramatically enhances the accumulation of phenolic acids in Salvia miltiorrhiza hairy root cultures. *Physiol. Plant.* **2009**, *137*, 1–9. [CrossRef] [PubMed]

10. Zhou, Y.J.; Gao, W.; Rong, Q.; Jin, G.; Chu, H.; Liu, W.; Yang, W.; Zhu, Z.; Li, G.; Zhu, G.; et al. Modular pathway engineering of diterpenoid synthases and the mevalonic acid pathway for miltiradiene production. *J. Am. Chem. Soc.* **2012**, *134*, 3234–3241. [CrossRef] [PubMed]
11. Avanci, N.C.; Luche, D.D.; Goldman, G.H.; Goldman, M.H. Jasmonates are phytohormones with multiple functions, including plant defense and reproduction. *Genet. Mol. Res.* **2010**, *9*, 484–505. [CrossRef] [PubMed]
12. Cheong, J.J.; Choi, Y.D. Methyl jasmonate as a vital substance in plants. *Trends Genet.* **2003**, *19*, 409–413. [CrossRef]
13. Creelman, R.A.; Mullet, J.E. Jasmonic acid distribution and action in plants: Regulation during development and response to biotic and abiotic stress. *Proc. Natl. Acad. Sci. USA* **1995**, *92*, 4114–4119. [CrossRef] [PubMed]
14. Creelman, R.A.; Mullet, J.E. Biosynthesis and Action of Jasmonates in Plants. *Annu. Rev. Plant Physiol. Plant Mol. Biol.* **1997**, *48*, 355–381. [CrossRef] [PubMed]
15. Creelman, R.A.; Mullet, J.E. Oligosaccharins, brassinolides, and jasmonates: Nontraditional regulators of plant growth, development, and gene expression. *Plant Cell* **1997**, *9*, 1211–1223. [CrossRef] [PubMed]
16. Beale, M.H.; Ward, J.L. Jasmonates: Key players in the plant defence. *Nat. Prod. Rep.* **1998**, *15*, 533–548. [CrossRef] [PubMed]
17. Preuss, A.; Augustin, C.; Figueroa, C.R.; Hoffmann, T.; Valpuesta, V.; Sevilla, J.F.; Schwab, W. Expression of a functional jasmonic acid carboxyl methyltransferase is negatively correlated with strawberry fruit development. *J. Plant Physiol.* **2014**, *171*, 1315–1324. [CrossRef] [PubMed]
18. Sasaki, Y.; Asamizu, E.; Shibata, D.; Nakamura, Y.; Kaneko, T.; Awai, K.; Amagai, M.; Kuwata, C.; Tsugane, T.; Masuda, T. Monitoring of Methyl Jasmonate-responsive Genes in Arabidopsis by cDNA Macroarray: Self-activation of Jasmonic Acid Biosynthesis and Crosstalk with Other Phytohormone Signaling Pathways. *DNA Res.* **2001**, *8*, 153–161. [CrossRef] [PubMed]
19. Delker, C.; Stenzel, I.; Hause, B.; Miersch, O.; Feussner, I.; Wasternack, C. Jasmonate biosynthesis in Arabidopsis thaliana—Enzymes, products, regulation. *Plant Biol.* **2006**, *8*, 297–306. [CrossRef] [PubMed]
20. Wang, B.; Wang, S.; Wang, Z. Genome-Wide Comprehensive Analysis the Molecular Phylogenetic Evaluation and Tissue-Specific Expression of SABATH Gene Family in Salvia miltiorrhiza. *Genes* **2017**, *8*, 365. [CrossRef] [PubMed]
21. Seo, H.S.; Song, J.T.; Cheong, J.J.; Lee, Y.H.; Lee, Y.W.; Hwang, I.; Lee, J.S.; Choi, Y.D. Jasmonic acid carboxyl methyltransferase: A key enzyme for jasmonate-regulated plant responses. *Proc. Natl. Acad. Sci. USA* **2001**, *98*, 4788–4793. [CrossRef] [PubMed]
22. Min, S.S.; Dong, G.K.; Sun, H.L. Isolation and characterization of a jasmonic acid carboxyl methyltransferase gene from hot pepper (*Capsicum annuum* L.). *J. Plant Biol.* **2005**, *48*, 292–297.
23. Zhao, N.; Yao, J.; Chaiprasongsuk, M.; Li, G.; Guan, J.; Tschaplinski, T.J.; Guo, H.; Chen, F. Molecular and biochemical characterization of the jasmonic acid methyltransferase gene from black cottonwood (*Populus trichocarpa*). *Phytochemistry* **2013**, *94*, 74–81. [CrossRef] [PubMed]
24. Qi, J.; Li, J.; Han, X.; Li, R.; Wu, J.; Yu, H.; Hu, L.; Xiao, Y.; Lu, J.; Lou, Y. Jasmonic acid carboxyl methyltransferase regulates development and herbivory-induced defense response in rice. *J. Integr. Plant Biol.* **2016**, *58*, 564–576. [CrossRef] [PubMed]
25. Wu, J.; Wang, L.; Baldwin, I.T. Methyl jasmonate-elicited herbivore resistance: Does MeJA function as a signal without being hydrolyzed to JA? *Planta* **2008**, *227*, 1161–1168. [CrossRef] [PubMed]
26. Sohn, H.B.; Lee, H.Y.; Seo, J.S.; Jung, C.; Jeon, J.H.; Kim, J.-H.; Lee, Y.W.; Lee, J.S.; Cheong, J.-J.; Choi, Y.D. Overexpression of jasmonic acid carboxyl methyltransferase increases tuber yield and size in transgenic potato. *Plant Biotechnol. Rep.* **2010**, *5*, 27–34. [CrossRef]
27. Gundlach, H.; Müller, M.J.; Kutchan, T.M.; Zenk, M.H. Jasmonic acid is a signal transducer in elicitor-induced plant cell cultures. *Proc. Natl. Acad. Sci. USA* **1992**, *89*, 2389–2393. [CrossRef] [PubMed]
28. Luo, H.; Zhu, Y.; Song, J.; Xu, L.; Sun, C.; Zhang, X.; Xu, Y.; He, L.; Sun, W.; Xu, H.; et al. Transcriptional data mining of Salvia miltiorrhiza in response to methyl jasmonate to examine the mechanism of bioactive compound biosynthesis and regulation. *Physiol. Plant.* **2014**, *152*, 241–255. [CrossRef] [PubMed]
29. Wasternack, C. Jasmonates: An Update on Biosynthesis, Signal Transduction and Action in Plant Stress Response, Growth and Development. *Ann. Bot.* **2007**, *100*, 681–697. [CrossRef] [PubMed]
30. Ge, Q.; Zhang, Y.; Hua, W.P.; Wu, Y.C.; Jin, X.X.; Song, S.H.; Wang, Z.Z. Combination of transcriptomic and metabolomic analyses reveals a JAZ repressor in the jasmonate signaling pathway of Salvia miltiorrhiza. *Sci. Rep.* **2015**, *5*, 14048. [CrossRef] [PubMed]

31. Gao, W.; Hillwig, M.L.; Huang, L.; Cui, G.; Wang, X.; Kong, J.; Yang, B.; Peters, R. A functional genomics approach to tanshinone biosynthesis provides stereochemical insights. *Org. Lett.* **2009**, *11*, 5170–5173. [CrossRef] [PubMed]
32. Xiao, Y.; Zhang, L.; Gao, S.; Saechao, S.; Di, P.; Chen, J.; Chen, W. The c4h, tat, hppr and hppd Genes Prompted Engineering of Rosmarinic Acid Biosynthetic Pathway in Salvia miltiorrhiza Hairy Root Cultures. *PLoS ONE* **2011**, *6*, e29713. [CrossRef] [PubMed]
33. Peng, D.; Lei, Z.; Chen, J.; Tan, H.; Ying, X.; Xin, D.; Xun, Z.; Chen, W. 13C Tracer Reveals Phenolic Acids Biosynthesis in Hairy Root Cultures of Salvia miltiorrhiza. *ACS Chem. Biol.* **2013**, *8*, 1537–1548.
34. Chen, C.; Zhang, Y.; Qiakefu, K.; Zhang, X.; Han, L.; Hua, W.; Yan, Y.; Wang, Z. Overexpression of tomato Prosystemin (LePS) enhances pest resistance and the production of tanshinones in Salvia miltiorrhiza Bunge. *J. Agric. Food Chem.* **2016**, *64*, 7760–7769. [CrossRef] [PubMed]
35. Attieh, J.; Djiana, R.; Koonjul, P.; Étienne, C.; Sparace, S.A.; Saini, H.S. Cloning and functional expression of two plant thiol methyltransferases: A new class of enzymes involved in the biosynthesis of sulfur volatiles. *Plant Mol. Biol.* **2002**, *50*, 511–521. [CrossRef] [PubMed]
36. Joshi, C.P.; Chiang, V.L. Conserved sequence motifs in plant S-adenosyl-L-methionine-dependent methyltransferases. *Plant Mol. Biol.* **1998**, *37*, 663–674. [CrossRef] [PubMed]
37. Kagan, R.M.; Clarke, S. Widespread occurrence of three sequence motifs in diverse S-adenosylmethionine-dependent methyltransferases suggests a common structure for these enzymes. *Arch. Biochem. Biophys.* **1994**, *310*, 417–427. [CrossRef] [PubMed]
38. Yang, N.; Zhou, W.; Su, J.; Wang, X.; Li, L.; Wang, L.; Cao, X.; Wang, Z. Overexpression of SmMYC2 Increases the Production of Phenolic Acids in Salvia miltiorrhiza. *Front. Plant Sci.* **2017**, *8*, 1804. [CrossRef] [PubMed]
39. Wenping, H.; Yuan, Z.; Jie, S.; Lijun, Z.; Zhezhi, W. De novo transcriptome sequencing in Salvia miltiorrhiza to identify genes involved in the biosynthesis of active ingredients. *Genomics* **2011**, *98*, 272–279. [CrossRef] [PubMed]
40. Kombrink, E. Chemical and genetic exploration of jasmonate biosynthesis and signaling paths. *Planta* **2012**, *236*, 1351–1366. [CrossRef] [PubMed]
41. Wasternack, C.; Hause, B. Jasmonates: Biosynthesis, perception, signal transduction and action in plant stress response, growth and development. An update to the 2007 review in Annals of Botany. *Ann. Bot.* **2013**, *111*, 1021–1058. [CrossRef] [PubMed]
42. Jung, C.K.; Lyou, S.H.; Koo, Y.J.; Song, J.T.; Choi, Y.D.; Cheong, J.J. Constitutive Expression of Defense Genes in Transgenic Arabidopsis Overproducing Methyl Jasmonate. *J. Appl. Biol. Chem.* **2003**, *46*, 52–57.
43. Baldwin, I.T. Jasmonate-induced responses are costly but benefit plants under attack in native populations. *Proc. Natl. Acad. Sci. USA* **1998**, *95*, 8113–8118. [CrossRef] [PubMed]
44. McConn, M.; Creelman, R.A.; Bell, E.; Mullet, J.E.; Browse, J. Jasmonate is essential for insect defense in Arabidopsis. *Proc. Natl. Acad. Sci. USA* **1997**, *94*, 5473–5477. [CrossRef] [PubMed]
45. Li, C.; Williams, M.M.; Loh, Y.T.; Lee, G.I.; Howe, G.A. Resistance of cultivated tomato to cell content-feeding herbivores is regulated by the octadecanoid-signaling pathway. *Plant Physiol.* **2002**, *130*, 494–503. [CrossRef] [PubMed]
46. Harms, K.; Atzorn, R.; Brash, A.; Kuhn, H.; Wasternack, C.; Willmitzer, L.; Penacortes, H. Expression of a Flax Allene Oxide Synthase cDNA Leads to Increased Endogenous Jasmonic Acid (JA) Levels in Transgenic Potato Plants but Not to a Corresponding Activation of JA-Responding Genes. *Plant Cell* **1995**, *7*, 1645–1654. [CrossRef] [PubMed]
47. Heilmann, B.; Hartung, W.; Gimmler, H. The Distribution of Abscisic Acid between Chloroplasts and Cytoplasm of Leaf Cells and the Permeability of the Chloroplast Envelope for Abscisic Acid. *Zeitschrift Für Pflanzenphysiologie* **1980**, *97*, 67–78. [CrossRef]
48. Castresana, J. Selection of conserved blocks from multiple alignments for their use in phylogenetic analysis. *Mol. Biol. Evol.* **2000**, *17*, 540–552. [CrossRef] [PubMed]
49. Huelsenbeck, J.P.; Ronquist, F. MRBAYES: Bayesian inference of phylogenetic trees. *Bioinformatics* **2001**, *17*, 754–755. [CrossRef] [PubMed]
50. Hall, B.G. Comparison of the accuracies of several phylogenetic methods using protein and DNA sequences. *Mol. Biol. Evol.* **2005**, *22*, 792–802. [CrossRef] [PubMed]
51. Abascal, F.; Zardoya, R.; Posada, D. ProtTest: Selection of best-fit models of protein evolution. *Bioinformatics* **2005**, *21*, 2104–2105. [CrossRef] [PubMed]

52. Zhai, Y.; Tchieu, J.; Saier, M.H., Jr. A web-based Tree View (TV) program for the visualization of phylogenetic trees. *J. Mol. Microbiol. Biotechnol.* **2002**, *4*, 69–70. [PubMed]
53. Depicker, A. Transfection and transformation of Agrobacterium tumefaciens. *Mol. Gen. Genet. MGG* **1978**, *163*, 181–187.
54. Yan, Y.P.; Wang, Z.Z. Genetic transformation of the medicinal plant Salvia miltiorrhiza by Agrobacterium tumefaciens -mediated method. *Plant Cell Tissue Organ Cult.* **2007**, *88*, 175–184. [CrossRef]
55. Hua, W.; Song, J.; Li, C.; Wang, Z. Molecular cloning and characterization of the promoter of SmGGPPs and its expression pattern in Salvia miltiorrhiza. *Mol. Biol. Rep.* **2012**, *39*, 5775–5783. [CrossRef] [PubMed]
56. Yang, Y.; Hou, S.; Cui, G.; Chen, S.; Wei, J.; Huang, L. Characterization of reference genes for quantitative real-time PCR analysis in various tissues of Salvia miltiorrhiza. *Mol. Biol. Rep.* **2010**, *37*, 507–513. [CrossRef] [PubMed]
57. Schmittgen, T.D.; Livak, K.J. Analyzing real-time PCR data by the comparative CT method. *Nat. Protoc.* **2008**, *3*, 1101–1108. [CrossRef] [PubMed]
58. Yikling, C.; Jookheng, G.; Yauyan, L. Assessment of in vitro antioxidant capacity and polyphenolic composition of selected medicinal herbs from Leguminosae family in Peninsular Malaysia. *Food Chem.* **2009**, *116*, 13–18.
59. Dewanto, V.; Wu, X.; Adom, K.K.; Liu, R.H. Thermal processing enhances the nutritional value of tomatoes by increasing total antioxidant activity. *Food Sci. Technol.* **2005**, *50*, 3010–3014. [CrossRef]
60. Cao, X.; Guo, X.; Yang, X.; Wang, H.; Hua, W.; He, Y.; Kang, J.; Wang, Z. Transcriptional Responses and Gentiopicroside Biosynthesis in Methyl Jasmonate-Treated Gentiana macrophylla Seedlings. *PLoS ONE* **2016**, *11*, e0166493. [CrossRef] [PubMed]
61. Kim, D.; Pertea, G.; Trapnell, C.; Pimentel, H.; Kelley, R.; Salzberg, S.L. TopHat2: Accurate alignment of transcriptomes in the presence of insertions, deletions and gene fusions. *Genome Biol.* **2013**, *14*, R36. [CrossRef] [PubMed]
62. Schulze, S.K.; Kanwar, R.; Gölzenleuchter, M.; Therneau, T.M.; Beutler, A.S. SERE: Single-parameter quality control and sample comparison for RNA-Seq. *BMC Genom.* **2012**, *13*, 524. [CrossRef] [PubMed]
63. Trapnell, C.; Williams, B.A.; Pertea, G.; Mortazavi, A.; Kwan, G.; Van, M.B.; Salzberg, S.L.; Wold, B.J.; Pachter, L. Transcript assembly and quantification by RNA-Seq reveals unannotated transcripts and isoform switching during cell differentiation. *Nat. Biotechnol.* **2010**, *28*, 511–515. [CrossRef] [PubMed]
64. Florea, L.; Song, L.; Salzberg, S.L. Thousands of exon skipping events differentiate among splicing patterns in sixteen human tissues. *F1000Research* **2013**, *2*, 188. [CrossRef] [PubMed]
65. Anders, S.; Huber, W. Differential expression analysis for sequence count data. *Genome Biol.* **2010**, *11*, R106. [CrossRef] [PubMed]
66. Alexa, A.; Rahnenführer, J.; Lengauer, T. Improved scoring of functional groups from gene expression data by decorrelating GO graph structure. *Bioinformatics* **2006**, *22*, 1600–1607. [CrossRef] [PubMed]
67. Kanehisa, M.; Goto, S.; Kawashima, S.; Okuno, Y.; Hattori, M. The KEGG resource for deciphering the genome. *Nucleic Acids Res.* **2004**, *32*, D277–D280. [CrossRef] [PubMed]
68. Mao, X.; Cai, T.; Olyarchuk, J.G.; Wei, L. Automated genome annotation and pathway identification using the KEGG Orthology (KO) as a controlled vocabulary. *Bioinformatics* **2005**, *21*, 3787–3793. [CrossRef] [PubMed]

8

Identification and Characterization of *EI* (*Elongated Internode*) Gene in Tomato (*Solanum lycopersicum*)

Xiaorong Sun †, Jinshuai Shu †, Ali Mohamed Ali Mohamed, Xuebin Deng, Xiaona Zhi, Jinrui Bai, Yanan Cui, Xiaoxiao Lu, Yongchen Du, Xiaoxuan Wang, Zejun Huang, Yanmei Guo, Lei Liu * and Junming Li *

Institute of Vegetables and Flowers, Chinese Academy of Agricultural Sciences, Key Laboratory of Biology and Genetic Improvement of Horticultural Crops, Ministry of Agriculture, 12 Zhongguancun Nandajie Street, Beijing 100081, China; sunxiaorong@caas.cn (X.S.); shujinshuai@caas.cn (J.S.); almuttaried@hotmail.com (A.M.A.M.); dengxuebin@caas.cn (X.D.); 18763825752@163.com (X.Z.); baijinrui@caas.cn (J.B.); 15621567802@163.com (Y.C.); luxiaoxiao914@163.com (X.L.); duyongchen@caas.cn (Y.D.); wangxiaoxuan@caas.cn (X.W.); huangzejun@caas.cn (Z.H.); guoyanmei@caas.cn (Y.G.)

* Correspondence: liulei02@caas.cn (L.L.); lijunming@caas.cn (J.L.)

† These authors equally contributed to this work.

Abstract: Internode length is an important agronomic trait affecting plant architecture and crop yield. However, few genes for internode elongation have been identified in tomato. In this study, we characterized an elongated internode inbred line P502, which is a natural mutant of the tomato cultivar 05T606. The mutant P502 exhibits longer internode and higher bioactive GA concentration compared with wild-type 05T606. Genetic analysis suggested that the elongated internode trait is controlled by quantitative trait loci (QTL). Then, we identified a major QTL on chromosome 2 based on molecular markers and bulked segregant analysis (BSA). The locus was designated as *EI* (*Elongated Internode*), which explained 73.6% genetic variance. The *EI* was further mapped to a 75.8-kb region containing 10 genes in the reference Heinz 1706 genome. One single nucleotide polymorphism (SNP) in the coding region of *solyc02g080120.1* was identified, which encodes gibberellin 2-beta-dioxygenase 7 (SlGA2ox7). SlGA2ox7, orthologous to AtGA2ox7 and AtGA2ox8, is involved in the regulation of GA degradation. Overexpression of the wild *EI* gene in mutant P502 caused a dwarf phenotype with a shortened internode. The difference of *EI* expression levels was not significant in the P502 and wild-type, but the expression levels of GA biosynthetic genes including *CPS, KO, KAO, GA20ox1, GA20ox2, GA20ox4, GA3ox1, GA2ox1, GA2ox2, GA2ox4*, and *GA2ox5*, were upregulated in mutant P502. Our results may provide a better understanding of the genetics underlying the internode elongation and valuable information to improve plant architecture of the tomato.

Keywords: tomato; *Elongated Internode* (*EI*); QTL; GA2ox7

---

1. Introduction

Plant height is an important component of plant architecture, and is highly correlated with the yield [1]. One of the decisive factors affecting plant height is internode length. The reduced plant height or internode length of semi-dwarf varieties has improved the harvest index and biomass production. The introduction of the "green revolution" semi-dwarf gene *SD1* in rice and *Rht* in wheat resulted in substantial increases in grain yields and helped to avert predicted food shortages in Asia during the 1960s and 1970s [2–5]. To explore the genetic potential, several genes or quantitative trait loci (QTL) controlling the internode length in rice [6–8], maize [9–11], wheat [12,13], and sorghum [14,15] have been identified. In a recent report, the semi-dwarf gene *SBI* was cloned, which could shorten the basal internode of rice. Moreover, the *SBI* allele-introduced varieties have great potential for improving lodging resistance and yield [16].

The molecular genetic analysis of dwarf and slender mutants revealed that most of the mutations are related to the biosynthesis pathways and signal transduction of phytohormones, mostly gibberellins (GAs) [2,9,17–19]. GAs are a group of tetracyclic diterpenoid that affect plant developmental processes such as stem elongation [20]. Nowadays, more than 130 GAs have been identified, but relatively few GAs (e.g., GA_1, GA_3, GA_4, and GA_7) have intrinsic biological activity [21]. The rice *SD1* encodes GA20ox2 and catalyzes the conversion of GA_{12}/GA_{53} to bioactive GA precursors GA_9/GA_{20}. The recessive semi-dwarf mutant *sd1* can be restored by exogenous GA_3 [22,23]. DELLA protein is a negative regulatory factor in the GA signaling pathway, which inhibits plant growth. The semi-dominant mutations that occurred in Arabidopsis *GAI*, maize *D8*, wheat *Rht*, and rice *GAI* were caused by the gain of function of DELLA protein, which led to the dwarfism [3,24–28]. In contrast, the loss-of-function of DELLA protein in barley *SLN1*, rice *SLR1*, and tomato *PRO* increased growth and caused a GA-constitutive response phenotype [29–31]. In addition, other phytohormones including brassinosteroid (BR), indole-3-acetic acid (IAA), and strigolactones (SLs) have also been proven to be involved in the regulation of internode length or plant height [7,14,32–34].

Tomato (*Solanum lycopersicum*) is the second most consumed vegetable crop and is widely grown around the world [35]. The tomato internode length not only affects the plant architecture, but also plays an important role in mechanized harvesting. However, there are few reports on the genetic regulation of internode length in tomato. Evidence has shown that the well-known dwarf cultivar Micro-Tom, which has the characteristics of extreme dwarfing, dark and wrinkled leaves, has at least two mutations (*d* and *mnt*) affecting internode length [36]. However, Micro-Tom was produced for ornamental purposes, and as a conventional model system for research due to its small size, rapid growth, and easy transformation. Due to its characteristics of extreme dwarfing, it seems very limited in a practical breeding program. Moreover, the tomato *br* locus contributes to a shorter internode and could be useful source for tomato short internode breeding. However, the *br* locus was only mapped to a 763.1-kb region on chromosome 1, and the gene has not been cloned yet [37]. Therefore, to clone new gene or locus that controls internode elongation is of great theoretical and practical significance. It is helpful to clarify the regulatory mechanism of tomato internode elongation and to provide the possibility of establishing a breeding approach. Furthermore, with the completion of tomato genome sequencing as well as the rapid development of sequencing, marker development has become easier, which has accelerated the speed of gene cloning [38].

In this study, we performed a phenotypic characterization of the mutant P502, which shows a significantly elongated internode compared with wild-type 05T606. Then, we report on the molecular identification of the *EI* (*Elongated Internode*) gene by map-based cloning. *EI*, which is expressed in the root, stem, and leaf, encodes the GA2ox7 enzyme involved in the GA metabolism pathway. Overexpressing of *EI* can cause a dwarf phenotype with short internode. Our results indicate that *EI* plays an important role in controlling the tomato internode elongation.

2. Results

2.1. Characterization of Elongated Internode Inbred Line P502

To compare the dynamic difference in the internode lengths of P502 and wild-type 05T606, we recorded the mean internode lengths of 20-day-old, 25-day-old, 30-day-old, 35-day-old, and 40-day-old seedlings. The results showed that P502 had a longer internode than the wild-type across the seedling stages (Figure 1a,b). Additionally, we compared individual internode lengths including the first, second, third, fourth, and fifth internodes of 40-day-old seedlings. The results indicated that each internode of P502 was significantly greater than the corresponding internode of the wild-type (Figure 1c,d). An analysis of the longitudinal sections of the third internode with a scanning electron microscope revealed that the cells were much longer in P502 than in the wild-type (Figure 1e,f). These results suggest that the mutant P502 phenotype is characterized by the elongated internode.

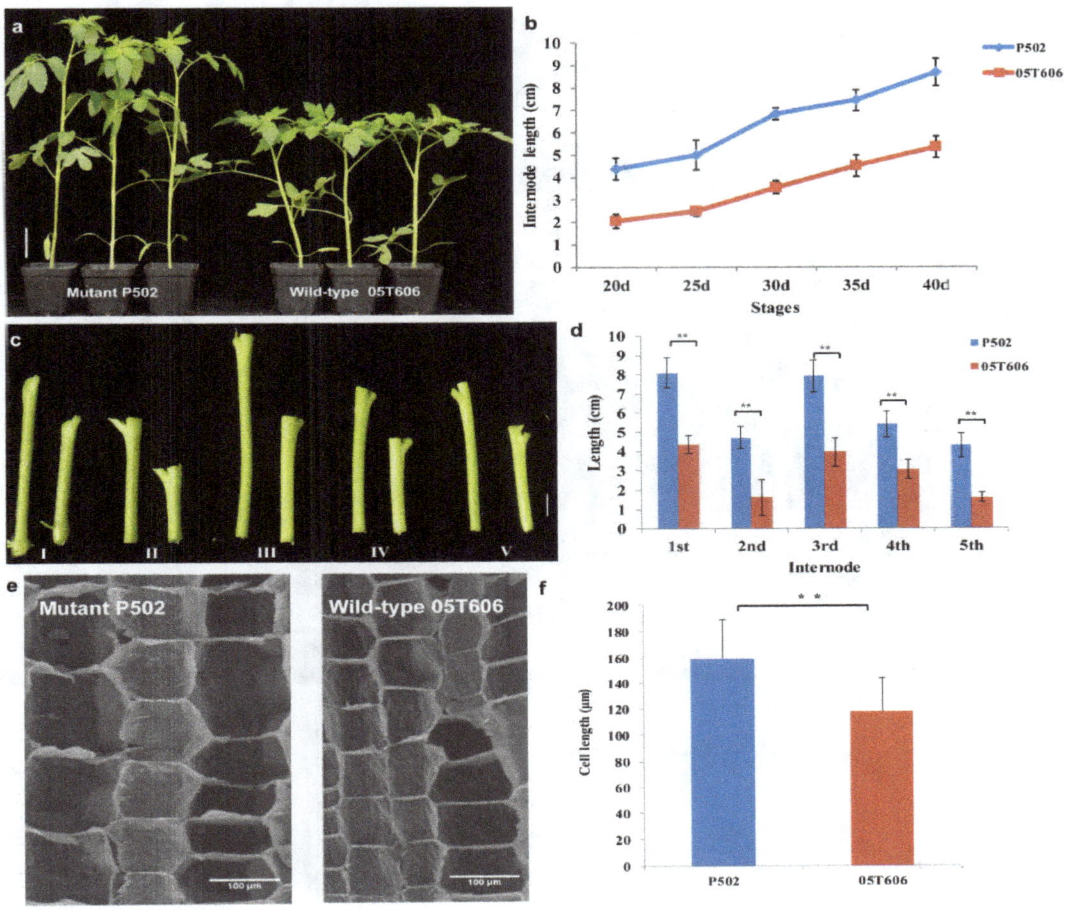

Figure 1. The phenotypic characterization of the mutant P502 and wild-type 05T606 at 40-day-old seedlings. Morphological phenotypes (**a**) and the statistical data of internode lengths (**b**) of the mutant P502 and wild-type 05T606. (**c**) Morphological phenotypes of the first five internode lengths of the mutant P502 (left) and wild-type (right). (**d**) Statistical data of internode length in (**c**). Longitudinal sections (**e**) and the statistical data (**f**) of the third internode pith cell length of P502 and wild-type. Scale bar is 5 cm in (**a**), 1 cm in (**c**), 100 μm in (**e**). A Student's t test indicated a significant difference ((**b**,**d**), $n = 30$ plants; (**f**) $n = 90$ cells) in (**d**,**f**). ** $p < 0.01$. All data are given as mean ± SD.

2.2. Elongated Internode Mutation Is Related to the GA Metabolic Pathway

The GAs stimulate cell elongation, and are effective internode elongation regulators. Paclobutrazol (PAC) inhibits the oxidation of *ent*-kaurene, an early step in GA biosynthesis, and can reduce endogenous GA level [39]. To examine the responses to GAs, we sprayed 20-day-old wild-type 05T606 and mutant P502 seedlings with exogenous GA_3 and PAC. For the 40-day-old seedlings, the plant height of 05T606 and P502 increased by 58.2% and 20.5%, respectively, after GA_3 treatment. However, the PAC treatment decreased the height of 05T606 by 57.6%, and decreased the height of P502 by 61.0% (Figure 2a–d). Next, we measured the endogenous GA concentration of the first five internodes in 30-day-old seedlings of P502 and 05T606 plants. There was an increase in the bioactive GA_1 in P502, and bioactive GA_4 was only detected in mutant P502 (Figure 2e). These results indicate that the elongated internode of mutant P502 is related to the GA metabolic pathway and was caused by a higher level of bioactive GAs.

Figure 2. The line P502 is a GA-sensitive mutant. (**a**,**b**) The morphological phenotypes of 30-day-old wild-type and P502 seedlings after treatment with GA_3 (10^{-5} M) and PAC (10^{-7} M, GA biosynthesis inhibitor), respectively. (**c**,**d**) The statistical data of wild-type and P502 plant height in different stages, respectively. (**e**) Concentration of endogenous bioactive GAs in the first five internodes of 30-day-old mutant P502 and wild-type seedlings. The water treatment was used as control and the scale bar is 5 cm in (**a**,**b**). Data for (**c**,**d**) are based on three replicates of eight plants per group. N.D. represents not detectable. Data for (**e**) are based on three independent biological replicates, and the asterisk indicates a statistically significant difference (Student's *t*-test, ** $p < 0.01$). All data are given as mean ± SD.

2.3. Genetic Analysis of the Elongated Internode Trait

To determine whether the elongated internode trait is controlled by a single gene or QTL, the first five average internode lengths of the P_1 (Heinz 1706), P_2 (P502), F_1 (Heinz 1706 × P502), and F_2 population were recorded in the spring of 2016, 2017, and 2018. The internode length frequency of the F_2 population exhibited a continuous and skewed distribution in different years (Figure 3), and the F_1 was biased toward the parent Heinz 1706. Moreover, the differences of P_1, P_2, and F_1 were significant (Table 1). These results indicate that the elongated internode length is a quantitative trait and that this population was ideal for elongated internode QTL analysis.

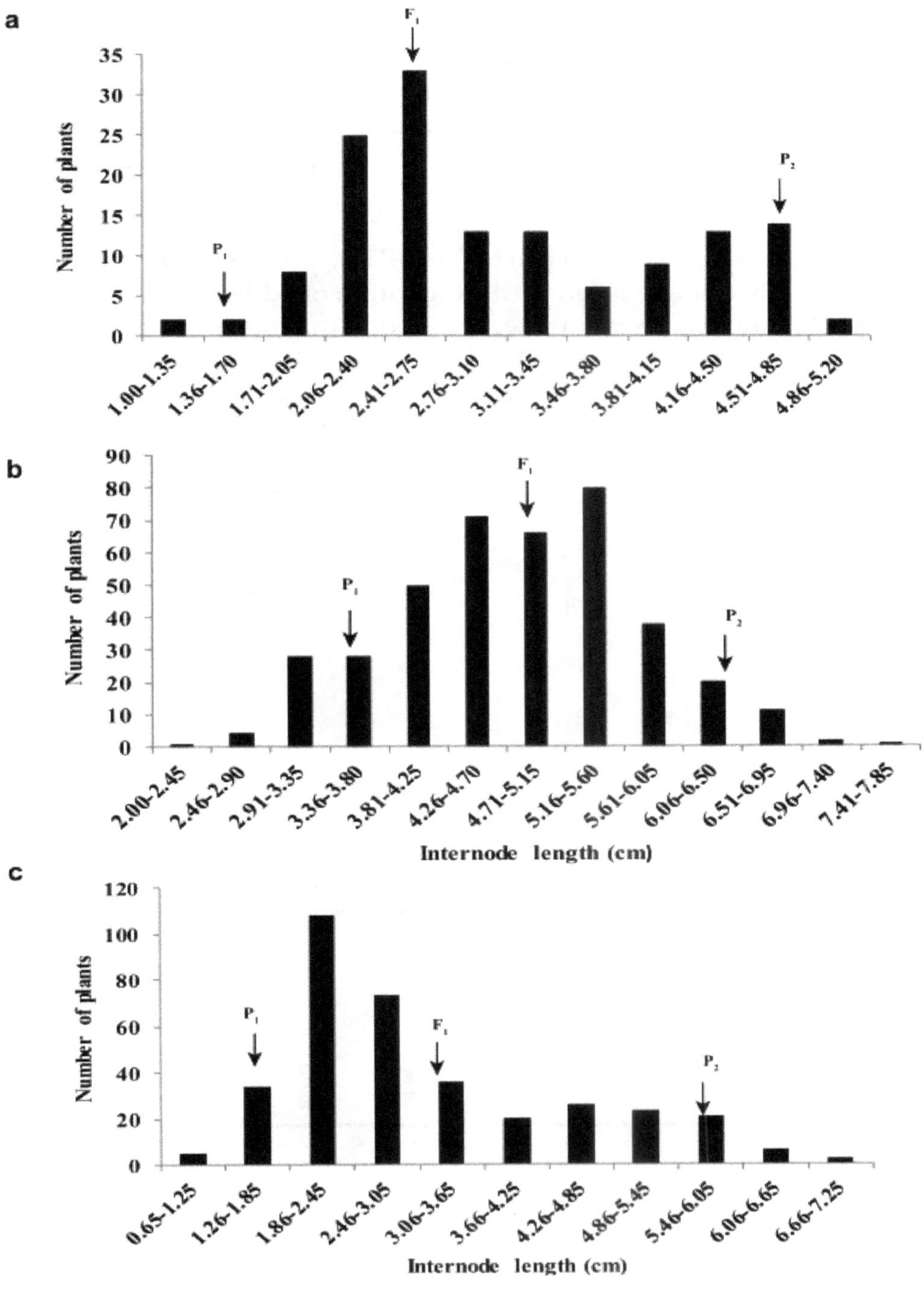

Figure 3. Internode length frequency distribution of F_2 individuals at 40-day-old seedlings. (**a**) 2016 ($n = 140$). (**b**) 2017 ($n = 400$). (**c**) 2018 ($n = 354$). The mean internode lengths were recorded from the first internode to the fifth internode (starting from the cotyledons).

Table 1. Internode lengths of P_1, P_2, and F_1 population plants.

Materials	2016 (cm) [a]	2017 (cm) [a]	2018 (cm) [a]
P_1 (Heinz 1706)	1.52 ± 0.35 a	1.48 ± 0.63 a	1.85 ± 0.30 a
P_2 (P502)	4.62 ± 0.19 c	4.66 ± 0.60 c	5.54 ± 0.48 c
F_1 (Heinz 1706 × P502)	2.74 ± 0.28 b	2.53 ± 0.47 b	3.10 ± 0.24 b

[a] Values followed by different letters (a, b, and c) are significantly different ($p < 0.01$). $n = 15$ plants.

2.4. Map-Based Cloning of the EI Gene

A total of 372 InDel markers distributed on 12 chromosomes were screened, and 90 were polymorphic between the parents. Of these 90 markers, four markers (D55, D57, D64, and D67) located on chromosome 2 were polymorphic between the E and N bulks. The bands of the E pool were consistent with those of the parent P502, whereas the bands of the N pool were heterozygous. These results suggest that the locus is present on chromosome 2.

The 354 F_2 individuals, derived from Heinz 1706 × P502, were used for QTL analysis. The results showed that there was only a single peak, with a logarithm of odds score of 102.4 explaining 73.6% phenotypic variance (Figure 4a). Therefore, we concluded that the locus named *elongated internode* (*ei*) was located between markers D64 and HP2509. Another 956 F_2 individuals were screened for recombinants with the flanking markers D64 and HP2509. The detected recombinants were analyzed with another seven CAPS and InDel markers between the flanking markers (Figure 4b). According to the genotypes of the recombinants and the phenotypes of F_3 individuals, we narrowed the *ei* locus to an interval between CAPS17 and InDel6 (Table 2), corresponding to a 75.8-kb region on chromosome 2 of the reference Heinz 1706 genome.

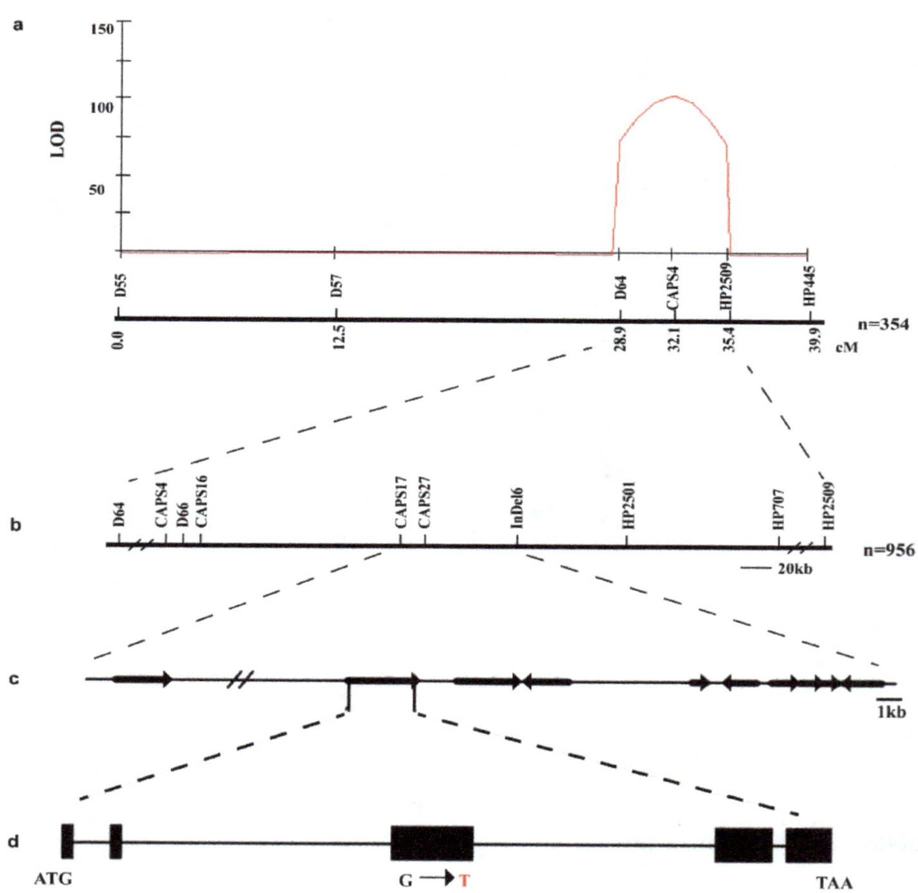

Figure 4. Map-based cloning of *ei* locus. (**a**) The genetic map of *ei* on chromosome 2, mapped using 354 F_2 individuals and six polymorphic markers. (**b**) The *ei* locus was fine-mapped to the interval between markers CAPS17 and InDel6. (**c**) Predicted genes in the region encompassing the *ei* locus. The arrows indicate the direction of transcription. (**d**) Gene structure and the mutation site. The black rectangles and black line indicate exons and introns, respectively. The red letter represents the mutation base in mutant P502.

Table 2. Genotypes of F_2 recombinants and phenotypes of $F_{2:3}$ individuals.

Recombinants	Genotype (F_2) [a]										Phenotype(F_3)	
	D64	CAPS4	D66	CAPS16	CAPS17	CAPS27	InDel6	HP2501	HP707	HP2509	N [b]	E [b]
6–11	b	h	h	h	h	h	h	h	h	h	-	-
13–5	b	b	b	h	h	h	h	h	h	h	44	15
7–48	b	b	b	b	h	h	h	h	h	h	45	15
7–71	b	b	b	b	h	h	h	h	h	h	43	14
15–543	b	b	b	b	b	h	h	h	h	h	45	14
17–18	b	b	b	b	b	h	h	h	h	h	45	15
15–741	h	h	h	h	h	h	b	b	b	b	46	14
1–2	h	h	h	h	h	h	b	b	b	b	45	15
1–72	h	h	h	h	h	h	h	b	b	b	44	16
6–21	h	h	h	h	h	h	h	h	h	b	-	-

[a] b in green backgroud: homozygous like P502; h in gray background: heterozygous; [b] N: the number of normal internode plants; E: the number of elongated internode plants; -: no data.

The Solanaceae Genomics Network website (SGN; http://solgenomics.net) searches [40] indicated that there were 10 genes in this region (Figure 4c, Table 3). By analyzing the sequenced P502 genome, we determined that the DNA sequence had no mutations in the other nine genes, whereas *solyc02g080120.1* contained a SNP (G–T) in the exon region (Figure 4d). Thus, *solyc02g080120.1* was amplified using genomic DNA extracted from 05T606 and P502 plants and eight primer pairs (A1, A2, A3, A4, A5, A6, A7, and A8; Table S1). The amplification results revealed that the *EI* gene consists of 4626 bp (with five exons and four introns). Moreover, the *EI* gene in mutant P502 includes a SNP mutation in the third exon (G2152T) (Figure S1). To confirm this mutation site, the *solyc02g080120.1* coding sequence (CDS) was amplified by RT-PCR (CDS-F and CDS-R primers) (Table S1). Sequences of the CDS further confirmed the presence of a SNP mutation in the coding region, which resulted in an amino acid mutation in mutant P502.

Table 3. Ten predicted genes in the 75.8-kb fine mapping interval according to the reference genome.

Gene ID [a]	Position	Functional Annotation
solyc02g080110.2	SL2.50ch02: 44414686..44417879 (+)	Unknown Protein (AHRD V1)
solyc02g080120.1	SL2.50ch02: 44432042..44436667 (+)	Gibberellin 2-beta-dioxygenase 7
solyc02g080130.2	SL2.50ch02: 44439349..44443883 (+)	Chaperone dnaj-like protein
solyc02g080140.2	SL2.50ch02: 44444670..44447584 (-)	cysteine-rich PDZ-binding protein
solyc02g080150.1	SL2.50ch02: 44458406..44458810 (+)	uncharacterized LOC101262168
solyc02g080160.2	SL2.50ch02: 44460578..44462486 (-)	probable xyloglucan endotransglucosylase/hydrolase protein 8
solyc02g080170.1	SL2.50ch02: 44471643..44473397 (+)	pentatricopeptide repeat-containing protein At4g21170
solyc02g080180.1	SL2.50ch02: 44474353..44475381 (+)	Probable dolichyl-diphosphooligosaccharide—protein glycosyltransferase subunit 3B
solyc02g080190.2	SL2.50ch02: 44477570..44478612 (+)	Nuclear transport factor 2
solyc02g080200.2	SL2.50ch02: 44478656..44480593 (-)	pectinesterase-like

[a] Genes were identified based on the tomato model (ITAG release 2.40, SL2.50) in SGN (https://solgenomics.net/) (Access on 21 June 2014).

To determine whether the G-to-T transition was directly associated with the elongated internode phenotype, a co-segregation analysis was conducted with a functional marker (KASP) developed from this SNP. The KASP marker (S-A1: GAAGGTGACCAAGTTCATGCTCACAAGCTTCACAAGAATGGGG; S-A2: GAAGGTCGGAGTCAACGGATTGCACAAGCTTCACAAGAATGGGT; S-C: GTGATACTCCATGGTTTACAACTTGGAA) was used to validate the genotypes of 354 F_2 individuals derived from the cross of Heinz 1706 × P502 hybridization. The results showed that this KASP marker was co-segregated with internode length (Figure S2), with a 100% accuracy rate. It further confirmed that SNP mutation is associated with the elongated internode.

2.5. Protein Sequence Alignment and Phylogenetic Analysis of SlGA2ox7

According to the gene annotation, *EI* encodes the SlGA2ox7 enzyme, which comprises 380 amino acids. The sequence alignment and phylogenetic analysis revealed that GA2ox7 and GA2ox8 are clustered in one group, which is separate from GA20oxs and GA3oxs (Figure 5a), indicating that GA2ox7 and GA2ox8 are conserved in tobacco, Arabidopsis, maize, and grape. Moreover, amino acid position 112 of wild-type SlGA2ox7 is a hydrophilic glycine, whereas it is a hydrophobic valine in P502.

The SlGA2ox7 at this position is located within a conserved region of the PcbC superfamily (Figure 5b), indicating that the altered hydrophobicity of the amino acid may affect the function of SlGA2ox7.

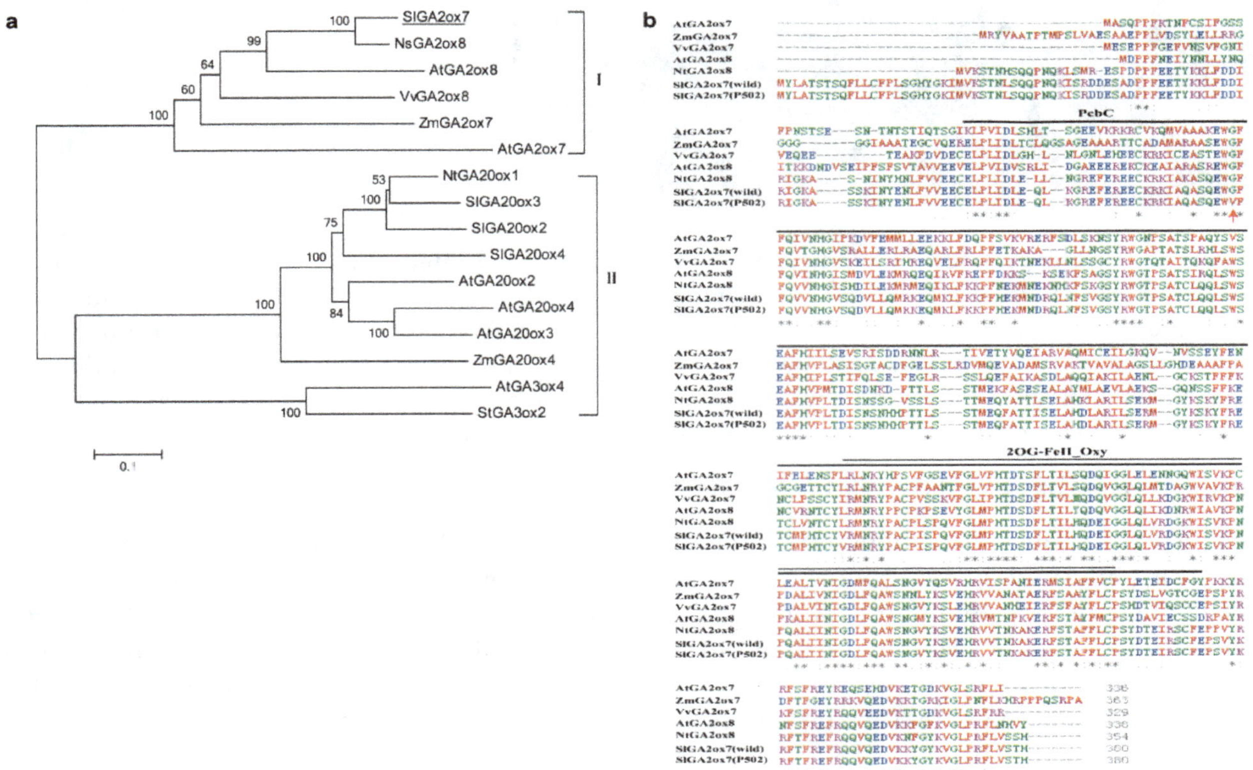

Figure 5. Phylogenetic analysis and sequence alignment of SlGA2ox7 with various species. (**a**) Phylogenetic analysis of SlGA2ox7. The phylogenetic tree was generated using the neighbor-joining method built in MEGA6.0, and the inferred phylogeny was tested by bootstrap analysis with 1000 replicate datasets. Numbers shown at the tree forks indicate the frequency of occurrence among all bootstrap iterations performed. (**b**) Alignment of the SlGA2ox7 sequences. The proteins were aligned with the ClustalW program. The bold black line and the thin black line indicate the PcbC domain (81–343) and 2OG-FeII_Oxy domain (237–332), respectively. The red arrow represents the amino acid at position 112 of SlGA2ox7.

2.6. *Overexpression of Wild-Type EI in P502 Resulted in Dwarfism*

To confirm the function of *EI*, the recombinant plasmid 35S: *EI* was introduced into the mutant P502. Transgenic plants were obtained after screening for regenerated shoots on selection medium containing kanamycin. The transgenic plants were analyzed further by PCR with primers NPTII-F and NPTII-R, and two positive transgenic plants (T_0-1 and T_0-2) were obtained. The *EI*-overexpressing transgenic T_1 homozygous lines (OE-1 and OE-2) exhibited dwarfism with shortened internodes (CK: 36.00 ± 1.32 cm; OE-1: 24.7 ± 1.14 cm; OE-2: 8.53 ± 0.91 cm) (Figure 6a). The *EI* expression level was 1.44-fold and 14-fold higher in the OE-1 and OE-2 plants, respectively, than in the P502 control plants (Figure 6b). These results indicated that overexpression of *EI* could result in dwarf phenotype with shortened internodes, and the degree of shortness is related to the expression level of *EI*.

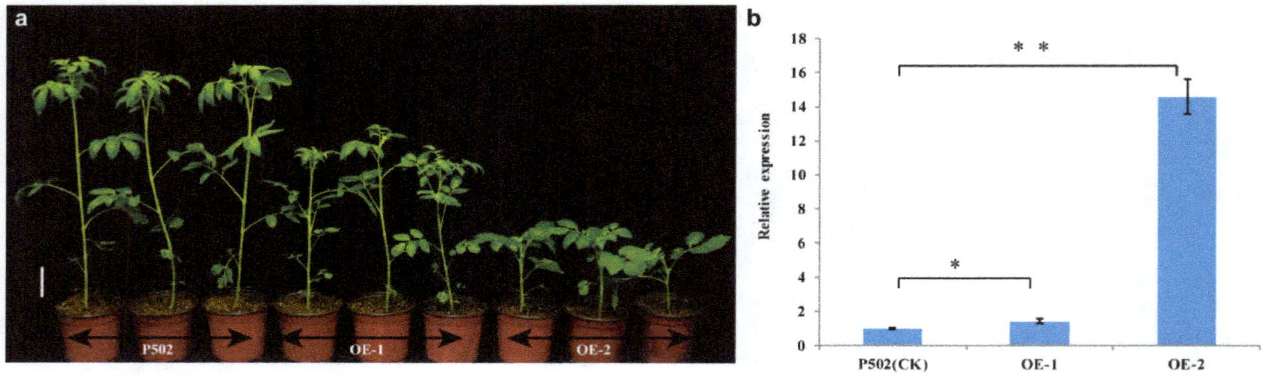

Figure 6. Overexpression of wild-type *EI* gene reduced the internode length of mutant P502. (**a**) Phenotype of P502 (CK) and T_1 transgenic plants (OE-1 and OE-2) at 40-day-old seedlings. (**b**) Expression level of *EI* in T_1 lines (OE-1 and OE-2). Total RNA was isolated from internode of P502 and transgenic T_1 plants at 40-day-old seedlings. Data represent mean ± SD based on three independent biological and three technical experiments. Scale bar is 5 cm in (**a**). Statistical significances were calculated based on two-tailed, two-sample Student's *t*-test at * $p < 0.05$ and ** $p < 0.01$.

2.7. The Expression Analysis of GA Metabolic Pathway-Related Genes

To study the spatiotemporal expression patterns of *EI*, total RNA was extracted from the leaves, stem (the third internode), and roots of 40-day-old wild-type and mutant P502 seedlings. The results of a qRT-PCR assay indicated that *EI* was expressed in the leaves, stem, and roots (Figure 7a), and the expression level in the leaves was more than 18-fold higher than that in the stem and roots. Interestingly, the differences of expression levels were not significant ($p > 0.05$) in the wild-type and mutant P502 (Figure 7a), indicating the G-to-T mutation does not alter the expression level of the *EI*.

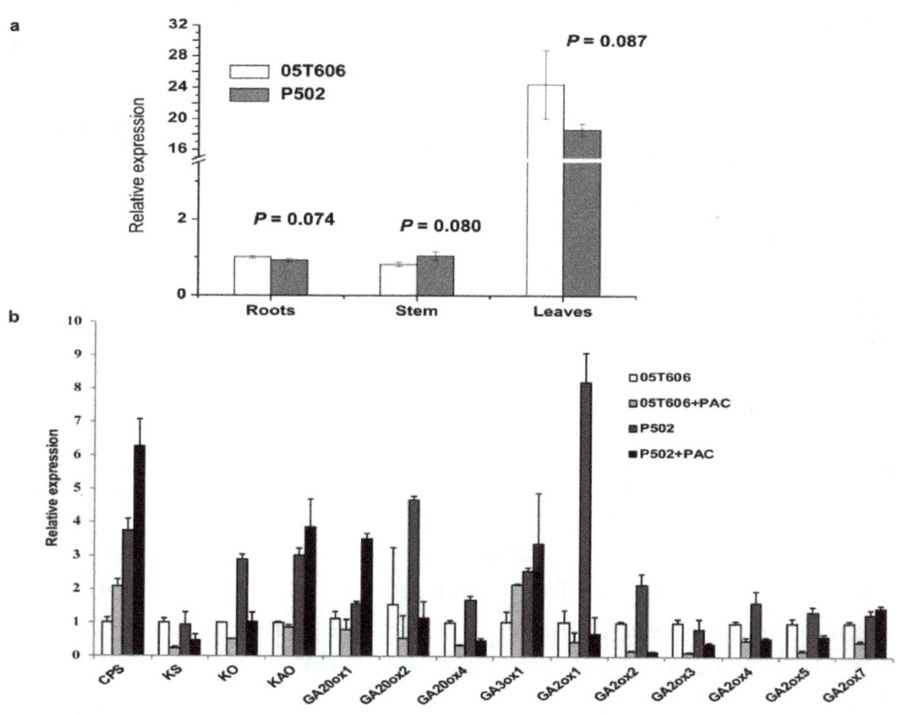

Figure 7. Expression analysis in the qRT-PCR assay. (**a**) Expression patterns of *EI* in the roots, stem, and leaves of P502 and the wild-type. Total RNA was isolated from P502 and wild-type at 40-day-old seedlings. A Student's *t* test was used for statistical analysis. (**b**) Expression levels of the GA biosynthetic genes after 10^{-7} M paclobutrazol (PAC, GA biosynthesis inhibitor) treatment. Data represent mean ± SD based on three independent biological and three technical experiments.

Many genes are involved in the GA biosynthetic pathway. CPS, KS, and KO are each encoded by a single gene in most plant species examined. However, the cytosol-localized GA20ox, GA3ox, and GA2ox each is encoded by a small gene family [41]. The qRT-PCR assay indicated that *CPS, KO, KAO, GA20ox1, GA20ox2, GA20ox4, GA3ox1, GA2ox1, GA2ox2, GA2ox4*, and *GA2ox5*, were more highly expressed in P502 than in the wild-type 05T606. PAC treatment decreased the expression levels of *KS, KO, GA20ox2, GA20ox4, GA2ox1, GA2ox2, GA2ox3, GA2ox4*, and *GA2ox5*, but increased the expression level of *CPS* and *GA3ox1*. However, the expression of *EI* (*GA2ox7*) in P502 was not significantly changed after PAC treatment (Figure 7b).

3. Discussion

The GA-related mutants have been categorized into GA-deficient (GA-sensitive) mutants and GA-insensitive mutants according to their response to exogenous GAs [42]. In GA-deficient dwarfs, the normal phenotype can be restored by the application of exogenous GAs and the mutations are usually due to a deficiency in the GA metabolic pathway [43]. In GA-insensitive types, the mutants are deficient in GA signaling and exhibit altered GA responses or the constitutive activation of GA responses [29–31,44]. In our study, the plant height of wild-type and the mutant P502 increased by 58.2% and 20.5%, respectively, after GA_3 treatment, whereas PAC treatment decreased the height of the wild-type by 57.6%, and the mutant P502 by 61.0% (Figure 2a–d). These results indicated that the mutant P502 was responsive to GA_3 and PAC, and the mutation was related to the GA metabolic pathway. In addition, it was found that mutant P502 was far less sensitive to GA_3 and more sensitive to PAC compared with the wild-type. This may be caused by a higher GA concentration of mutant P502 (Figure 2e), which reduced its sensitivity to GA_3 and increased its sensitivity to PAC.

GA2ox members are thought to disable GA functions by hydroxylating the C-2 position of active GAs or their precursors. The genes encoding 2-oxidases have been isolated from Arabidopsis, rice, spinach, and pea [45–48]. However, few GA2-oxidase genes have been isolated from tomato. In our study, we isolated the tomato *EI* gene by map-based cloning. The *EI* gene encodes SlGA2ox7, which belongs to GA2-oxidase. A point mutation in the exon region of *EI* gene resulted in the amino acid mutation (glycine to valine) of SlGA2ox7. GA2ox7 or GA2ox8 is conserved in various species (Figure 5a), and amino acid mutation occurs in the conserved domain of the PcbC superfamily (Figure 5b). Overexpression of *EI* inhibited the internode elongation of mutant P502, leading to dwarfism with a shortened internode (Figure 6a). These results are consistent with the research reported by Schomburg et al. [45], who revealed that the overexpression of *AtGA2ox7* and *AtGA2ox8* induced a dwarf phenotype of Arabidopsis and tobacco. Similar results have been obtained in transgenic plants overexpressing GA 2-oxidases from rice (*O. sativa*) [49]. Taken together, the elongated internode is caused by the loss-of-function of SlGA2ox7.

So far, three different kinds of GA deactivation have been identified. One type of GA2oxs hydroxylates the C-2 of active C_{19}-GAs (GA_1 and GA_4) or C_{19}-GA precursors (GA_{20} and GA_9) to produce biologically inactive GAs (GA_8, GA_{34}, GA_{29}, and GA_{51}, respectively) [18]. Another type of GA2oxs including AtGA2ox7, AtGA2ox8, OsGA2ox5, OsGA2ox6, and SoGA2ox3 (*Spinacia oleracea*) accept C_{20}-GAs (GA_{12} and GA_{53}) as their substrates to produce GA_{110} and GA_{97}, respectively [45,50]. In addition, the recombinant SoGA2ox1 can work on both C_{19}-GA and C_{20}-GA substrates [47]. Although SlGA2ox7 is orthologous to NsGA2ox8 and AtGA2ox8, whether they catalyze the same substrate needs further research.

Bioactive GA (GA_1, GA_3, GA_4, and GA_7) concentrations are maintained mainly by the balanced activities of GA 3-oxidases (GA3oxs) and GA 20-oxidases (GA20oxs), essential enzymes regulating GA biosynthesis, and GA 2-oxidases (GA2oxs) necessary for GA inactivation [51]. In our study, the expression of *EI* (*GA2ox7*) did not significantly change in the mutant P502 compared with the wild-type 05T606 (Figure 7a), indicating that the mutation of *EI* did not affect its transcript level. However, the expression of GA biosynthesis pathway genes including *CPS, KO, KAO, GA20ox1, GA20ox2, GA20ox4, GA3ox1, GA2ox1, GA2ox2, GA2ox4* and *GA2ox5* were increased (Figure 7b).

Similarly, the *dw* mutant of the soybean had lower expression levels of *CPS* and *GA20oxs* than the wild-type [19]. These results indicated that the mutations of genes related to GA synthesis may regulate the expression of other genes involved in the GA biosynthesis pathway by changing the GA concentrations. After exogenous PAC treatment, the growths of the mutant P502 and wild-type 05T606 were blocked. The expression levels of *KS*, *KO*, *GA20ox2*, *GA20ox4*, *GA2ox1*, *GA2ox2*, *GA2ox3*, *GA2ox4*, and *GA2ox5*, were downregulated (Figure 7b), revealing that these genes may be involved in the homeostatic maintenance of bioactive GA levels. Interestingly, the expression of the *EI* (*GA2ox7*) gene in the mutant P502 was not significantly changed after PAC treatment. We speculated that the mutation of *EI* may result in more complex regulation of GA homeostasis.

Tomato cultivars with determinate growth habit, compact, and short internode have been developed for commercial use [35]. The *sp* gene controlling determinate growth habit was cloned and the introduction of the *sp* allele into tomato cultivars has transformed the industry by creating a major modification in plant architecture [52]. However, few genes controlling internode elongation have been cloned. In our study, the *EI* gene for internode elongation was identified by map-based cloning and this gene encodes SlGA2ox7. Increased expression of *EI* caused different degrees of dwarfism, which may provide a useful resource for improving the plant architecture in tomato. Meanwhile, the co-segregated KASP marker developed in our study might be useful for high throughput maker assistant selection (MAS) in short internode breeding programs.

4. Materials and Methods

4.1. Plant Materials

Three determinate tomato inbred lines, P502, 05T606, and Heinz 1706, were used in this study. The mutant P502 shows an elongated internode, derived from a natural mutant of tomato cultivar 05T606. These were generated at the Institute of Vegetables and Flowers, Chinese Academy of Agricultural Sciences. Heinz 1706 was obtained from the Tomato Genetics Resource Center and displays a normal internode length, which is significantly shorter than the mutant P502 [53].

The mutant P502 (as the male parent) and Heinz 1706 (as the female parent) were hybridized to obtain an F_1 generation, and F_1 plants were self-pollinated to generate the F_2 population, which were used for inheritance analysis and fine mapping. All of the plant materials were grown in a greenhouse in Beijing, China. The average day and night temperatures were set at 25 °C and 20 °C, respectively.

4.2. Scanning Electron Microscopy Observation

To measure the cell lengths, the third internode was collected from 40-day-old 05T606 and P502 seedlings. The internode was cut into 5-mm segments and fixed in 3.5% glutaraldehyde for 24 h at room temperature. After washing in 0.1 M phosphate buffer (pH 7.4), the samples were fixed in 1% osmic acid for 2 h, dehydrated in a graded ethanol series, and dried in a Leica-EM CPD 300 desiccator (Leica, Frankfurt, Germany). Longitudinal sections were prepared by cutting the middle of the internode, which was then coated with a gold film. The pith cells at approximately the center of the stem were visualized and photographed with the Hitachi SU 8010 scanning election microscope (Hitachi, Tokyo, Japan). The cell length was measured with IMAGEJ software [54].

4.3. Exogenous GA_3 Treatments and Endogenous GA Quantification

To assess the response of P502 and 05T606 to GAs, the aerial parts of the 20-day-old seedlings were separately sprayed with 10^{-5} M GA_3 (Sigma, St. Louis Missouri, USA) and 10^{-7} M paclobutrazol (PAC, GA biosynthesis inhibitor; Biotopped, Beijing, China) [55] solutions containing 0.02% Tween-20 at an interval of one day. Control plants were sprayed with water. We sprayed ten times, and stopped at 40-day-old seedlings. The effects of GA_3 and PAC on stem expansion (from the cotyledons to the uppermost internode) were evaluated every four days. Each treatment was completed with three replicates (each with eight plants).

To determine the concentration of endogenous GAs, the first five internodes from the 30-day-old seedlings of P502 and wild-type 05T606 were collected into three biological replicates. Each biological replicate contained 1 g of tissue fresh weight. Tissue was immediately frozen in liquid nitrogen, and then was stored at −80 °C. The phytohormone extraction and quantitative profiling of GAs (GA_1, GA_4, GA_9, GA_{19}, and GA_{20}) were performed by HPLC-MS/MS [56].

4.4. DNA Extraction and Molecular Marker Development

A single young leaf was collected from each plant at 2-week-old seedlings. Genomic DNA was extracted according to a modified CTAB method [57] and then diluted to a concentration of 100–150 ng/μL in RNase (10 mg/mL) H_2O (1:100). To develop new markers, the elongated internode line P502 was sequenced with the Illumina HiSeq PE150 system, with a 50× genome coverage (Sequence Read Archive accession number: PRJNA540748). According to differences with the reference genome sequence (Heinz 1706), insertion and deletion (InDel) and competitive allele specific PCR (KASP) markers were designed using Primer Premier 5.0 software [58], and cleaved amplified polymorphic sequence (CAPS) markers were designed by dCAPS Finder 2.0 [59].

4.5. Mapping Strategy

The mean internode lengths from the first internode to the fifth internode (starting from the cotyledons) of 40-day-old seedlings were recorded for phenotypic analysis [36,53]. The bulked segregant analysis (BSA) strategy was used for the quick identification of molecular markers linked with the target locus [60]. Two DNA bulks, the elongated internode bulk (E bulk) and the normal internode bulk (N bulk), were generated by pooling equal amounts of DNA from ten elongated internode and ten normal F_2 plants, respectively.

To screen for polymorphic makers, the two parents were genotyped with 372 InDel markers across 12 tomato chromosomes (unpublished). All polymorphic markers were used to analyze the two bulked DNA samples. The target chromosome was identified based on the BSA results. Subsequently, QTL mapping was conducted according to the internode lengths of 354 F_2 individuals and genotypes of ideal markers on the target chromosome by using JoinMap 4.0 and MapQTL 6.0 [61,62]. After flanking markers were identified, another 956 F_2 individuals derived from the same cross were used for selecting recombinants. The F_3 individuals from eight F_2 recombinants were obtained for the confirmation of the progeny phenotype. Each $F_{2:3}$ contained 60 plants, and the internode lengths were evaluated on 40-day-old seedlings. Details regarding the polymorphic markers are provided in Table 4.

Table 4. Markers used for mapping of *EI* gene.

Primer Name	Forward Primer (5′-3′)	Reverse primer (5′-3′)	Type	Enzyme
D55	AATGACTTACCTACTGGAAAGC	GATTGATCACCCTTTGGATA	InDel	
D57	GAGACATCACTTTGCCTTTC	AAAAGTCTCTCCGCCTATGT	InDel	
D64	TTGTTACCGCTTACTTTGGT	CACAGCTGTTGATTTCTTCA	InDel	
CAPS4	GCATTGCAACCTATTCTCAC	TCTGTAGTTTCCGTCTTCTT	CAPS	*HaeIII*
D66	CGTTGTCTAGGTCAATAGCC	AGGTGTTACACTTTCTACGTCT	InDel	
CAPS16	AGAGAAGGAGGATTCGGGTT	ATAGGGGCATTATCAAAAGG	CAPS	*BsrDI*
CAPS17	TAAGTTAGCCATATAAAAC	AAATGACACAGCGAGACA	CAPS	*MboII*
CAPS27	GAGAAAATTATTTGGGATAC	ATTAAAACTTTGATGCCTAC	CAPS	*MfeI*
InDel6	ACAATCCCAGTTTATGTGAT	ATATTTGGTGTTTTCTGTTT	InDel	
HP2501	CTTTTCACAAAACTAACACAGG	TGACAATATAAGCATTTGTCGC	InDel	
HP707	TCCGATGTAACATCACGCAA	GTTGATCACCTTCAGACAGC	InDel	
D67	AGCTTTTATAGCACGTACCG	CCATACTCTACTTATGCTGCAA	InDel	
HP2509	ACCTCGACACTGGTTCACTC	GTGACTCATATACACCCTTACCTA	InDel	
HP445	GAGAACATCTGTACCAGCCT	CAAGTATCTATATGCCTGACAAC	InDel	

4.6. RNA Extraction and qRT-PCR

Total RNA was extracted using an RNA pure kit (Aidlab, Beijing, China) following the user manual. First-strand cDNA was synthesized using TransScript One-Step gDNA Removal and cDNA Synthesis SuperMix (Transgene, Beijing, China). A quantitative real-time polymerase chain reaction (qRT-PCR) assay was conducted using a SYBR Green reagent (Yeasen, Shanghai, China) and the LightCycler 480 Real-Time detection system (Roche, Basel, Switzerland). Housekeeping gene *actin* (*solyc03g078400*) was used as an internal control to normalize the data. Details regarding the qRT-PCR primers are listed in Table S1 [63]. The qRT-PCR data for each sample were validated with three biological and three technical replicates. The relative expression levels were quantified according to the $2^{-\Delta\Delta Ct}$ method [64].

4.7. Protein Sequence Alignment and Phylogenetic Analysis

The sequence was retrieved from the National Center for Biotechnology Information database (NCBI). The BLASTP program [65] was used for homology searches in GenBank. A multiple protein sequence alignment was performed by the ClustalW program, and the phylogenetic tree was constructed according to the neighbor-joining method of the MEGA 6.0 program with 1000 bootstrap replicates [66].

4.8. Plasmid Construction and Transformation

The genomic clone including the whole *EI* coding region was amplified from full-length cDNA of wild-type 05T606 with primers OE-F (5′-CACGGGGACTCTAGAATGTACTTAGCCACCTCCA-3′) and OE-R (5′-GATCGGGGAAATTCGAGCTCTTAGTGAGTTGAGACAAGAAAC-3′). The amplified fragment was cloned into the *Xba*I and *Sac*I sites of the binary vector *pBI*121 by using an In-Fusion HD Cloning Kit (Takara, Dalian, China) to generate an *EI* transformation plasmid under the control of the CaMV35S promoter. The plasmid mediated by *Agrobacterium tumefaciens* strain GV3101 was transformed into the mutant P502 as described by the method of Sharma et al. [67]. After screening for regenerated shoots on the selection medium, the transgenic plants were further confirmed by PCR using NPTII-F (5′-GACAATCGGCTGCTCTGA-3′) and NPTII-R (5′-AACTCCAGCATGAGATCC-3′) primers. The positive transgenic plants were selected and the T_1 generation was obtained for phenotypic observation and gene expression analysis.

Accession numbers: SlGA2ox7 (XP_004232746), NsGA2ox8 (NP_001289506.1), AtGA2ox8 (NP_193852.2), AtGA2ox7 (NP_175509.1), VvGA2ox8 (NP_001268435.1), ZmGA2ox7 (NP_001148252.2), NtGA20ox1 (NP_001313089.1), SlGA20ox3 (NP_001234579.1), SlGA20ox2 (NP_001234628.2), SlGA20ox4 (NP_001234363.1), AtGA20ox2 (NP_199994.1), AtGA20ox4 (NP_176294.1), AtGA20ox3 (NP_196337.1), ZmGA20ox4 (NP_001308615.1) StGA3ox2 (NP_001275412.1), and AtGA3ox4 (NP_178149.1).

Author Contributions: J.L., L.L., J.S., and X.S. conceived and designed the experiments; X.S., J.S., A.M.A.M., and X.D. performed the experiments; X.S. analyzed the data and prepared the manuscript; J.S. improved the manuscript; X.Z., J.B., Y.C., and X.L. assisted in the field experiment; J.L., L.L., J.S., Y.D., X.W., Z.H., and Y.G. contributed reagents and materials; J.L. provided guidance on the whole study and revised the manuscript. All authors read and approved the final manuscript.

References

1. Salas Fernandez, G.S.; Becraft, P.W.; Yin, Y.; Lübberstedt, T. From dwarves to giants? Plant height manipulation for biomass yield. *Trends Plant Sci.* **2009**, *14*, 454–461. [CrossRef] [PubMed]

2. Itoh, H.; Tatsumi, T.; Sakamoto, T.; Otomo, K.; Toyomasu, T.; Kitano, H.; Ashikari, M.; Ichihara, S.; Matsuoka, M. A rice semi-dwarf gene, *Tan-Ginbozu (D35)*, encodes the gibberellin biosynthesis enzyme, *ent*-kaurene oxidase. *Plant Mol. Biol.* **2004**, *54*, 533–547. [CrossRef]
3. Peng, J.; Richards, D.E.; Hartley, N.M.; Murphy, G.P.; Devos, K.M.; Flintham, J.E.; Beales, J.; Fish, L.J.; Worland, A.J.; Pelica, F.; et al. Green revolution' genes encode mutant gibberellin response modulators. *Nature* **1999**, *400*, 256–261. [CrossRef]
4. Hedden, P. The genes of the Green Revolution. *Trends Genet.* **2003**, *19*, 5–9. [CrossRef]
5. Khush, G.S. Green revolution: The way forward. *Nat. Rev. Genet.* **2001**, *2*, 815–822. [CrossRef] [PubMed]
6. Md. Babul, A.; Piao, R.; Reflinur; Md. Lutfor, R.; Yunjoo, L.; Jeonghwan, S.; Backki, K.; Hee-Jong, K. Characterization and mapping of *d13*, a dwarfing mutant gene, in rice. *Genes Genom.* **2015**, *37*, 893–903.
7. Hong, Z.; Ueguchi-Tanaka, M.; Umemura, K.; Uozu, S.; Fujioka, S.; Takatsuto, S.; Yoshida, S.; Ashikari, M.; Kitano, H.; Matsuoka, M. A rice brassinosteroid-deficient mutant, *ebisu dwarf* (*d2*), is caused by a loss of function of a new member of cytochrome P450. *Plant Cell* **2003**, *15*, 2900–2910. [CrossRef]
8. Zhou, H.; He, S.; Cao, Y.; Chen, T.; Du, B.; Chu, C.; Zhang, J.; Chen, S. OsGLU1, a putative membrane-bound endo-1,4-ß-D-glucanase from rice, affects plant internode elongation. *Plant Mol. Biol.* **2006**, *60*, 137–151. [CrossRef]
9. Chen, Y.; Hou, M.; Liu, L.; Wu, S.; Shen, Y.; Ishiyama, K.; Kobayashi, M.; McCarty, D.R.; Tan, B.C. The maize *DWARF1* encodes a gibberellin 3-oxidase and is dual localized to the nucleus and cytosol. *Plant Physiol.* **2014**, *166*, 2028–2039. [CrossRef]
10. Li, X.; Zhou, Z.; Ding, J.; Wu, Y.; Zhou, B.; Wang, R.; Ma, J.; Wang, S.; Zhang, X.; Xia, Z.; et al. Combined linkage and association mapping reveals QTL and candidate genes for plant and ear height in maize. *Front Plant Sci.* **2016**, *7*, 833. [CrossRef]
11. Teng, F.; Zhai, L.; Liu, R.; Bai, W.; Wang, L.; Huo, D.; Tao, Y.; Zheng, Y.; Zhang, Z. *ZmGA3ox2*, a candidate gene for a major QTL, *qPH3.1*, for plant height in maize. *Plant J.* **2013**, *73*, 405–416. [CrossRef] [PubMed]
12. Tian, X.; Wen, W.; Xie, L.; Fu, L.; Xu, D.; Fu, C.; Wang, D.; Chen, X.; Xia, X.; Chen, Q.; et al. Molecular mapping of reduced plant height gene *Rht24* in Bread Wheat. *Front. Plant Sci.* **2017**, *8*, 1379. [CrossRef]
13. Zhai, H.; Feng, Z.; Li, J.; Liu, X.; Xiao, S.; Ni, Z.; Sun, Q. QTL analysis of spike morphological traits and plant height in Winter Wheat (*Triticum aestivum* L.) using a high-density SNP and SSR-based linkage map. *Front Plant Sci.* **2016**, *7*, 1617. [CrossRef] [PubMed]
14. Multani, D.S.; Briggs, S.P.; Chamberlin, M.A.; Blakeslee, J.J.; Murphy, A.S.; Johal, G.S. Loss of an MDR transporter in compact stalks of maize *br2* and sorghum *dw3* mutants. *Science* **2003**, *302*, 81–84. [CrossRef]
15. Yamaguchi, M.; Fujimoto, H.; Hirano, K.; Araki-Nakamura, S.; Ohmae-Shinohara, K.; Fujii, A.; Tsunashima, M.; Song, X.J.; Ito, Y.; Nagae, R.; et al. Sorghum *Dw1*, an agronomically important gene for lodging resistance, encodes a novel protein involved in cell proliferation. *Sci. Rep.* **2016**, *6*, 28366. [CrossRef] [PubMed]
16. Liu, C.; Zheng, S.; Gui, J.; Fu, C.; Yu, H.; Song, D.; Shen, J.; Qin, P.; Liu, X.; Han, B.; et al. *Shortened Basal Internodes* encodes a gibberellin 2-oxidase and contributes to lodging resistance in rice. *Mol. Plant* **2018**, *11*, 288–299. [CrossRef] [PubMed]
17. Helliwell, C.A.; Sheldon, C.C.; Olive, M.R.; Walker, A.R.; Zeevaart, J.A.; Peacock, W.J.; Dennis, E.S. Cloning of the Arabidopsis *ent*-kaurene oxidase gene GA_3. *Proc. Natl. Acad. Sci. USA* **1998**, *95*, 9019–9024. [CrossRef]
18. Sakamoto, T.; Miura, K.; Itoh, H.; Tatsumi, T.; Ueguchi-Tanaka, M.; Ishiyama, K.; Kobayashi, M.; Agrawal, G.K.; Takeda, S.; Abe, K.; et al. An overview of gibberellin metabolism enzyme genes and their related mutants in rice. *Plant Physiol.* **2004**, *135*, 1642–1653. [CrossRef] [PubMed]
19. Li, Z.; Guo, Y.; Ou, L.; Hong, H.; Wang, J.; Liu, Z.; Guo, B.; Zhang, L.; Qiu, L. Identification of the dwarf gene *GmDW1* in soybean (*Glycine max* L.) by combining mapping-by-sequencing and linkage analysis. *Theor. Appl. Genet.* **2018**, *131*, 1001–1016. [CrossRef]
20. Richards, D.E.; King, K.E.; Ait-Ali, T.; Harberd, N.P. How gibberellin regulates plant growth and development: A molecular genetic analysis of gibberellin signaling. *Annu. Rev. Plant Physiol. Plant Mol. Biol.* **2001**, *52*, 67–88. [CrossRef]
21. Hedden, P.; Thomas, S.G. Gibberellin biosynthesis and its regulation. *Biochem. J.* **2012**, *444*, 11–25. [CrossRef]

22. Sasaki, A.; Ashikari, M.; Ueguchi-Tanaka, M.; Itoh, H.; Nishimura, A.; Swapan, D.; Ishiyama, K.; Saito, T.; Kobayashi, M.; Khush, G.S.; et al. Green revolution: A mutant gibberellin-synthesis gene in rice. *Nature* **2002**, *416*, 701–702. [CrossRef] [PubMed]
23. Spielmeyer, W.; Ellism, M.H.; Chandler, P.M. *Semidwarf* (*sd-1*), 'green revolution' rice, contains a defective gibberellin 20-oxidase gene. *Proc. Natl. Acad. Sci. USA* **2002**, *99*, 9043–9048. [CrossRef] [PubMed]
24. Koorneef, M.; Elgersma, A.; Hanhart, C.J.; Van Loenen Martinet, E.P.; Van Rijn, L.; Zeevaart, J.A. A gibberellin insensitive mutant of *Arabidopsis thaliana*. *Physiol Plant.* **2010**, *65*, 33–39. [CrossRef]
25. Peng, J.; Carol, P.; Richards, D.E.; King, K.E.; Cowling, R.J.; Murphy, G.P.; Harberd, N.P. The Arabidopsis *GAI* gene defines a signaling pathway that negatively regulates gibberellin responses. *Gene Dev.* **1997**, *11*, 3194–3205. [CrossRef] [PubMed]
26. Harberd, N.P.; Freeling, M. Genetics of dominant gibberellin-insensitive dwarfism in maize. *Genetics* **1989**, *121*, 827–838.
27. Winkler, R.G.; Freeling, M. Physiological genetics of the dominant gibberellin-nonresponsive maize dwarfs, *Dwarf 8* and *Dwarf 9*. *Planta* **1994**, *193*, 341–348. [CrossRef]
28. Ogawa, M.; Kusano, T.; Katsumi, M.; Sano, H. Rice gibberellin-insensitive gene homolog, *OsGAI*, encodes a nuclear-localized protein capable of gene activation at transcriptional level. *Gene* **2000**, *245*, 21–29. [CrossRef]
29. Fu, X.D.; Richards, D.E.; Ait-Ali, T.; Hynes, L.W.; Ougham, H.; Peng, J.; Harberd, N.P. Gibberellin-mediated proteasome-dependent degradation of the barley DELLA protein SLN1 repressor. *Plant Cell* **2002**, *14*, 3191–3200. [CrossRef]
30. Ikeda, A.; Ueguchi-Tanaka, M.; Sonoda, Y.; Kitano, H.; Koshioka, M.; Futsuhara, Y.; Matsuoka, M.; Yamaguchi, J. *Slender* rice, a constitutive gibberellin response mutant, is caused by a null mutation of the *SLR1* gene, an ortholog of the height-regulating gene *GAI/RGA/RHT/D8*. *Plant Cell* **2001**, *13*, 999–1010. [CrossRef]
31. Carrera, E.; Ruiz-Rivero, O.; Peres, L.E.; Atares, A.; Garcia-Martinez, J.L. Characterization of the *procera* tomato mutant shows novel functions of the SlDELLA protein in the control of flower morphology, cell division and expansion, and the auxin-signaling pathway during fruit-set and development. *Plant Physiol.* **2012**, *160*, 1581–1596. [CrossRef]
32. Yamamuro, C.; Ihara, Y.; Wu, X.; Noguchi, T.; Fujioka, S.; Takatsuto, S.; Ashikari, M.; Kitano, H.; Matsuoka, M. Loss of function of a rice *brassinosteroid insensitive1* homolog prevents internode elongation and bending of the lamina joint. *Plant Cell* **2000**, *12*, 1591–1606. [CrossRef]
33. Lin, H.; Wang, R.; Qian, Q.; Yan, M.; Meng, X.; Fu, Z.; Yan, C.; Jiang, B.; Su, Z.; Li, J.; et al. DWARF27, an iron-containing protein required for the biosynthesis of strigolactones, regulates rice tiller bud outgrowth. *Plant Cell* **2009**, *21*, 1512–1525. [CrossRef]
34. Jiang, L.; Liu, X.; Xiong, G.; Liu, H.; Chen, F.; Wang, L.; Meng, X.; Liu, G.; Yu, H.; Yuan, Y.; et al. *DWARF 53* acts as a repressor of strigolactone signalling in rice. *Nature* **2013**, *504*, 401–405. [CrossRef]
35. Foolad, M.R. Genome mapping and molecular breeding of tomato. *Int. J. Plant Genom.* **2007**, *2007*, 64358. [CrossRef] [PubMed]
36. Martí, E.; Gisbert, C.; Bishop, G.J.; Dixon, M.S.; Garcíamartínez, J.L. Genetic and physiological characterization of tomato cv. Micro-Tom. *J. Exp. Bot.* **2006**, *57*, 2037–2047. [CrossRef]
37. Barabaschi, D.; Tondelli, A.; Desiderio, F.; Volante, A.; Vaccino, P.; Valè, G.; Cattivelli, L. Next generation breeding. *Plant Sci.* **2016**, *242*, 3–13. [CrossRef]
38. Tong, G.L.; Samuel, F.H. Fine mapping of the *brachytic* locus on the tomato genome. *J. Amer. Soc. Hort. Sci.* **2018**, *143*, 239–247.
39. Cowling, R.J.; Kamiya, Y.; Seto, H.; Harberd, N.P. Gibberellin dose-response regulation of GA_4 gene transcript levels in Arabidopsis. *Plant Physiol.* **1998**, *117*, 1195–1203. [CrossRef] [PubMed]
40. Mueller, L.A.; Solow, T.H.; Taylor, N.; Skwarecki, B.; Buels, R.; Binns, J.; Lin, C.; Wright, M.H.; Ahrens, R.; Wang, Y.; Herbst, E.V.; Keyder, E.R.; Menda, N.; Zamir, D.; Tanksley, S.D. The SOL Genomics Network: A comparative resource for Solanaceae biology and beyond. *Plant Physiol.* **2005**, *138*, 1310–1317. [CrossRef]
41. Olszewski, N.; Sun, T.P.; Gubler, F. Gibberellin signaling: Biosynthesis, catabolism, and response pathways. *Plant Cell* **2002**, *14*, S61–S80. [CrossRef]
42. Mitsunaga, S.; Tashiro, T.; Yamaguchi, J. Identification and characterization of gibberellin-insensitive mutants selected from among dwarf mutants of rice. *Theor. Appl. Genet.* **1994**, *87*, 705–712. [PubMed]
43. Hedden, P.; Phillips, A.L. Gibberellin metabolism: New insights revealed by the genes. *Trends. Plant Sci.* **2000**, *5*, 523–530. [CrossRef]

44. Gomi, K.; Matsuoka, M. Gibberellin signalling pathway. *Curr. Opin. Plant Biol.* **2003**, *6*, 489–493. [CrossRef]
45. Schomburg, F.M.; Bizzell, C.M.; Lee, D.J.; Zeevaart, J.A.; Amasino, R.M. Overexpression of a novel class of gibberellin 2-oxidases decreases gibberellin levels and creates dwarf plants. *Plant Cell* **2003**, *15*, 151–163. [CrossRef]
46. Lo, S.F.; Yang, S.Y.; Chen, K.T.; Hsing, Y.I.; Zeevaart, J.A.; Chen, L.J.; Yu, S.M. A novel class of gibberellin 2-oxidases control semidwarfism, tillering, and root development in rice. *Plant Cell* **2008**, *20*, 2603–2618. [CrossRef] [PubMed]
47. Lee, D.J.; Zeevaart, J.A. Differential Regulation of RNA Levels of Gibberellin Dioxygenases by Photoperiod in Spinach. *Plant Physiol.* **2002**, *130*, 2085–2094. [CrossRef] [PubMed]
48. Martin, D.N.; Proebsting, W.M.; Hedden, P. The *SLENDER* gene of pea encodes a gibberellin 2-oxidase. *Plant Physiol.* **1999**, *121*, 775–781. [CrossRef] [PubMed]
49. Huang, J.; Tang, D.; Shen, Y.; Qin, B.; Hong, L.; You, A.; Li, M.; Wang, X.; Yu, H.; Gu, M.; et al. Activation of gibberellin 2-oxidase 6 decreases active gibberellin levels and creates a dominant semi-dwarf phenotype in rice (*Oryza sativa* L.). *J. Genet. Genom.* **2010**, *37*, 23–36. [CrossRef]
50. Lee, D.J.; Zeevaart, J.A. Molecular Cloning of GA 2-Oxidase3 from Spinach and Its Ectopic Expression in Nicotiana sylvestris. *Plant Physiol.* **2005**, *138*, 243–254. [CrossRef]
51. Sun, T.P. Gibberellin metabolism, perception and signaling pathways in Arabidopsis. *Arabidopsis Book* **2008**, *6*, e0103. [CrossRef]
52. Pnueli, L.; Carmel-Goren, L.; Hareven, D.; Gutfinger, T.; Alvarez, J.; Ganal, M.; Zamir, D.; Lifschitz, E. The *SELF-PRUNING* gene of tomato regulates vegetative to reproductive switching of sympodial meristems and is the ortholog of CEN and TFL1. *Development* **1998**, *125*, 1979–1989.
53. Sun, X.R.; Liu, L.; Zhi, X.N.; Bai, J.R.; Cui, Y.N.; Shu, J.S.; Li, J.M. Genetic analysis of tomato internode length via mixed major gene plus polygene inheritance model. *Sci. Hortic.* **2019**, *246*, 759–764. [CrossRef]
54. Collins, T.J. ImageJ for microscopy. *BioTechniques* **2007**, *43*, 25–30. [CrossRef] [PubMed]
55. García-Hurtado, N.; Carrera, E.; Ruiz-Rivero, O.; López-Gresa, M.P.; Hedden, P.; Gong, F.; García-Martínez, J.L. The characterization of transgenic tomato overexpressing gibberellin 20-oxidase reveals induction of parthenocarpic fruit growth, higher yield, and alteration of the gibberellin biosynthetic pathway. *J. Exp. Bot.* **2012**, *63*, 5803–5813. [CrossRef]
56. Dong, C.F.; Gu, H.R.; Ding, C.L.; Xu, N.X.; Zhang, W.J. Effects of gibberellic acid on forage quality of rice (*Oryza sativa*) straw. *Acta Pratacult. Sin.* **2016**, *25*, 94–102.
57. Saghaimaroof, M.A.; Soliman, K.M.; Jorgensen, R.A.; Allard, R.W. Ribosomal DNA spacer-length polymorphisms in barley: Mendelian inheritance, chromosomal location, and population dynamics. *Proc. Natl. Acad. Sci. USA* **1984**, *81*, 8014–8018. [CrossRef]
58. Lalitha, S. Primer Premier 5. *Biotech Softw. Internet Report.* **2000**, *1*, 270–272. [CrossRef]
59. Neff, M.M.; Turk, E.; Kalishman, M. Web-based primer design for single nucleotide polymorphism analysis. *Trends Genet.* **2002**, *18*, 613–615. [CrossRef]
60. Michelmore, R.W.; Paran, I.; Kesseli, R.V. Identification of markers linked to disease-resistance genes by bulked segregant analysis: A rapid method to detect markers in specific genomic regions by using segregating populations. *Proc. Natl. Acad. Sci. USA* **1991**, *88*, 9828–9832. [CrossRef]
61. Van Ooijen, J.W. *JoinMap® 4.0. Software for the Calculation of Genetic Linkage Maps in Experimental Populations*; Kyazma: Wageningen, The Netherlands, 2006.
62. Van Ooijen, J.W. *MapQTL® 6.0, Software for the Mapping of Quantitative trait Loci in Experimental Populations*; Kyazma: Wageningen, The Netherlands, 2009.
63. Li, J.; Sima, W.; Ouyang, B.; Wang, T.; Ziaf, K.; Luo, Z.; Liu, L.; Li, H.; Chen, M.; Huang, Y.; et al. Tomato *SlDREB* gene restricts leaf expansion and internode elongation by downregulating key genes for gibberellin biosynthesis. *J. Exp. Bot.* **2012**, *63*, 6407–6420. [CrossRef]
64. Livak, K.J.; Schmittgen, T.D. Analysis of relative gene expression data using real-time quantitative PCR and the $2^{-\Delta\Delta CT}$ method. *Methods* **2001**, *25*, 402–408. [CrossRef] [PubMed]
65. Altschul, S.; Madden, T.; Schäffer, A. Gapped BLAST and PSI-BLAST: A new generation of protein database search programs. *Nucleic Acids Res.* **1997**, *25*, 3389–3402. [CrossRef]
66. Tamura, K.; Stecher, G.; Peterson, D.; Filipski, A.; Kumar, S. MEGA6: Molecular evolutionary genetics analysis version 6.0. *Mol. Biol. Evol.* **2013**, *30*, 2725–2729. [CrossRef]
67. Sharma, M.K.; Solanke, A.U.; Jani, D.; Singh, Y.; Sharma, A.K. A simple and efficient *Agrobacterium*-mediated procedure for transformation of tomato. *J. Biosci.* **2009**, *34*, 423. [CrossRef] [PubMed]

9

Enhanced Expression of QTL *qLL9/DEP1* Facilitates the Improvement of Leaf Morphology and Grain Yield in Rice

Xue Fu †, Jing Xu †, Mengyu Zhou, Minmin Chen, Lan Shen, Ting Li, Yuchen Zhu, Jiajia Wang, Jiang Hu, Li Zhu, Zhenyu Gao, Guojun Dong, Longbiao Guo, Deyong Ren, Guang Chen, Jianrong Lin, Qian Qian * and Guangheng Zhang *

State Key Laboratory of Rice Biology, China National Rice Research Institute, Hangzhou 310006, China; fuxuezsy@163.com (X.F.); xujing87@126.com (J.X.); zhoumy@mail.sustc.edu.cn (M.Z.); jiashf@126.com (M.C.); baishushenlan@126.com (L.S.); 13258376009@163.com (T.L.); zyc1205926704@sina.com (Y.Z.); wgwangjiajia@163.com (J.W.); hujiang588@163.com (J.H.); zhuli05@caas.cn (L.Z.); zygao2000@hotmail.com (Z.G.); dongguojun@caas.cn (G.D.); guolongbiao@caas.cn (L.G.); rendeyongsd@163.com (D.R.); chenguang0066@126.com (G.C.); ljr1970@hotmail.com (J.L.)

* Correspondence: qianqian188@hotmail.com (Q.Q.); zhangguangheng@126.com (G.Z.)

† These authors contributed equally to this work.

Abstract: In molecular breeding of super rice, it is essential to isolate the best quantitative trait loci (QTLs) and genes of leaf shape and explore yield potential using large germplasm collections and genetic populations. In this study, a recombinant inbred line (RIL) population was used, which was derived from a cross between the following parental lines: hybrid rice Chunyou84, that is, *japonica* maintainer line Chunjiang16B (CJ16); and *indica* restorer line Chunhui 84 (C84) with remarkable leaf morphological differences. QTLs mapping of leaf shape traits was analyzed at the heading stage under different environmental conditions in Hainan (HN) and Hangzhou (HZ). A major QTL *qLL9* for leaf length was detected and its function was studied using a population derived from a single residual heterozygote (RH), which was identified in the original population. *qLL9* was delimitated to a 16.17 kb region flanked by molecular markers C-1640 and C-1642, which contained three open reading frames (ORFs). We found that the candidate gene for *qLL9* is allelic to *DEP1* using quantitative real-time polymerase chain reaction (qRT-PCR), sequence comparison, and the clustered regularly interspaced short palindromic repeat-associated Cas9 nuclease (CRISPR/Cas9) genome editing techniques. To identify the effect of *qLL9* on yield, leaf shape and grain traits were measured in near isogenic lines (NILs) NIL-*qLL9*^{CJ16} and NIL-*qLL9*^{C84}, as well as a chromosome segment substitution line (CSSL) CSSL-*qLL9*^{KASA} with a Kasalath introgressed segment covering *qLL9* in the Wuyunjing (WYJ) 7 backgrounds. Our results showed that the flag leaf lengths of NIL-*qLL9*^{C84} and CSSL-*qLL9*^{KASA} were significantly different from those of NIL-*qLL9*^{CJ16} and WYJ 7, respectively. Compared with NIL-*qLL9*^{CJ16}, the spike length, grain size, and thousand-grain weight of NIL-*qLL9*^{C84} were significantly higher, resulting in a significant increase in yield of 15.08%. Exploring and pyramiding beneficial genes resembling *qLL9*^{C84} for super rice breeding could increase both the source (e.g., leaf length and leaf area) and the sink (e.g., yield traits). This study provides a foundation for future investigation of the molecular mechanisms underlying the source–sink balance and high-yield potential of rice, benefiting high-yield molecular design breeding for global food security.

Keywords: *Oryza sativa* L.; leaf shape; yield trait; molecular breeding; hybrid rice

1. Introduction

Rice leaf morphogenesis and its spatial extension posture are important components of ideal plant architecture, which play a significant role in the photosynthetic efficiency and grain yield [1,2]. During the grain-filling stage, the top three leaves, particularly the flag leaf, are the main carbohydrate sources that were transported to panicles and grains for yield formation. The establishment of better source-to-sink biomass allocation would greatly contribute to the improvement of rice yield potential [3,4]. Therefore, using molecular genetic techniques to modulate the top three leaf morphology and improve the photosynthesis rate so as to balance the relationship with the grain sink will effectively achieve a high yield in rice.

The polarity development of leaves along the adaxial–abaxial, the medial–lateral, and the apical–basal axis determines the construction of the three-dimensional spatial morphology of the leaf. About 40 genes related to leaf morphogenesis have been cloned, and studies have mainly focused on leaf width, length, and rolling. Rice leaf width is mainly related to the number of veins and the distance between veins, which are regulated by the following aspects: microRNA shear-related genes, such as *OsDCL1* [5] and *GIF1* [6]; cell division-related genes, such as *SRL2* [7], *SLL1/RL9* [8,9], *OsCCC1* [10], and *OsWOX4* [11]; *NAL2 /NAL3* [12], *SLL1* [8], *OsCD1* [13], and genes related to coding transcription factors and cellulases; *NAL1/LSCHL4* [14,15], *NAL7* [16], *TDD1* [17], *OsCOW1* [18], *OsARF19* [19], and genes associated with auxin synthesis and metabolism; and genes including *NAL9* encoding an ATP-dependent Clp protease proteolytic subunit [20]. In a series of cloned rolling genes in rice, the type I genes, such as *ADL1* [21], *OsAGO7* [22], *OsAGO1a* [23], *SLL1* [8], and *RL9* [9], regulate the unbalanced development of different tissues in the adaxial/abaxial side, which affects the curl degree of blades. The type II genes are associated with the development of bulliform cells in the adaxial side, and changes in the size or amount affect the curl degree of the blade. Adaxially and abaxially rolled leaves appear as favored by various genes, such as *ACL1* [24], *LC2/OsVIL3* [25,26], *OsCOW1/NAL7* [16,18], *OsCD1/NRL1/ND1/sle1/DNL1* [13,14,27–29], *OsHox32* [30], *OsMYB103L* [31], *OsZHD1* [32], *REL1* [33], *REL2* [34], *RL14* [35], *ROC5* [36], *CLD1* [37], *SRL1* [38] *SFL1* [39], *SLL2* [40], *YABBY1* [41], and *LRRK1* [42]. The type III genes, such as *SLL1* [8], *RL9* [9], *SRL2* [7], *AVB* [43], and *OsSND2* [44], control the development of sclerenchyma in the abaxial side and affect the curl degree of blades. The genes of type IV, such as *CFL1* [45], include those with an abnormal cuticle development, leading to leaf curl. The genes of type V, such as *CVD1*, regulate commissural veins (CVs), and the lack of CV in *cvd1* mutant is the main cause of leaf curl [46].

The length, width, and area are the three traits determining the shape and size of a leaf, which are quantitatively inherited. Using a DH (doubled haploid) population, Li et al. detected two major QTLs for the flag leaf length, which were located on chromosome 4 and chromosome 8, respectively [47]. Yan et al. studied the genotype–environment interaction of eight plant morphological traits using a DH population and mapped seven QTLs on chromosomes 1, 2, 3, 4, 6, 9, and 10 related to the length of the flag leaf [48]. Farooq et al. identified three leaf length QTLs on chromosomes 1, 2, and 4 using IR64 derived introgression lines [49]. Although the important role of leaf traits in plant ideotype in rice has attracted great attention, the cloning of QTLs for leaf length is rarely documented.

The coordinated balance of source and sink is an essential component to ensure a high yield in rice. Notably, genetic populations used for leaf traits were generally among those used for yield traits, where QTLs for leaf traits were frequently located in regions in which QTLs for yield traits were detected [50–58]. The major leaf width QTL *qFLW4/LSCHL4/SPIKE* allelic to *NAL1* is related to both leaf morphology and yield traits [15,58,59]. Similarly, the pleiotropic effect on leaf morphology regulation was also found in the cloning of rice grain-shaped QTL. Large-grain alleles in the *GS2* locus simultaneously increase leaf length [60,61]. However, in rice high-yield breeding, studies have yet to be conducted on the influence of the leaf shape alleles/QTLs from different donors in the interaction of molecular level between the regulation of leaf morphogenesis and yield formation. As reported, up to 50% of the variation in panicle weight depended on the variation in leaf size [50]. The co-location of QTLs/genes for source–sink traits in rice could increase the source while expanding the sink, provided that the QTLs have the same effect direction for both traits, which will be invaluable genetic resources for breeding high-yield varieties [62].

In this study, QTL mapping for leaf length and width of the top three leaves in rice was performed at two different environments using a recombinant inbred lines (RILs) set derived from the cross between maintainer line and restorer line of the *indica–japonica* super hybrid rice Chunyou 84, followed by validation and fine mapping. The target major QTL *qLL9* controlling leaf length was delimitated into a 16.17 kb interval between markers C-1640 and C-1642 on chromosome 9, using a residual heterozygote identified from the RIL population, which were segregated at *qLL9* with high homogenous backgrounds. Then, gene cloning, functional analysis, and breeding potential assessments of *qLL9* were conducted.

2. Results

2.1. Analysis of Leaf Morphology in the RIL Population and Their Parental Cultivars

The morphology of the top three leaves of RIL parents Chunjiang 16B (CJ16) and Chunhui 84 (C84) was investigated and significant differences were found for leaf length, width, and area (Figure 1a–c). Compared with CJ16, the top three leaves length of C84 were 42.2%, 40.2%, and 46.6% longer, respectively (Figure 1d). Similarly, the leaf width of C84 was significantly wider than that of CJ16 (Figure 1e). Thus, a much larger leaf area was found in C84, that is, 2.7 times the flag leaf and 2.2 times both the second and third leaves (Figure 1f). In RIL population, the leaf traits of the top three leaves were all continuously distributed with large variations and transgressive segregation, showing a typical pattern of quantitative variation at both Hangzhou and Hainan experimental sites (Figure 2), which were suitable for QTL mapping. Furthermore, we observed that the leaf traits of RILs in Hangzhou generally tended to higher values, while those in Hainan tended to lower ones.

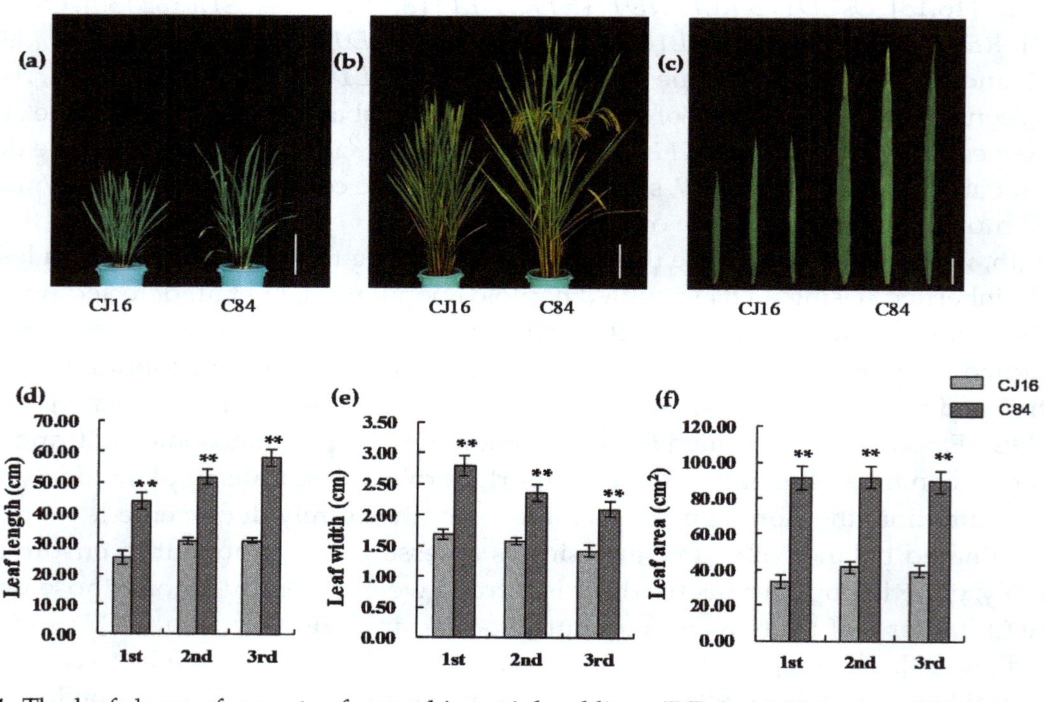

Figure 1. The leaf shape of parents of recombinant inbred lines (RILs). (**a**) Plant morphology at tillering stage; bar = 18 cm. (**b**) Plant morphology at heading stage; bar = 18 cm. (**c**) The top three leaves' shape of CJ16 and C84. From left to right are the first leaf, the second leaf, and the third leaf, respectively; bar = 5 cm. (**d**) Comparison of the top three leaves' length between CJ16 and C84. (**e**) Comparison of the top three leaves' width between CJ16 and C84. (**f**) Comparison of the top three leaves' area between CJ16 and C84. Data are represented as mean ± SD (n = 11). Asterisks represent significant difference determined by Student's t-test at p-value <0.01 (**), p-value <0.05 (*).

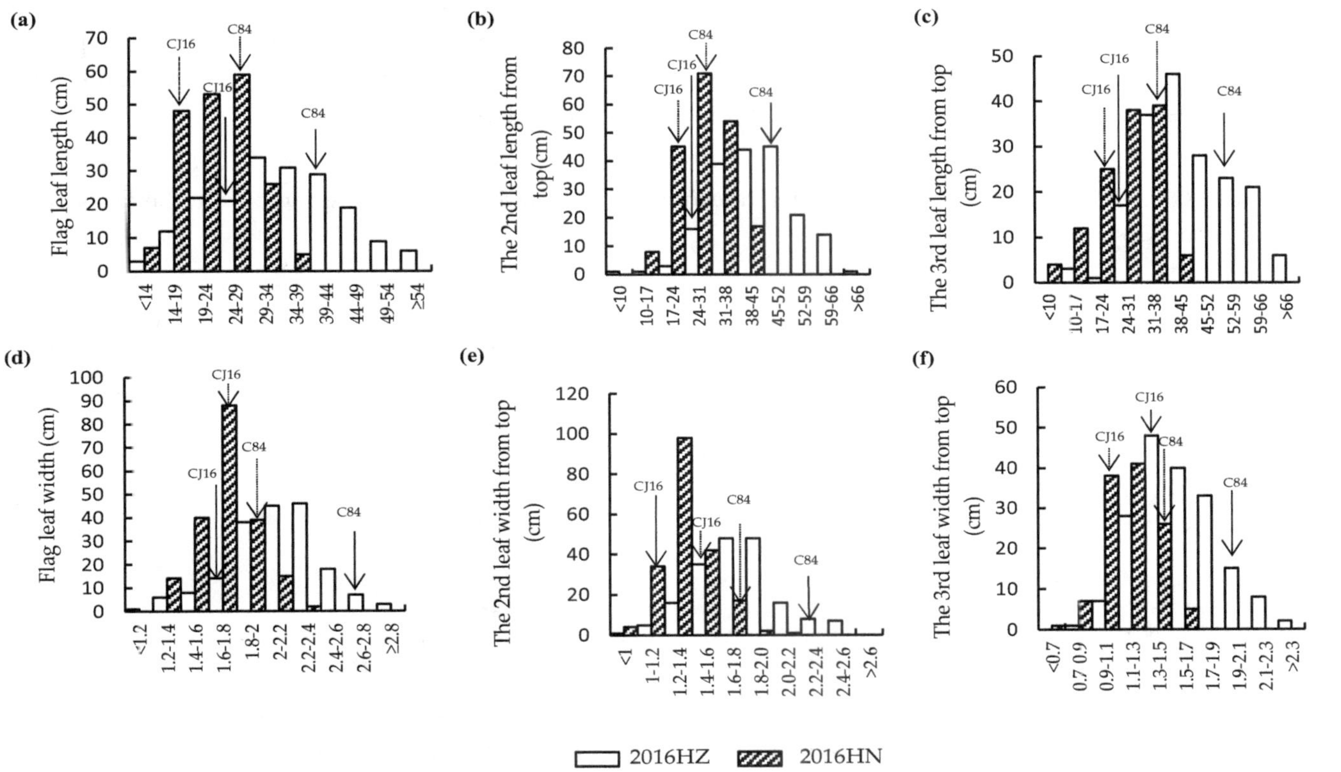

Figure 2. Frequency distributions of leaf traits in CJ16/C84 RILs. (**a**) Flag leaf length; (**b**) the second leaf length from top; (**c**) the third leaf length from top; (**d**) flag leaf width; (**e**) the second leaf width from top; (**f**) the third leaf width from top. HZ: Hangzhou; HN: Hainan.

2.2. Correlation Analysis and QTL Mapping

The correlation analysis for leaf traits in the RIL population showed a low correlation between Hangzhou and Hainan for the length of the second and the third leaf. However, the flag leaf length and width, and the second and third leaf width, were significantly positively correlated between the two environments. This could be the result of the *indica–japonica* subspecies differentiation of the parents and the distinct temperature and light conditions of Hangzhou and Hainan (Table 1).

QTL mapping was performed for the leaf length and width of the top three leaves (Table 2 and Figure 3). The results showed that a total of 27 QTLs were detected in the two environments, which were distributed on chromosomes 1, 2, 3, 5, 6, 9, 10, and 11. In Hangzhou, nine QTLs were detected, including one QTL for the flag leaf length, three QTLs for the second leaf length, and five QTLs for the third leaf length, which explained phenotypic variance in the range of 7.76%–32.41 %. In Hainan, 18 QTLs were detected, including two QTLs for the flag length, two QTLs for the flag leaf width, five QTLs for the second leaf length, four QTLs for the second leaf width, two QTLs for the third leaf length, and three QTLs for the third leaf width, which explained phenotypic variance in the range of 1.85%–30.63%. Among them, three leaf length QTLs, namely, the flag leaf length QTL *qFLL9*, the second leaf length QTL *qSLL9*, and the third leaf length QTL *qTLL9*, were simultaneously mapped in RM3700-B9-11 interval on chromosome 9 across the two environments, with the enhancing alleles all from C84, explaining phenotypic variance ranging from 19.19% to 32.41%, which agreed with the significantly positive correlation of flag leaf length between Hangzhou and Hainan (Table 1). Meanwhile, even though no significant correlation was found for the leaf length of the second and third leaf, consistent QTLs across both locations were also detected, that is, *qSLL6* and *qTLL2*, which could be because of their larger genetic effects and/or less sensitivity to environmental variation. In addition, we noted QTLs for the width of the second and third leaf were also detected in the interval between RM3700 and B9-11, but with the increasing alleles coming from CJ16. These results prompt us to mainly focus on the major region flanked by RM3700 and B9-11 on chromosomes 9, which showed stable effects on leaf morphological development, and named *qLL9*.

Table 1. Correlation analysis on leaf traits in recombinant inbred lines (RILs) derived from the cross of CJ16 and C84.

	HZ						HN				
	FLL	FLW	SLL	SLW	TLL	TLW	FLL	FLW	SLL	SLW	TLL
HZ-FLL											
HZ-FLW	−0.078										
HZ-SLL	0.833 **	−0.013									
HZ-SLW	−0.178 *	0.627 **	−0.072								
HZ-TLL	0.525 **	0.000	0.746 **	−0.023							
HZ-TLW	−0.203 **	0.597 **	−0.073	0.699 **	0.086						
HN-FLL	0.341 **	−0.317 **	0.271 **	−0.202 *	0.336 **	−0.252 **					
HN-FLW	−0.250 **	0.389 **	−0.298 **	0.382 **	−0.287 **	0.246 **	0.076				
HN-SLL	0.225 **	−0.268 **	0.135	−0.137	0.195 *	−0.226 **	0.873 **	0.229 **			
HN-SLW	−0.418 **	0.390 **	−0.510 **	0.441 **	−0.395 **	0.361 **	−0.021	0.754 **	0.160		
HN-TLL	0.000	−0.182 *	-0.098	−0.087	0.007	−0.132	0.642 **	0.254 **	0.813 **	0.313 **	
HN-TLW	−0.405 **	0.246 **	−0.477 **	0.337 **	−0.355 **	0.300 **	0.088	0.624 **	0.206 *	0.760 **	0.358 **

FLL: flag leaf length; FLW: flag leaf width; SLL: the second leaf length; SLW: the second leaf width; TLL: the third leaf length; TLW: the third leaf width. Data are represented as mean ± SD (n = 3). Asterisks represent significant difference determined by Student's t-test at p-value < 0.01 (**), p-value < 0.05 (*).

Table 2. Locations of quantitative trait loci (QTLs) for leaf traits in the RIL population.

Trait	QTL	Interval		Peak Position		Additive Effect		Explained Phenotypic Variance (%)	
		Marker 1	Marker 2	HZ	HN	HZ	HN	HZ	HN
FLL	*qFLL9*	RM3700	B9-11	61.1	56.87	−7.1707	−6.8682	32.41	30.63
	qFLL10	RM1375	H10-3		20.09		3.3926		5.09
FLW	*qFLW3*	H3-11	H3-12		152.3		0.0648		6.6
	qFLW6	RM3496	H6-3		27.14		−0.0839		16.07
SLL	*qSLL2-1*	2-30.2-a	2-42.1-a	36.2		−2.8643		8.98	
	qSLL2-2	H2-1	2-30.2-a		23.9		−1.3914		5.29
	qSLL6	wgw1	RM3183	59.4	60.1	3.5234	−4.2842	14.97	17.29
	qSLL9	RM3700	B9-11	60.1	57.6	−6.4659	−6.5431	29.62	20.85
	qSLL9-2	H9-3	H9-4		12.18		−0.7557		8.8
	qSLL10	RM1375	H10-3		27.11		3.9058		5.91
SLW	*qSLW1*	RM1247	H1-1		24.2		−0.0759		10.69
	qSLW6	wgw1	RM3183		59.35		−0.0727		10.31
	qSLW9	RM3700	B9-11		59.2		0.0839		18.39
	qSLW11	H11-9	H11-10		71.4		−0.0329		1.85
TLL	*qTLL2*	H2-1	2-30.2-a	13.0	19.44	−3.6555	−0.9935	13.10	8.99
	qTLL5	RM430	RM18751	70.3		−2.5759		8.43	
	qTLL6	wgw1	RM3183	35.2		−4.8389		15.65	
	qTLL9	RM3700	B9-11	59.1	61.2	−6.0826	−4.8326	21.87	19.19
	qTLL10	RM5689	RM1375	28.6		2.3523		7.76	
	qTLW1	H1-1	RM8111		22.41		−0.0669		10.79
TLW	*qTLW6*	wgw1	RM3183		59.35		−0.0684		13.84
	qTLW9	RM3700	B9-11		61.31		0.0834		16.28

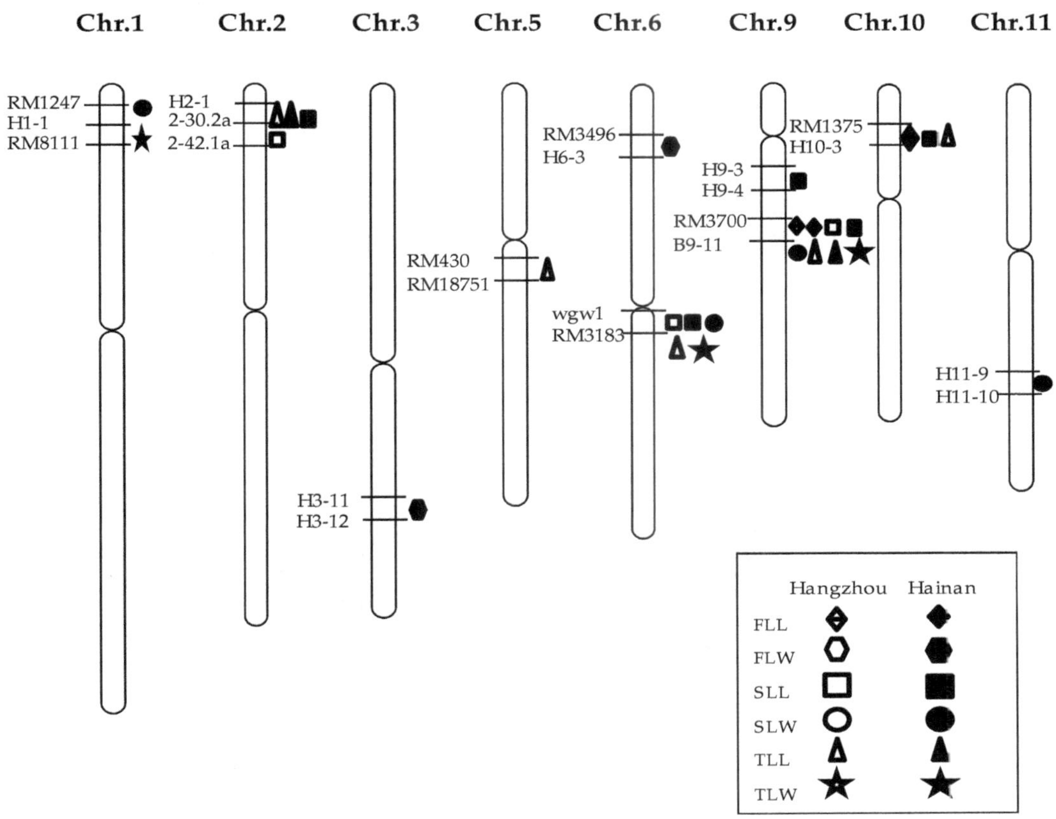

Figure 3. Locations of quantitative trait loci (QTLs) for leaf traits in the genetic map. FLL: flag leaf length; FLW: flag leaf width; SLL: the second leaf length; SLW: the second leaf width; TLL: the third leaf length; TLW: the third leaf width.

2.3. *Fine Mapping and Leaf Shape Characterization of qLL9*

According to the flanking markers RM3700 and B9-11 on chromosome 9, one residual heterozygote with a heterozygous segment covering the interval was identified from the RILs, from which a large population was derived. Nine representative near isogenic lines (NILs) with introgressions covering different portions of the target region were identified using DNA markers in the target interval for QTL fine mapping (Table 3; Figure 4a). Then, combined with their flag leaf length, *qLL9* was delimitated into a 16.17 kb region between markers C9-1640 and C9-1642 (Figure 4a).

To further clarify the effect of *qLL9* on leaf morphology, we compared the leaf traits in a near-isogenic line set of NIL-$qLL9^{CJ16}$ and NIL-$qLL9^{C84}$, and a chromosome segment substitution line (CSSL)-$qLL9^{KASA}$ and its recurrent parent Wuyunjing (WYJ) 7. In both groups, significant differences for the flag leaf were only found for the length, but not the width (Figure 5), while both the leaf length and width variation were significant for the second and third leaf from the top (Figure S1), which were highly consistent with the findings in our QTL primary mapping (Table 2). Obviously, compared with NIL-$qLL9^{CJ16}$, the longer flag leaf length of NIL-$qLL9^{C84}$ resulted in a larger flag leaf area (Figure 5d–f). However, the opposite allelic effects on leaf length and width of the second and third leaf from the top even made the leaf area variation not significant (Supplementary Figure 1). Similarly, after introgression of the Kasalath allele at *qLL9* in WYJ 7, CSSL-$qLL9^{KASA}$ resulted in a 1.5 times longer flag leaf from 29.4–43.5 cm^2 (Figure 5g–i).

On the other hand, it is noted that the flag leaf epidermal cells showed no significant difference in either the size or number in unit area between NIL-$qLL9^{CJ16}$ and NIL-$qLL9^{C84}$, which suggested the variation of flag leaf length could be mainly attributed to the increase of the total cell number (Figure 6). In summary, *qLL9* influenced the leaf length and area variation by regulating the development of leaf cells.

Table 3. Primers for QTL fine mapping, quantitative real-time polymerase chain reaction (qRT-PCR), vector construction, and gene editing.

Primer	Forward (5′-3′)	Reverse (5′-3′)	Experiment
RM3700	AAATGCCCCATGCACAAC	TTGTCAGATTGTCACCAGGG	
C9-1594	CCTGTACACTGTAGGCCTGT	GGTGTCAAAGTACATAGGCCC	
C9-1635	GGTGGAAAGGAAGGAGAGCT	CTAGCCCTGCCTCGTTGTAA	
C9-1638	GTGTGTGTGTGTGTGTGTGT	TCATAGTACATGCCCTCCGT	
C9-1640	ATAAGTCCATATTGCCCACCTC	AAGCTTCTGGATCGTTAACAGG	Fine mapping
C9-1642	GTACCCTCCTCCGATGACAC	TTGTGGAGGACGAGAAGGTG	
C9-1715	GGTGGCGAGAAGAATTTGCA	TTTCGCCTCTCACTGACCTT	
B9-11	TCTTACGAATAGGCCCTTGG	AGAGCCCACAACACTTGTGC	
Actin	ATCCATCTTGGCATCTCTCAGC	CACAATGGATGGGCCAGACT	
LOC_Os09g26960	CTGAGCCTCGCCAATCTG	CGAAGATCTCCTCCATGCTC	
LOC_Os09g26970	CAAACATCTGGGCTTGGTCT	TCTAAGCAACCTGCCCAATC	qRT-PCR
LOC_Os09g26980	ATTGATGTGAAAGGGCAAGACT	CACCTTAAGCCCAAGGTTGTAG	
LOC_Os09g26999	GTAGCTGCAAGCCAAGCTG	TTGAAGCAGCTGGAGCAAC	
POs26999	GGCCAGTGCCAAGCTTAAGGGAAGTTGGCCGCCTGCC	AGGGTCTTGCAGATCTCTCCACACGCAGCACGCCAAC	Vector construction
Os09g26999-g++/g–	GGCAGGTGGTGATGGAGGCGCCG	AAACCGGCGCCTCCATCACCACC	
Os09g26999-JC	CGGCGATTTATACCCACCAC	CGCTCACCTTGAGGAACGT	
Hyg-F1	GCTGTTATGCGGCCATTGTC	GACGTCTGTCGAGAAGTTTC	Detection of target mutations
Cas9-F2/pC1300-R2	ACCAGACACGAGACGACTAA	ATCGGTGCGGGCCTCTTC	
T3	ATCGGTGCGGGCCTCTTC		

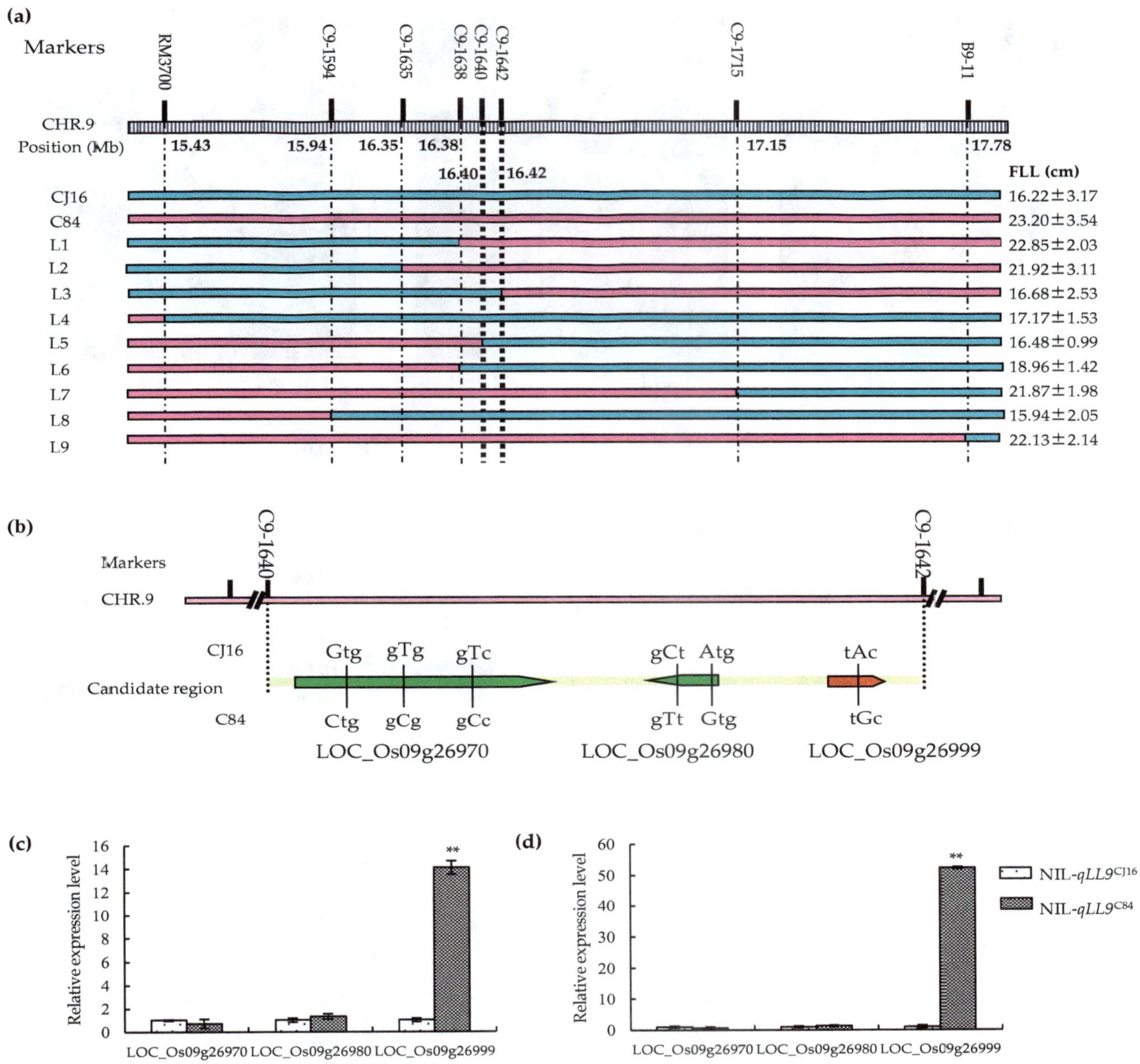

Figure 4. Map-based cloning of *qLL9* and expression analysis of candidate genes. (**a**) High-resolution mapping of *qLL9*. Numbers on the map indicate the physical distance on chromosome 9. Nine recombinant plants were used to refine the candidate region to 16.17 kb region by substitution mapping, in which green and pink rectangles indicate the homozygous CJ16 genotype and homozygous C84 genotype, respectively. Flag leaf length (FLL) values were obtained from the corresponding selfed progenies and represented as mean ± SD (n = 12). (**b**) Predicted open reading frames and sequence difference of the *qLL9* between CJ16 and C84 are shown. Relative expression of three candidate genes *LOC_09g26970*, *LOC_09g26980*, and *LOC_09g26999* of near isogenic line (NIL)-$qLL9^{CJ16}$ and NIL-$qLL9^{C84}$ at tillering stage (**c**) and heading stage (**d**) by quantitative real-time polymerase chain reaction (qRT-PCR). Data are mean ± SD (n = 3). Asterisks represent significant difference determined by Student's t-test at p-value < 0.01 (**).

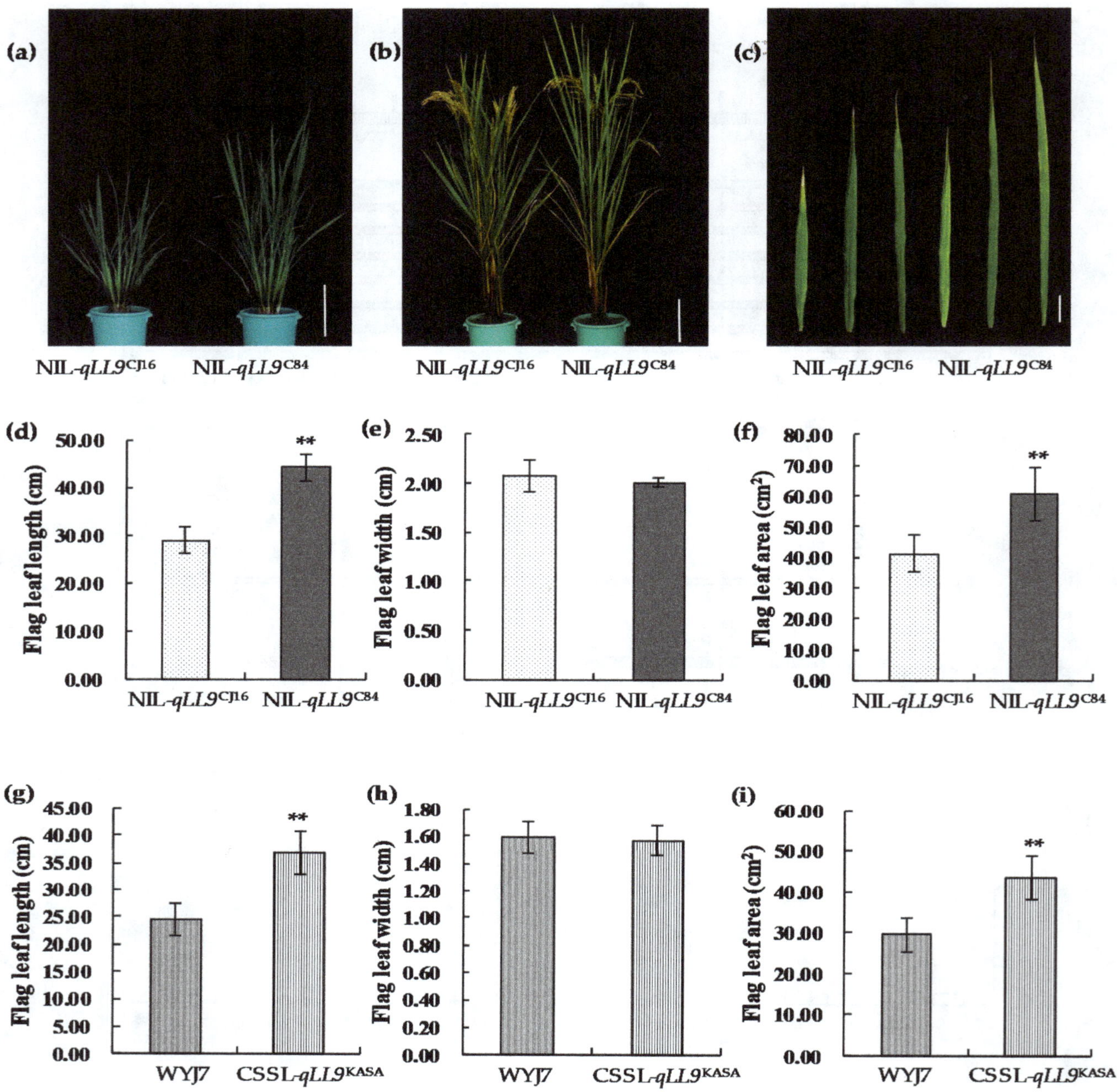

Figure 5. Phenotypes of near-isogenic lines and chromosome segment substitution lines. (**a**) Plant morphology at tillering stage; bar = 18 cm. (**b**) Plant morphology at heading stage; bar = 18 cm. (**c**) The top three leaf shape of NIL-*qLL9*^{CJ16} and NIL-*qLL9*^{C84}. From left to right are the first leaf, the second leaf, and the third leaf, respectively; bar = 5 cm. Comparison between the NIL-*qLL9*^{CJ16} and NIL-*qLL9*^{C84} for the flag leaf length (**d**), the flag leaf width (**e**), and the flag leaf area (**f**). Comparison between Wuyunjing (WYJ) 7 and chromosome segment substitution line (CSSL)-*qLL9*^{KASA} for the flag leaf length (**g**), the flag leaf width (**h**), and the flag leaf area (**i**). Data are mean ± SD (n = 15). Asterisks represent significant difference determined by Student's t-test at p-value < 0.01 (**).

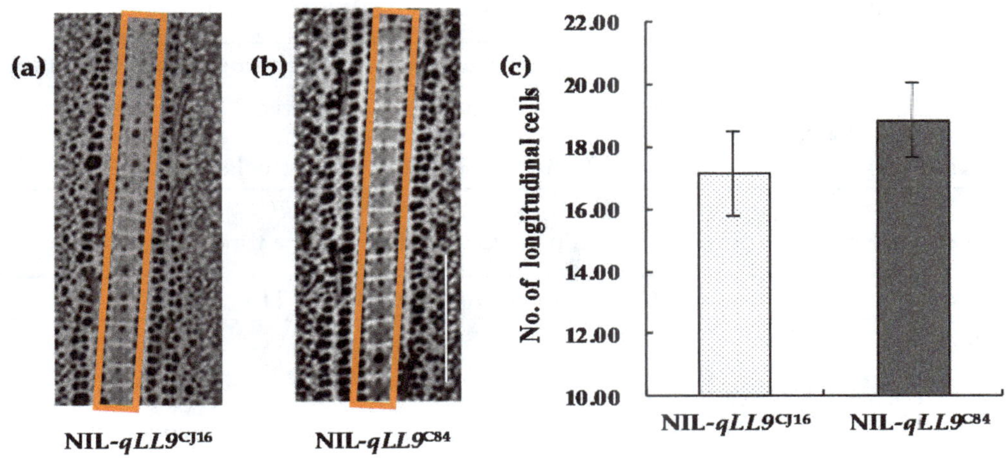

Figure 6. Histological cell morphology of NIL-*qLL9*^{CJ16} and NIL-*qLL9*^{C84}. (**a**,**b**) Comparison of cytological morphological characteristics in the orange boxes between the two genotypes of the NIL set; bar = 200 μm. (**c**) Number of longitudinal cells in NIL-*qLL9*^{CJ16} and NIL-*qLL9*^{C84}. Data are mean ± SD (n = 15) and no significant difference was found.

2.4. Determination of the Candidate Gene

According to the database of Rice Genome Annotation Project (http://rice.plantbiology.msu.edu), three open reading frames (ORFs) were predicted in the target region defined by C9-1640 and C9-1642 (Figure 4b and Table 4), namely, LOC_Os09g26970 encoding cytochrome P450 protein, LOC_Os09g26980 encoding retrotransposon protein, and LOC_Os09g26999 encoding G protein γ subunit (a cloned gene *DEP1*) [63]. By sequencing and qRT-PCR for the three candidates, six non-synonymous single nucleotide polymorphisms (SNPs) were found across the three candidate genes (Figure 4b and Table 5). No significant difference was detected in the LOC_Os09g26970 and LOC_Os09g26980 expression between NIL-*qLL9*^{CJ16} and NIL-*qLL9*^{C84} at both the tillering and heading stages, but the expression of LOC_Os09g26999 in NIL-*qLL9*^{C84} was significantly increased by 14.1 times and 52.3 times, respectively, compared with NIL-*qLL9*^{CJ16} at the two stages (Figure 4c,d). Therefore, LOC_Os09g26999 could be the best potential candidate for *qLL9*. In addition, we found that *qLL9* was allelic to the known gene *DEP1* [63]. We inferred that the difference of amino acid of *qLL9* would have an influence on the gene expression and would thus affect the leaf morphological development.

It has been reported that *DEP1* is mainly related to panicle development [63]. To further verify its correlation with leaf development, the CRISPR/Cas9 gene knockout technique was adopted in the genetic background of Nipponbare (Figure 7). Two knockout plants of LOC_Os09g26999, namely mutant 1 (MT-1) and mutant 2 (MT-2), were screened in T0 generation (Figure 8; Figure S2). After continuous selfing and marker assay, homozygous positive transgenic lines in T2 generation were obtained and used for investigating the flag leaf morphology. The results showed no significant difference in the flag leaf width between the knockout and wild-type plants. However, the flag leaf length and area were significantly decreased in both knockout lines. (Figure 8b–e). The results showed that *qLL9* was mainly responsible for the leaf length development, and the loss-function of *qLL9* could lead to a reduction in leaf length, and thereby leaf area.

To further explore the causal factor for differential expression of *LOC_Os09g26999* observed in the NIL set, we compared the promoter sequence of 2.0 kb upstream of the ATG between them, and found nine variations (Table 6). Then, the activity difference of the two types of promoters was compared by dual luciferase reporter assay. The result showed that the LUC-C84 promoter had 3.9 times higher expression of reporter genes than the LUC-CJ16 promoter (Figure 9). That is to say, the promoter activity of *qLL9* also played a significant role in the gene expression and leaf-trait variation.

Table 4. Annotated genes included in the 16.17kb region for *qLL9*.

Gene ID	Annotation from the Rice Genome Annotation Project
LOC_Os09g26970	Retrotransposon protein, Putative, Unclassified, Expressed
LOC_Os09g26980	Cytochrome P450, Putative, Expressed
LOC_Os09g26999	Gγ subunit; Dense and Erect Panicle1; DENSE PANICLE 1

Table 5. The position of SNP and amino acid variation for the three candidate genes.

Locus Name	Position on chr.9	Position on gene	CJ16	C84	CJ16	C84
LOC_Os09g26970	16393266	397	G	C	Val	Leu
LOC_Os09g26970	16393717	848	T	C	Val	Ala
LOC_Os09g26970	16395923	3054	T	C	Val	Ala
LOC_Os09g26980	16403850	4146	G	A	Ala	Val
LOC_Os09g26980	16404094	3902	T	C	Met	Cys
LOC_Os09g26999	16414735	3182	A	G	Tyr	Cys

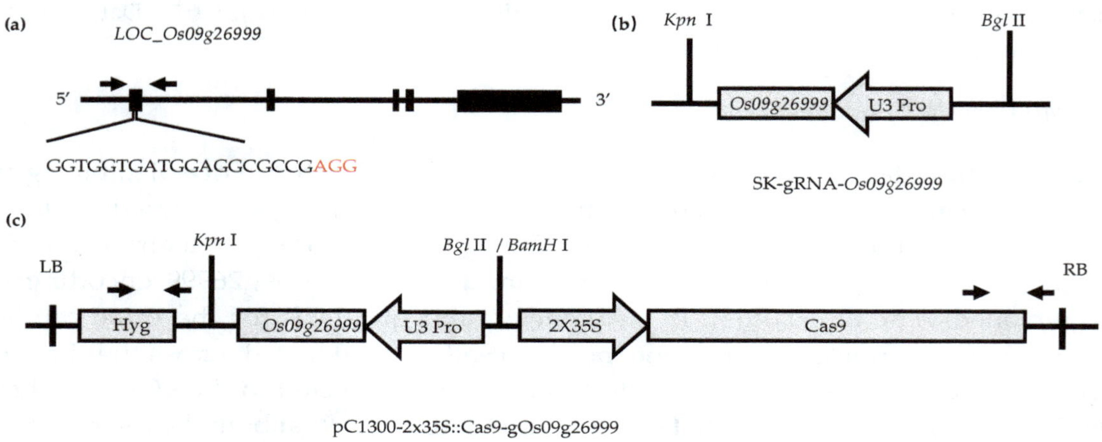

Figure 7. CRISPR/Cas9 vector construction. (**a**) Primer sequence on *LOC_Os09g26999*. (**b**) The intermediate vector SK-gRNA contains the U3 promotor and sgRNA scaffold. (**c**) Binary vector pC1300-Cas9 contains the 2 × 35S promotor and a Cas9 protein. SK-gRNA-*Os09g26999* are digested with *Kpn* I and *Bgl* II, respectively, and cloned into pC1300-Cas9 (digested with *Kpn* I and *BamH* I) by a one-step ligation.

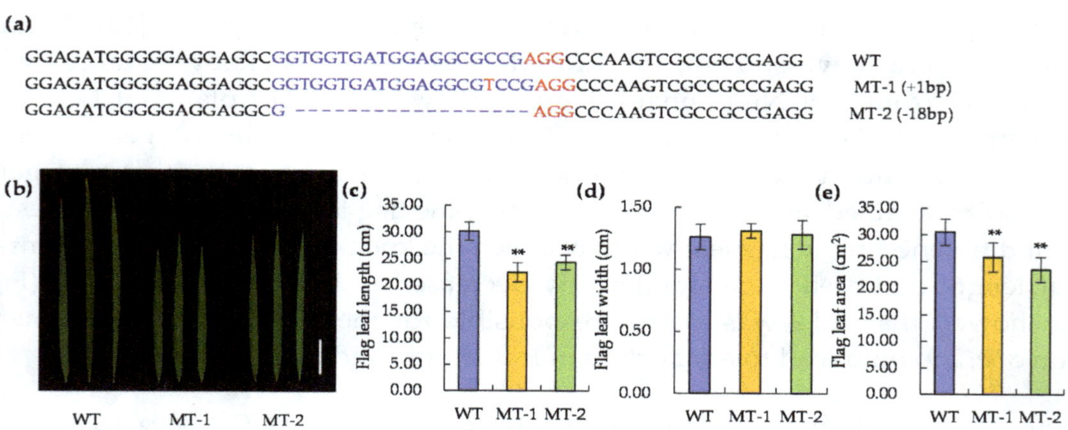

Figure 8. Sequence of target loci in CRISPR/Cas9 transgenic plants and phenotypic comparison between the knockout and wild-type (WT) plants. (**a**) Sequence variation of target loci in two transgenic plants. (**b**) Flag leaf variation; bar = 5 cm. Difference in (**c**) flag leaf length (cm), (**d**) flag leaf width (cm), and (**e**) flag leaf area. WT, Nipponbare; MT-1, Mutant 1; MT-2, Mutant 2. Data are represented as mean ± SD (n = 15). Asterisks represent significant difference determined by Student's t-test at p-value <0.01 (**).

Table 6. The position of nine SNPs in the promoter of LOC_Os09g26999.

Position	CJ16	C84
−1341	A	C
−1255	G	C
−951	C	G
−906	T	G
−629	G	T
−503	A	C
−493	G	C
−486	T	G
−31	TG	T

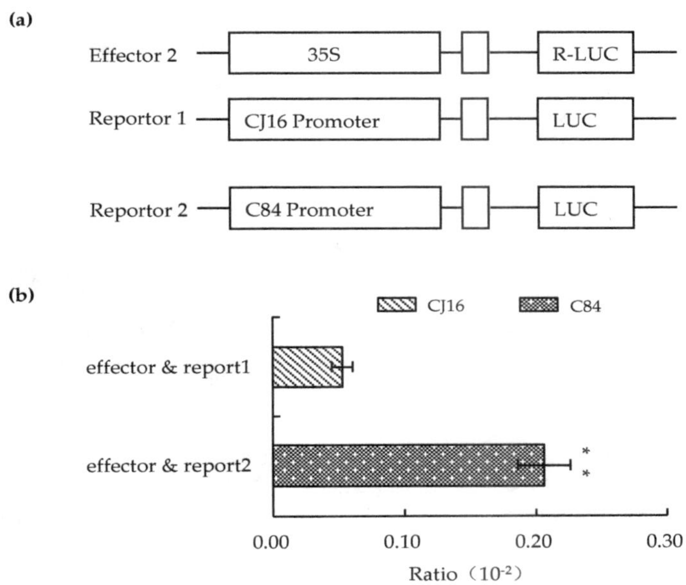

Figure 9. Comparison of the promoter activity of LOC_Os09g26999 between CJ16 and C84. (**a**) Diagram of carrier construction. (**b**) Ratio of promoter activity. Data are represented as mean ± SD (n = 3). Asterisks represent significant difference determined by Student's t-test at p-value <0.01 (**).

2.5. *qLL9* Affecting the Yield Traits

To verify whether *qLL9* affected the yield formation, we firstly investigated its effects on grain shape using NIL-*qLL9*^{CJ16} and NIL-*qLL9*^{C84} plants (Figure 10). We found that the grain length, width, and thickness of NIL-*qLL9*^{C84} are all slightly larger than those of NIL-*qLL9*^{CJ16} by 6.66%, 2.55%, and 3.37%, respectively.

Then, we compared the other yield component traits between the NIL-*qLL9*^{CJ16} and NIL-*qLL9*^{C84} plants. There was no significant difference between them in the number of productive panicles per plant, the number of primary branches per panicle, the number of second branches per panicle, and the number of grains per panicle (Table 7). However, the thousand-grain weight of NIL-*qLL9*^{C84} was significantly higher than that of NIL-*qLL9*^{CJ16} (22.99 g and 20.79 g, respectively), while the seed setting rate of NIL-*qLL9*^{C84} was lower (69.71% and 75.80%, respectively). Even so, the yield per plant of NIL-*qLL9*^{C84} was 16.59% higher than that of NIL-*qLL9*^{CJ16}. That is, the increase in the yield per plant was mainly attributed to the increase in thousand-grain weight. Further, yield measurement in plots also showed that the NIL-*qLL9*^{C84} could yield more grains than NIL-*qLL9*^{CJ16}, increasing by 15.08%. As for the actual yield per hectare, NIL-*qLL9*^{C84} could increase by 991.67 kg compared with NIL-*qLL9*^{CJ16}. These results showed that the C84 allele at *qLL9* significantly increases the grain size, thousand-grain weight, and grain yield in the field production.

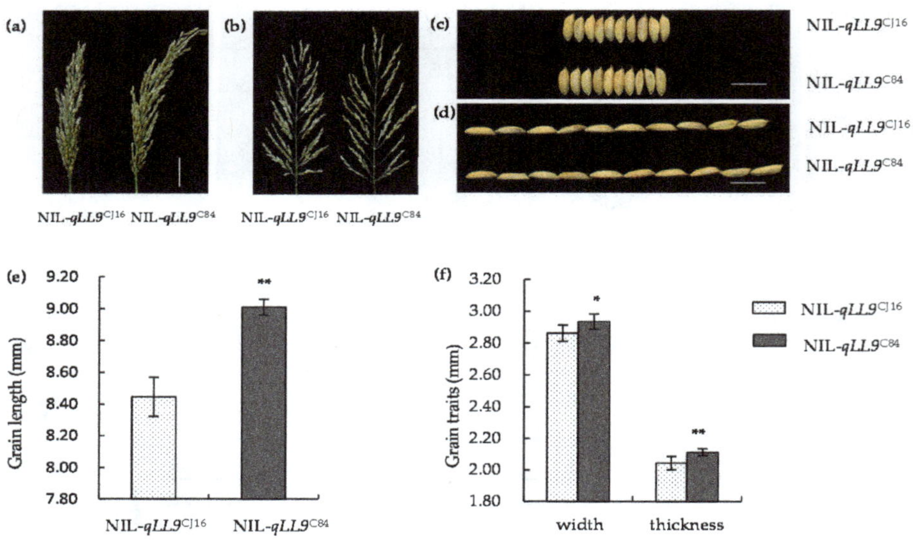

Figure 10. Panicle and grain morphology of NIL-*qLL9*^{CJ16} and NIL-*qLL9*^{C84}. (**a**,**b**) Panicle morphology at heading stage; bars = 4 cm. (**c**,**d**) Grain morphology at repining stage; bar = 1 cm. (**e**) Grain length. (**f**) Grain width and thickness. Data are represented as mean ± SD (*n* = 10). Asterisks represent significant difference determined by Student's *t*-test at *p*-value <0.01 (**), *p*-value <0.05 (*).

Table 7. Yield traits of NIL-*qLL9*^{CJ16} and NIL-*qLL9*^{C84}.

Trait	NIL-*qLL9*^{CJ16}	NIL-*qLL9*^{C84}
Panicle length (cm)	20.44 ± 0.84	26.7 ± 0.83 **
Panicles per plant	9.25 ± 1.42	9.42 ± 1.26
Number of primary branches	22.11 ± 2.18	21.4 ± 2.42
Number of secondry branches	80.00 ± 15.51	77.60 ± 7.84
Grains per panicle	404.11 ± 71.70	355.17 ± 34.09
1000-grain weight (g)	20.79 ± 0.74	22.99 ± 0.45 **
Seed setting rate (%)	75.80 ± 0.02	69.71 ± 0.02 **
Yield per plant (g)	27.93 ± 4.41	32.56 ± 8.19
Actual yield per plot (kg/48 m^2)	31.52 ± 3.20	36.28 ± 3.27 *
Actual yield change (%)	–	15.08

Data are represented as mean ± SD. Asterisks represent significant difference determined by Student's *t*-test at *p*-value <0.01(**), *p*-value <0.05 (*).

3. Discussion

In this study, the main effect QTL *qLL9* related to the leaf length was positioned by RIL population. *LOC_Os09g26999* was identified as the target gene through sequence comparison, expression analysis, and CRISPR-Cas9 gene editing technology. It is allelic to the known spike-shaped gene *DEP1/EP/qPE9-1* [63–65]. The plants of NILs and CSSLs carrying different alleles of *qLL9* showed great differences in leaf, spike, and yield. All results indicated that QTL *qLL9* has a pleiotropic function in rice. In addition to regulating spike development, *qLL9* is also plays a key role in the development of leaf morphology, grain shape, yield, and other traits.

Previous studies have demonstrated that QTL *qLL9* candidate gene *LOC_Os09g26999* encodes a cysteine-rich region [66]. Huang et al found that the replacement of a 637 bp stretch from the fifth exon of *LOC_Os09g26999* with a 12 bp sequence results in erect panicle architecture because of the early termination of translation [63]. Due to the differences of *LOC_Os09g26999* relative expression and promoter activity in CJ16 and C84, the promoter sequences and regulatory elements were compared by website (https://sogo.dna.affrc.go.jp). We found that nine nucleotide differences were observed among the parental lines, eight of which were located in cis-acting elements possibly by mediating enhancer activity depending on upstream region G/C mutation at 1255 bp upstream of ATG, which is consistent with site II transcriptional core sequence regulatory elements (TGGGCC^{CJ16} to TGGCCC^{C84}). It plays an important role in the specific expression of *proliferating cell nuclear antigen* (*PCNA*) gene in rice meristem [67,68]. We speculate that G/C mutation may be through mediation of enhancer activity dependent on far upstream regions. Comparison of CDS showed that a single base substitution event is observed from A to G at 3,182 bp downstream of ATG in *LOC_Os09g26999* between CJ16 and C84, leading to the substitution of amino acids from tyrosine to cysteine. This event might lead to changes in the structure of the γ subunit, affecting the signal transduction of G protein. Further work is needed to determine whether the difference expression of *LOC_Os09g26999* is due to changes in protein structure or differences in promoter activity.

The leaf morphological development in rice is regulated by the size and number of leaf epidermal cells. In *rot4-1D* mutant, the reduction in the number of leaf longitudinal cells induces a short leaf [69]. Tsuge et al revealed that the *Arabidopsis thaliana rotundifolia3* leaf mutant has the same number of cells as the wild type, but with reduced cell elongation in the leaf-length direction [70]. Rice flag leaf width QTL *qFLW7*, homologous to Arabidopsis *LNG1*, regulates the longitudinal growth of cells and its overexpression results in elongated leaves [71]. In this study, no significant difference was observed in the cell size per unit area of the leaf epidermis between NIL-*qLL9*^{C84} and NIL-*qLL9*^{CJ16}, whereas the total number of cells increased leading to leaf longer. The expression levels of eight cell cycle-related genes in NILs were analyzed at the heading stage (Figure 11 and Supplementary Table S1), which showed that the expression of *MCM5* plays an important role in the initiation and extension of DNA replication in the G1 phase, which was significantly up-regulated in NIL-*qLL9*^{C84} compared to NIL-*qLL9*^{CJ16} [72]. Plant Class A cyclin (Cyclin) *CYCA2;3* and *CYCA2;2* [73,74] could identify and interact with different cyclin-dependent kinases, which were also remarkably up-regulated in NIL-*qLL9*^{C84}. Therefore, *qLL9* from *indica* rice C84 could improve the DNA replication efficiency of leaf tissue cells, accelerate cell division, and promote leaf elongation, which also corresponded to the fact that the morphology of the leaf epidermal cells of NIL-*qLL9*^{C84} was unchanged while the total number of the cells increased. These findings indicated that *qLL9* may affect plant morphological development by participating in the regulation of cell division cycle. Notably, the regulatory mechanisms of different genes on leaf morphological development are different, thus their interactions in leaf morphogenesis should be studied in details in the future.

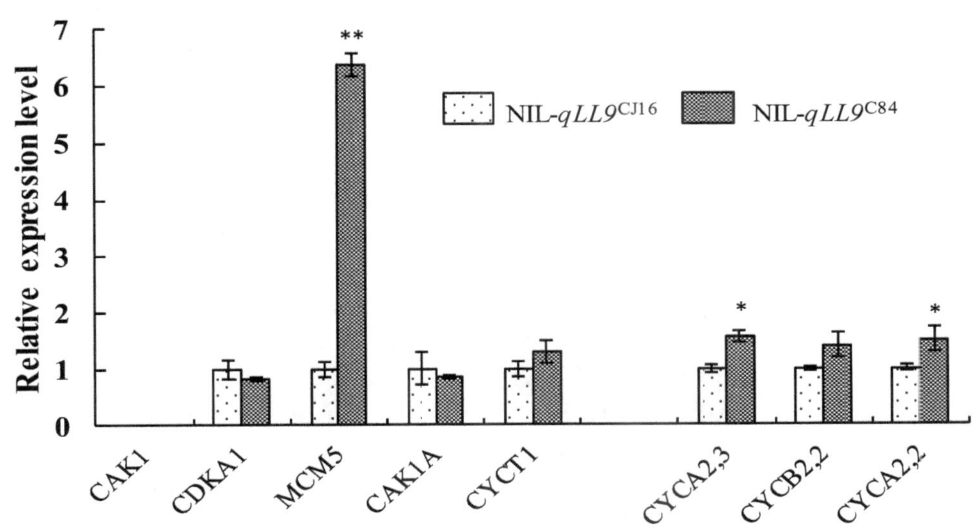

Figure 11. Comparison of eight cell cycle related gene expression between NIL-*qLL9*^{CJ16} and NIL-*qLL9*^{C84} at the heading stage by qRT-PCR, and the data are represented as mean ± SD (n = 3); asterisks represent significant difference determined by Student's t-test at p-value <0.01(**), p-value <0.05 (*).

4. Materials and Methods

4.1. RIL Population and Field Trial

The RIL population consisting of 188 lines was derived from the cross of the *japonica* rice Chunjiang 16B (CJ16) as the female parent and the *indica* rice Chunhui84 (C84) as the male parent, which are the maintainer and restorer lines of the commercial intersubspecific hybrid rice Chunyou84. The rice population was tested at experimental fields of the China Rice Research Institute located in Hangzhou, Zhejiang, and Lingshui, Hainan, during May–October 2016 and November 2016–April 2017, respectively. Twenty-five-day-old seedlings were transplanted at a hill spacing of 20 cm × 20 cm with three replications. In each replication, one line was grown in three-row plots with six plants per row. The block was managed in accordance with conventional field management, and diseases, insects, and weeds were controlled [75]. The leaf morphology and yield traits were investigated at the heading stage and mature stage. In each trial, data of the three replications were averaged for each line and used for data analysis.

4.2. Statistical and Genetic Analysis

The linkage map of the RILs consisted of sixty-nine simple sequence repeats (SSR) and eighty-nine sequence tagged site (STS) DNA markers. QTL analysis was conducted using QTL Network 2.1. Critical F values for genome-wise type I error were calculated with 1000 permutation tests and used for claiming a significant event. A significant level of $p < 0.005$ was set for candidate interval selection, putative QTL detection, and QTL effect estimation. The proportion of phenotypic variance (R^2) explained by a single main QTL for a given trait in a given population was calculated by Markov Chain Monte Carlo algorithm. In the genome scan, a testing window of 10 centimorgan (cM), filtration window of 10 cM, and walk speed of 1 cM were chosen. The naming of QTLs was based on the method of McCouch et al. [75].

4.3. Map-Based Cloning and Candidate-Gene Promoter Activity for *qLL9*

Using flanking markers RM3700 and B9-11 of *qLL9* obtained from QTL mapping, residual heterozygotes (RHs) were screened from the RIL population, which were segregated at *qLL9* with high homogenous genetic background. Six pairs of molecular markers were developed in this target interval (Table 3). By substitution mapping, we investigated the flag leaf length and the new developed marker genotypes of 2307 RH-derived introgression lines, and then the *qLL9* was fine mapped.

4.4. RNA Extraction and qRT-PCR

Total RNA of NIL-*qLL9*^{CJ16} and NIL-*qLL9*^{C84} was extracted from the penultimate leaves at the tillering stage and the flag leaf at heading stage to analyze the expression difference of candidate genes (Tables 3 and 4). We used the AxyPrepTM Multisource Total RNA Miniprep Kit (Axygen) to extract total RNA, which was then retro-transcribed using PrimeScriptTM RT Reagent Kit with gDNA Eraser (Takara, Dalian, China). Quality and concentration of the RNA extracted were checked with electrophoresis on 1% agarose gel and measured using the Nanodrop ND-2000 spectrophotometer (NanoDrop Technologies, Wilmington, CA, USA). Concentration of the RNA samples used for cDNA synthesis was normalized by dilution with RNase-free ultra-pure water. qRT-PCR assays of 20 μL reaction volumes, which contained 0.5 μL of synthesized cDNA, 0.4 μM of gene-specific primers, and 10 μL of SYBR® Premix Ex Taq™ (Takara), were conducted using ABI 7500 Real-time PCR System (Applied Biosystems, Foster, CA, USA). Following the manufacturer's instruction, the qRT-PCR conditions were set up as follows: denaturing at 95 °C for 30 s, then 40 cycles of 95 °C for 5 s, 55 °C for 30 s, and 72 °C for 30 s. To standardize the quantification of gene expression, we used the rice *Ubiquitin* (*UBQ*) gene (Os03g0234200, http://rapdb.dna.affrc.go.jp/) as an internal control.

According to the sequence shown in http://rice.plantbiology.msu.edu/index.shtml, the primer POs26999-F/R (Table 3) was used to amplify LOC_Os09g26999 promoter of CJ16 and C84, which were constructed into *pGreenII0800-LUC* using homologous recombination [76]. Positive clones were screened by colony PCR and sequencing and were named as *proCJ16-LUC* and *proC84-LUC*. The plasmids and the internal reference (R-LUC) were transformed into the protoplasts of rice variety 93-11 by 40% PEG-3350 (Sigma, St. Louis, MO, USA) solution-mediated transformation [77]. The dual-luciferase reporter gene detection kit (Promega Company, Madison, WI, USA) was used for detection and analysis of the promoter activity.

4.5. CRISPR/Cas9 Transgene Analysis

The target sequence of the potential candidate LOC_Os09g26999 refers to a previous study [78]. Synthetic primers Os26999-g++ and Os26999-g−−were used to make the target site adapter (Table 3). This adapter was connected to an SK-gRNA carrier, and positive intermediate carrier SK-gRNA-Os26999 was screened out. pC1300-2×35S::Cas9-gOs26999 final expression carrier was constructed by means of enzyme digestion-joining method. *Agrobacterium tumefaciens* EHA105 was transformed through positive cloning. *Agrobacterium tumefaciens*-mediated gene transfer experiments were carried out with the background of Nipponbare [79]. The result was analyzed via the sequential decoding method (http://dsdecode.scgene.com/) to identify transgenic positive plants. At heading stage, 15 wild-type and 15 mutant plants were selected to measure the length, width, and area of the flag leaf.

4.6. Construction of Near Isogenic Lines and Chromosome Segment Substitution Lines and Trait Measurement

Using flanking markers C-1640 and C-1642 (Table 3), a set of near isogenic lines (NILs) for *qLL9* was identified from RH-derived segregating populations with a higher homogenous background, named NIL-$qLL9^{CJ16}$ and NIL-$qLL9^{C84}$. At the same time, one chromosome segment substitution line (CSSL) for *qLL9* was obtained with Kasalath (KASA) as the donor parent and Wuyunjing 7 (WYJ 7) as the recurrent parent through one cross followed by six continuous backcrosses and one self-crossing, named CSSL-$qLL9^{KASA}$.

The leaf length, width, and area of the NIL and CSSL set were investigated at the heading stage, while the panicle length, the number of primary and secondary branches, the number of grains per panicle, grain length, grain width, grain thickness, thousand-grain weight, and setting percentage were scored from 10 randomly selected main panicles at maturity. Then, Student's *t*-test was adopted to analyze the phenotypic difference between the two genotypic groups in each set.

The grain yield was measured using the NIL set NIL-$qLL9^{CJ16}$ and NIL-$qLL9^{C84}$, of which each was planted in three 48 m^2 plots, at a hill spacing of 20 cm × 20 cm. At maturity, three points in each plot were randomly harvested, with 30 hills per point, to determine the grain yield, which is then converted into the grain yield per hectare.

4.7. Morphological Observation on Leaf Epidermal Cytology

The commercially available transparent nail polish without color is selected, which is conducive to the transparency of microscopic materials. The flag leaves of tested rice plants at the heading stage were sampled and painted with the nail polish evenly at the same part for 10 min air-dry. When an open mouth exists at the end of the coating layer of the nail polish, the dried coating is torn with a transparent tape and placed on the fragment. Then, the cover slip was covered. The filter paper was covered by the blunt end of the dissecting needle, and the fragment was gently pressed to make a temporary filling piece. Under the electron microscopy, the cell size and number of 15 leaves of a single plant were analyzed [80].

5. Conclusions

Our work identified the genetic contribution of *qLL9* to both flag leaf morphologic development and yield formation. Improving the photosynthetic efficiency and coordinating the source–sink interaction through leaf morphogenesis are the premise to increase the rice yield and to establish the ideal plant type. Identification and utilization of the QTLs with beneficially pleiotropism for both leaf shape and grain yield would greatly contribute to molecular breeding of superior rice. For example, the QTL *qLSCHL4/NAL1* is correlated with leaf shape and involved in regulating the development of chlorophyll content, grain number and grain weight [15]. QTL *qFLW7*^{9311} related to flag leaf width, could improve leaf shape and grain traits, remarkably increase rice yield in the field [71]. Similarly, *GS2* could increase both grain weight/size and leaf length and thus grain yield [60,61]. According to the main objectives of rice molecular design breeding, future studies should systematically analyze the source-sink relationship and the genetic network of plant type establishment, directional polymerize the *qLL9, NAL1, qFLW7, GS2,* and other beneficial genes from *indica* and *japonica* subspecies of rice to improve the yield potential and establish the ideal plant architecture in molecular breeding of superior rice.

Author Contributions: Data curation, X.F., J.X., M.Z., T.L., Y.Z., and J.W.; Formal analysis, X.F., T.L., and D.R.; Funding acquisition, G.Z.; Investigation, X.F., J.X., M.Z., M.C., L.S., Y.Z., J.W., and J.H.; Methodology, J.H., L.Z., Z.G., L.G., D.R., G.C., and G.Z.; Project administration, Q.Q. and G.Z.; Resources, G.D. and J.L.; Software, M.Z. and Z.G.; Supervision, Q.Q. and G.Z.; Writing—original draft, X.F. and J.X.; Writing—review & editing, G.Z.

References

1. Heath, D.V.; Gregory, F.G. The constancy of the mean net assimilation rate and its ecological importance. *Ann. Bot. Lond.* **1938**, *2*, 811–818. [CrossRef]
2. Donald, C.M. The breeding of crop ideotypes. *Euphytica* **1968**, *17*, 385–403. [CrossRef]
3. Mason, T.G.; Maskell, E.J. Studies on the transport of carbohydrates in the cotton plant. I. A study of diurnal variation in the carbohydrates of leaf, bark, and wood, and of the effects of ringing. *Ann. Bot.-Lond.* **1928**, *42*, 189–253. [CrossRef]
4. Hasson, A.; Blein, T.; Laufs, P. Leaving the meristem behind: The genetic and molecular control of leaf patterning and morphogenesis. *C. R. Biol.* **2010**, *333*, 350–360. [CrossRef] [PubMed]
5. Liu, B.; Li, P.C.; Li, X.; Liu, C.Y.; Cao, S.Y.; Chu, C.C.; Cao, X.F. Loss of function of *OsDCL1* affects micro RNA accumulation and causes developmental defects in rice. *Plant Physiol.* **2005**, *139*, 296–305. [CrossRef] [PubMed]
6. Wang, E.; Xu, X.; Zhang, L.; Zhang, H.; Lin, L.; Wang, Q.; Li, Q.; Ge, S.; Lu, B.R.; Wang, W.; et al. Duplication and independent selection of cell-wall invertase genes *GIF1* and *OsCIN1* during rice evolution and domestication. *BMC Evol. Biol.* **2010**, *10*, 108–120. [CrossRef] [PubMed]
7. Liu, X.F.; Li, M.; Liu, K.; Tang, D.; Sun, M.F.; Li, Y.F.; Shen, Y.; Du, G.J.; Cheng, Z.K. *Semi-Rolled Leaf2* modulates rice leaf rolling by regulating abaxial side cell differentiation. *J. Exp. Bot.* **2016**, *67*, 2139–2150. [CrossRef] [PubMed]
8. Zhang, G.H.; Xu, Q.; Zhu, X.D.; Qian, Q.; Xue, H.W. *SHALLOT-LIKE1* is a KANADI transcription factor that modulates rice leaf rolling by regulating leaf abaxial cell development. *Plant Cell* **2009**, *21*, 719–735. [CrossRef]
9. Yan, C.J.; Yan, S.; Zhang, Z.q.; Liang, G.H.; Lu, J.F.; Gu, M.H. Genetic analysis and gene fine mapping for a rice novel mutant ($rl_{9(t)}$), with rolling leaf character. *Chin. Sci. Bull.* **2006**, *51*, 63–69. [CrossRef]
10. Kong, X.Q.; Gao, X.H.; Sun, W.; An, J.; Zhao, Y.X.; Zhang, H. Cloning and functional characterization of a cation–chloride cotransporter gene *OsCCC1*. *Plant Mol. Biol.* **2011**, *75*, 567–578. [CrossRef]
11. Tian, X.; Zhang, D.B. cDNA cloning and expression analysis of *OsWOX4* in rice. *J. Shanghai Jiaotong Univ. Agric. Sci.* **2011**, *29*, 8–14.
12. Cho, S.H.; Yoo, S.C.; Zhang, H.T.; Pandeya, D.; Koh, H.J.; Hwang, J.Y.; Kim, G.T.; Paek, N.C. The rice narrow leaf2 and narrow leaf3 loci encode WUSCHEL-related homeobox 3A *OsWOX3A*, and function in leaf, spikelet, tiller and lateral root development. *New Phytol.* **2013**, *198*, 1071–1084. [CrossRef] [PubMed]
13. Luan, W.J.; Liu, Y.Q.; Zhang, F.X.; Song, Y.L.; Wang, Z.Y.; Peng, Y.K.; Sun, Z.X. *OsCD1* encodes a putative member of the cellulose synthase-like D sub-family and is essential for rice plant architecture and growth. *Plant Biotechnol. J.* **2010** *9*, 513–524. [CrossRef] [PubMed]

14. Hu, J.; Zhu, Li.; Zeng, D.L.; Gao, Z.Y.; Guo, L.B.; Fang, Y.X.; Zhang, G.H.; Dong, G.J.; Yan, M.X.; Liu, J.; Qian, Q. Identification and characterization of *NARROW AND ROLLED LEAF 1*, a novel gene regulating leaf morphology and plant architecture in rice. *Plant Mol. Biol.* **2010**, *73*, 283–292. [CrossRef] [PubMed]
15. Zhang, G.H.; Li, S.Y.; Wang, L.; Ye, W.J.; Zeng, D.L.; Rao, Y.C.; Peng, Y.L.; Hu, J.; Yang, Y.L.; Xu, J.; et al. *LSCHL4* from japonica cultivar, which is allelic to *NAL1*, increases yield of indica super rice 93-11. *Mol. Plant* **2014**, *7*, 1350–1364. [CrossRef] [PubMed]
16. Fujino, K.; Matsuda, Y.; Ozawa, K.; Nishimura, T.; Koshiba, T.; Fraaije, M.W.; Sekiguchi, H. *NARROW LEAF 7* controls leaf shape mediated by auxin in rice. *Mol. Genet. Genom.* **2008**, *279*, 499–507. [CrossRef] [PubMed]
17. Sazuka, T.; Kamiya, N.; Nishimura, T.; Ohmae, K.; Sato, Y.; Imamura, K.; Nagato, Y.; Koshiba, T.; Nagamura, Y.; Ashikari, M.; et al. A rice *tryptophan deficient dwarf* mutant, *tdd1*, contains a reduced level of indole acetic acid and develops abnormal flowers and organless embryos. *Plant J.* **2009**, *60*, 227–241. [CrossRef]
18. Woo, Y.M.; Park, H.J.; Su'udi, M.; Yang, J.I.; Park, J.J.; Back, K.; Park, Y.M.; An, G. *Constitutively wilted 1*, a member of the rice YUCCA gene family, is required for maintaining water homeostasis and an appropriate root to shoot ratio. *Plant Mol. Biol.* **2007**, *65*, 125–136. [CrossRef]
19. Zhang, S.Z.; Wu, T.; Liu, S.J.; Liu, X.; Jiang, L.; Wan, J.M. Disruption of *OsARF19* is Critical for Floral Organ Development and Plant Architecture in Rice (*Oryza sativa* L.). *Plant Mol. Biol. Rep.* **2016**, *34*, 748–760. [CrossRef]
20. Li, W.; Wu, C.; Hu, G.C.; Xing, Li.; Qian, W.J.; Si, H.M.; Sun, Z.X.; Wang, X.C.; Fu, Y.P.; Liu, W.Z. Characterization and fine mapping of a novel rice narrow leaf mutant *nal9*. *J. Integr. Plant Biol.* **2013**, *55*, 1016–1025. [CrossRef]
21. Hibara, K.; Obara, M.; Hayashida, E.; Abe, M.; Ishimaru, T.; Satoh, H.; Itoh, J.; Nagato, Y. The *ADAXIALIZED LEAF1* gene functions in leaf and embryonic pattern formation in rice. *Dev. Biol.* **2009**, *334*, 345–354. [CrossRef] [PubMed]
22. Shi, Z.Y.; Wang, J.; Wan, X.S.; Shen, G.Z.; Wang, X.; Zhang, J.L. Over-expression of rice *OsAGO7* gene induces upward curling of the leaf blade that enhanced erect-leaf habit. *Planta* **2007**, *226*, 99–108. [CrossRef] [PubMed]
23. Li, L.; Xue, X.; Zou, S.M.; Chen, Z.X.; Zhang, Y.F.; Li, Q.Q.; Zhu, J.K.; Ma, Y.Y.; Pan, X.B.; Pan, C.H. Suppressed expressed of AGO1a leads to adaxial leaf rolling in rice. *Chin. J. Rice Sci.* **2013**, *273*, 223–230.
24. Li, L.; Shi, Z.Y.; Li, L.; Shen, G.Z.; Wang, X.Q.; An, L.S.; Zhang, J.L. Overexpression of *ACL1 abaxially curled leaf 1*, increased bulliform cells and induced abaxial curling of leaf blades in rice. *Mol. Plant* **2010**, *35*, 807–817. [CrossRef] [PubMed]
25. Wang, J.; Hu, J.; Qian, Q.; Xue, H.W. LC2 and OsVIL2 promote rice flowering by photoperoid-induced epigenetic silencing of *OsLF*. *Mol. Plant* **2013**, *62*, 514–527. [CrossRef] [PubMed]
26. Zhao, S.Q.; Hu, J.; Guo, L.B.; Qian, Q.; Xue, H.W. *Rice leaf inclination 2*, a VIN3-like protein, regulates leaf angle through modulating cell division of the collar. *Cell Res.* **2010**, *20*, 935–947. [CrossRef] [PubMed]
27. Li, M.; Xiong, G.Y.; Li, R.; Cui, J.J.; Tang, D.; Zhang, B.C.; Pauly, M.; Cheng, Z.K.; Zhou, Y.H. Rice cellulose synthase-like D4 is essential for normal cell-wall biosynthesis and plant growth. *Plant J.* **2009**, *60*, 1055–1069. [CrossRef]
28. Yoshikaw, T.; Eiguchi, M.; Hibara, K.I.; Ito, J.I.; Nagato, Y. Rice *SLENDER LEAF 1* gene encodes cellulose synthase-like D4 and is specifically expressed in M-phase cells to regulate cell proliferation. *J. Exp. Bot.* **2013**, *647*, 2049–2061. [CrossRef]
29. Ding, Z.Q.; Lin, Z.F.; Li, Q.; Wu, H.; Xiang, C.Y.; Wang, J.F. *DNL1*, encodes cellulose synthase-like D4, is a major QTL for plant height and leaf width in rice (*Oryza sativa* L.). *Biochem. Biophys. Res. Commun.* **2014**, *457*, 133–140. [CrossRef]
30. Li, Y.Y.; Shen, A.; Xiong, W.; Sun, Q.L.; Luo, Q.; Song, T.; Li, Z.L.; Luan, W.J. Overexpression of *OsHox32* results in pleiotropic effects on plant type architecture and leaf development in rice. *Rice* **2016**, *9*, 46. [CrossRef]
31. Yang, C.H.; Li, D.Y.; Liu, X.; Ji, C.J.; Hao, L.L.; Zhao, X.F.; Li, X.B.; Chen, C.Y.; Cheng, Z.K.; Zhu, L.H. OsMYB103L, an R2R3-MYB transcription factor, influences leaf rolling and mechanical strength in rice (*Oryza sativa* L.). *BMC Plant Biol.* **2014**, *14*, 158. [CrossRef] [PubMed]
32. Xu, Y.; Wang, Y.H.; Long, Q.Z.; Huang, J.X.; Wang, Y.L.; Zhou, K.N.; Zheng, M.; Sun, J.; Chen, H.; Chen, S.H.; et al. Overexpression of *OsZHD1*, a zinc finger homeodomain class homeobox transcription factor, induces

abaxially curled and drooping leaf in rice. *Planta* **2009**, *239*, 803–816. [CrossRef] [PubMed]

33. Chen, Q.L.; Xie, Q.J.; Gao, J.; Wang, W.Y.; Sun, B.; Liu, B.H.; Zhu, H.T.; Peng, H.F.; Zhao, H.B.; Liu, C.H.; et al. Characterization of *Rolled and Erect Leaf 1* in regulating leave morphology in rice. *J. Exp. Bot.* **2015**, *66*, 6047–6058. [CrossRef] [PubMed]
34. Yang, S.Q.; Li, W.Q.; Miao, H.; Gan, P.F.; Qiao, L.; Chang, Y.L.; Shi, C.H.; Chen, K.M. *REL2*, A Gene Encoding An Unknown Function Protein which Contains DUF630 and DUF632 Domains Controls Leaf Rolling in Rice. *Rice* **2016**, *9*, 37. [CrossRef] [PubMed]
35. Fang, L.K.; Zhao, F.M.; Cong, Y.F.; Sang, X.C.; Du, Q.; Wang, D.Z.; Li, Y.F.; Ling, Y.H.; Yang, Z.L.; He, G.H. Rolling-leaf14 is a 2OG-Fe II, oxygenase family protein that modulates rice leaf rolling by affecting secondary cellwall formation in leaves. *Plant Biotechnol. J.* **2012**, *10*, 524–532. [CrossRef] [PubMed]
36. Zou, L.P.; Sun, X.H.; Zhang, Z.G.; Liu, P.; Wu, J.X.; Tian, C.J.; Qiu, J.L.; Lu, T.G. Leaf Rolling Controlled by the Homeodomain Leucine Zipper Class IV Gene *Roc5* in Rice. *Plant Physiol.* **2011**, *156*, 1589–1602. [CrossRef] [PubMed]
37. Li, W.Q.; Zhang, M.J.; Gan, P.F.; Qiao, L.; Yang, S.Q.; Miao, H.; Wang, G.F.; Zhang, M.M.; Liu, W.T.; Li, H.F.; et al. *CLD1/SRL1* modulates leaf rolling by affecting cell wall formation, epidermis integrity and water homeostasis in rice. *Plant J.* **2017**, *92*, 904–923. [CrossRef]
38. Xiang, J.J.; Zhang, G.H.; Qian, Q.; Xue, H.W. *Semi-rolled leaf1* encodes a putative glycosylphosphatidylinositol-anchored protein and modulates rice leaf rolling by regulating the formation of bulliform cells. *Plant Physiol.* **2012**, *159*, 1488–1500. [CrossRef]
39. Alamin, M.; Zeng, D.D.; Qin, R.; Sultana, H.; Jin, X.L.; Shi, C.H. Characterization and fine mapping of *SFL1*, a gene controlling *Screw Flag Leaf* in Rice. *Plant Mol. Biol. Rep.* **2017**, *35*, 491–503. [CrossRef]
40. Zhang, J.J.; Wu, S.Y.; Jiang, L.; Wang, J.L.; Zhang, X.; Guo, X.P.; Wu, C.Y.; Wan, J.M. A detailed analysis of the leaf rolling mutant *sll2* reveals complex nature in regulation of bulliform cell development in rice (*Oryza sativa* L.). *Plant Biol.* **2015**, *17*, 437–448. [CrossRef]
41. Dai, M.Q.; Zhao, Y.; Ma, Q.; Hu, Y.F.; Hedden, P.; Zhang, Q.F.; Zhou, D.X. The rice *YABBY1* gene is involved in the feedback regulation of gibberellin metabolism. *Plant Physiol.* **2007**, *144*, 121–133. [CrossRef] [PubMed]
42. Zhou, Y.; Wang, D.; Wu, T.; Yang, Y.; Liu, C.; Yan, L.; Tang, D.; Zhao, X.; Zhu, Y.; Lin, J.; et al. LRRK1, a receptor-like cytoplasmic kinase, regulates leaf rolling through modulating bulliform cell development in rice. *Mol. Breed.* **2018**, *38*, 48. [CrossRef]
43. Ma, L.; Sang, X.C.; Zhang, T.; Yu, Z.Y.; Li, Y.F.; Zhao, F.M.; Wang, Z.W.; Wang, Y.T.; Yu, P.; Wang, N.; et al. *ABNORMAL VASCULAR BUNDLES* regulates cell proliferation and procambium cell establishment during aerial organ development in rice. *New Phytol.* **2017**, *213*, 275–286. [CrossRef] [PubMed]
44. Ye, Y.F.; Wu, K.; Chen, J.F.; Liu, Q.; Wu, Y.J.; Liu, B.M.; Fu, X.D. OsSND2, a NAC family transcription factor, is involved in secondary cell wall biosynthesis through regulating MYBs expression in rice. *Rice* **2018**, *11*, 36. [CrossRef] [PubMed]
45. Wu, R.H.; Li, S.B.; He, S.; Waßmann, F.; Yu, C.; Qin, G.; Schreiber, L.; Qu, L.J.; Gu, H. CFL1, a WW domain protein, regulates cuticle development by modulating the function of HDG1, a class IV homeodomain transcription factor, in rice and *Arabidopsis*. *Plant Cell* **2011**, *23*, 3392–3411. [CrossRef] [PubMed]
46. Jing, W.; Cao, C.J.; Shen, L.K.; Zhang, H.S.; Jing, G.Q.; Zhang, W.H. Characterization and fine mapping of a rice leaf-rolling mutant deficient in commissural veins. *Crop Sci.* **2017**, *57*, 2595–2604. [CrossRef]
47. Li, S.G.; He, P.; Wang, Y.P.; Li, H.Y.; Chen, Y.; Zhou, K.D.; Zhu, L.H. Genetic analysis and gene mapping of the leaf traits in rice *Oryza sativa* L. *Acta Agron. Sin.* **2000**, *26*, 261–265.
48. Yan, J.Q.; Zhu, J.; He, C.X.; Benmoussa, M.; Wu, P. Molecular marker-assisted dissection of genotype × environment interaction for plant type traits in rice (*Oryza sativa* L.). *Crop Sci.* **1999**, *39*, 538–544. [CrossRef]
49. Farooq, M.; Tagle, A.G.; Santos, R.E.; Ebron, L.A.; Fujita, D.; Kobayashi, N. Quantitative trait loci mapping for leaf length and leaf width in rice cv. IR64 derived lines. *J. Integr. Plant Biol.* **2010**, *52*, 578–584. [CrossRef]
50. Li, Z.K.; Pinson, S.R.M.; Stansel, J.W.; Paterson, A. Genetic dissection of the source-sink relationship affecting fecundity and yield in rice (*Oryza sativa* L.). *Mol. Breed.* **1998**, *4*, 419–426. [CrossRef]
51. Mei, H.W.; Luo, L.J.; Ying, C.S.; Wang, Y.P.; Yu, X.Q.; Guo, L.B.; Paterson, A.H.; Li, Z.K. Gene actions of QTLs affecting several agronomic traits resolved in a recombinant inbred rice population and two testcross populations. *Theor. Appl. Genet.* **2005**, *110*, 649–659. [CrossRef] [PubMed]
52. Cui, K.H.; Peng, S.; Xing, Y.; Yu, S.; Xu, C.; Zhang, Q. Molecular dissection of the genetic relationships of source, sink and transport tissue with yield traits in rice. *Theor. Appl. Genet.* **2003**, *106*, 649–658. [CrossRef]

[PubMed]
53. Yue, B.; Xue, W.Y.; Luo, L.J.; Xing, Y.Z. QTL Analysis for flag leaf characteristics and their relationships with yield and yield traits in rice. *Acta Agron. Sin.* **2006**, *33*, 824–832. [CrossRef]
54. Jiang, S.k.; Zhang, X.J.; Wang, J.Y.; Chen, W.F.; Xu, Z.J. Fine mapping of the quantitative trait locus *qFLL9* controlling flag leaf length in rice. *Euphytica* **2010**, *176*, 341–347. [CrossRef]
55. Wang, P.; Zhou, G.L.; Cui, K.H.; Li, Z.K.; Yu, S.B. Clustered QTL for source leaf size and yield traits in rice (*Oryza sativa* L.). *Mol. Breed.* **2012**, *29*, 99–113. [CrossRef]
56. Wang, P.; Zhou, G.L.; Yu, H.H.; Yu, S.B. Fine mapping a major QTL for flag leaf size and yield-related traits in rice. *Theor. Appl. Genet.* **2011**, *123*, 1319–1330. [CrossRef]
57. Ding, X.P.; Li, X.K.; Xiong, L.Z. Evaluation of near-isogenic lines for drought resistance QTL and fine mapping of a locus affecting flag leaf width, spikelet number, and root volume in rice. *Theor. Appl. Genet.* **2011**, *123*, 815–826. [CrossRef]
58. Chen, M.L.; Luo, J.; Shao, G.N.; Wei, X.J.; Tang, S.Q.; Sheng, Z.H.; Song, J.; Hu, P.S. Fine mapping of a major QTL for flag leaf width in rice, *qFLW4*, which might be caused by alternative splicing of *NAL1*. *Plant Cell Rep.* **2012**, *31*, 863–872. [CrossRef]
59. Fujita, D.; Trijatmiko, K.R.; Tagle, A.G.; Sapasap, M.V.; Koide, Y.; Sasaki, K.; Tsakirpaloglou, N.; Gannaban, R.B.; Nishimura, T.; Yanagihara, S.; et al. *NAL1* allele from a rice landrace greatly increases yield in modern indica cultivars. *Proc. Natl. Acad. Sci. USA* **2013**, *110*, 20431–20436. [CrossRef]
60. Hu, J.; Wang, Y.X.; Fang, Y.X.; Zeng, L.J.; Xu, J.; Yu, H.P.; Shi, Z.Y.; Pan, J.J.; Zhang, D.; Kang, S.J.; et al. A Rare allele of *GS2* enhances grain size and grain yield in rice. *Mol. Plant* **2015**, *8*, 1455–1465. [CrossRef]
61. Duan, P.; Ni, S.; Wang, J.M.; Zhang, B.L.; Xu, R.; Wang, Y.X.; Chen, H.Q.; Zhu, X.D.; Li, Y.H. Regulation of *OsGRF4* by OsmiR396 controls grainsize and yield in rice. *Nat. Plants* **2015**, *2*, 15203. [CrossRef] [PubMed]
62. Quarrie, S.A.; Pekic, S.; Radosevic, Q.; Rancic, R.D.; Kaminska, A.; Barnes, J.D.; Leverington, M.; Ceoloni, C.; Dodig, D. Dissecting a wheat QTL for yield present in a range of environments: From the QTL to candidate genes. *J. Exp. Bot.* **2006**, *57*, 2627–2637. [CrossRef] [PubMed]
63. Huang, X.Z.; Qian, Q.; Liu, Z.B.; Sun, H.Y.; He, S.Y.; Luo, D.; Xia, G.M.; Chu, C.C.; Li, J.Y.; Fu, X.D. Natural variation at the *DEP1* locus enhances grain yield in rice. *Nat. Genet.* **2009**, *41*, 494–497. [CrossRef] [PubMed]
64. Wang, J.Y.; Nakazaki, T.; Chen, S.Q.; Chen, W.F.; Saito, H.; Tsukiyama, T.; Okumoto, Y.; Xu, Z.J.; Tanisaka, T. Identification and characterization of the erect-pose panicle gene *EP* conferring high grain yield in rice (*Oryza sativa* L.). *Theor. Appl. Genet.* **2009**, *119*, 85–91. [CrossRef] [PubMed]
65. Zhou, Y.; Zhu, J.Y.; Li, Z.Y.; Yi, C.D.; Liu, J.; Zhang, H.G.; Tang, S.Z.; Gu, M.H.; Liang, G.H. Deletion in a quantitative trait gene *qpe9-1* associated with panicle erectness improves plant architecture during rice domestication. *Genetics* **2009**, *183*, 315–324. [CrossRef] [PubMed]
66. Mao, S.; Lu, G.H.; Yu, K.H.; Bo, Z.; Chen, J.H. Specific protein detection using thermally reduced graphene oxide sheet decorated with gold nanoparticle-antibody conjugates. *Adv. Mater* **2010**, *22*, 3521–3526. [CrossRef] [PubMed]
67. Kosugi, S.; Suzuka, I.; Ohashi, Y. Two of three promoter elements identified in a rice gene for proliferating cell nuclear antigen are essential for meristematic tissue-specific expression. *Plant J.* **1995**, *7*, 877–886. [CrossRef]
68. Zhao, M.Z.; Sun, J.; Xiao, Z.Q.; Cheng, F.; Xu, H.; Tang, L.; Chen, W.F.; Xu, Z.J.; Xu, Q. Variations in *DENSE AND ERECT PANICLE 1* (*DEP1*) contribute to the diversity of the panicle trait in high-yielding japonica rice varieties in northern China. *Breed. Sci.* **2016**, *66*, 599–605. [CrossRef]
69. Narita, N.N.; Moore, S.; Horiguchi, G.; Kubo, M.; Demura, T.; Fukuda, H.; Goodrich, J.; Tsukaya, H. Overexpression of a novel small peptide ROTUNDIFOLIA4 decreases cell proliferation and alters leaf shape in Arabidopsis thaliana. *Plant J.* **2004**, *38*, 399–713. [CrossRef]
70. Tsuge, T.; Tsukaya, H.; Uchimiya, H. Two independent and polarized processes of cell elongation regulate leaf blade expansion in *Arabidopsis thaliana* (L.) Heynh. *Development* **1996**, *122*, 1589–1600.
71. Xu, J.; Wang, L.; Wang, Y.X.; Zeng, D.L.; Zhou, M.Y.; Fu, X.; Ye, W.J.; Hu, J.; Zhu, L.; Ren, D.Y.; et al. Reduction of *OsFLW7* expression enhanced leaf area and grain production in rice. *Sci. Bull.* **2017**, *62*, 1631–1633. [CrossRef]
72. Snyder, M.; He, W.; Zhang, J.J. The DNA replication factor MCM5 is essential for stat1-mediated transcriptional activation. *Proc. Natl. Acad. Sci. USA* **2005**, *102*, 14539–14544. [CrossRef] [PubMed]
73. Imai, K.K.; Ohashi, Y.H.; Tsuge, T.; Yoshizumi, T.; Matsui, M.; Oka, A.; Aoyama, T. The a-type cyclin CYCA2;3

is a key regulator of ploidy levels in *Arabidopsis* endoreduplication. *Plant Cell* **2006**, *18*, 382–396. [CrossRef] [PubMed]

74. Boudolf, V.; Lammens, T.; Boruc, J.; Leene, J.V.; Daele, H.V.D.; Maes, S.; Isterdael, G.V.; Russinova, E.; Kondorosi, E.; Witters, E.; et al. CDKB1;1 Forms a functional complex with CYCA2;3 to suppress endocycle onset. *Plant Physiol.* **2009**, *150*, 1482–1493. [CrossRef] [PubMed]
75. Zhou, M.Y.; Song, X.W.; Xu, J.; Fu, X.; Li, T.; Zhu, Y.C.; Xiao, X.Y.; Mao, Y.J.; Zeng, D.L.; Hu, J.; et al. Construction of genetic map and mapping and verification of grain traits QTLs using recombinant inbred lines derived from a cross between indica C84 and japonica CJ16B. *Chin. J. Rice Sci.* **2018**, *32*, 207–218.
76. Shen, L.; Hua, Y.F.; Fu, Y.P.; Li, J.; Liu, Q.; Jiao, X.Z.; Xin, G.W.; Wang, J.J.; Wang, X.C.; Yan, C.J.; et al. Rapid generation of genetic diversity by multiplex CRISPR/Cas9 genome editing in rice. *Sci. China Life Sci.* **2017**, *60*, 506–515. [CrossRef] [PubMed]
77. Ma, X.L.; Zhang, Q.Y.; Zhu, Q.L.; Liu, W.; Chen, Y.; Qiu, R.; Wang, B.; Yang, Z.F.; Li, H.Y.; Lin, Y.R.; et al. A robust CRISPR/Cas9 system for convenient, high-efficiency multiplex genome editing in monocot and dicot plants. *Mol. Plant* **2015**, *8*, 1274–1284. [CrossRef]
78. Hellens, R.P.; Allan, A.C.; Friel, E.N.; Bolitho, K.; Grafton, K.; Templeton, M.D.; Karunairetnam, S.; Gleave, A.P.; Laing, W.A. Transient expression vectors for functional genomics, quantification of promoter activity and RNA silencing in plants. *Plant Methods* **2005**, *1*, 13. [CrossRef]
79. You, M.K.; Lim, S.H.; Kim, M.J.; Jeong, Y.S.; Lee, M.G.; Ha, S.H. Improvement of the fluorescence intensity during a flow cytometric analysis for rice protoplasts by localization of a green fluorescent protein into chloroplasts. *Int. J. Mol. Sci.* **2015**, *16*, 788–804. [CrossRef]
80. Ishimaru, K.; Shirota, K.; Higa, M.; Kawamitsu, Y. Identification of quantitative trait loci for adaxial and abaxial stomatal frequencies in *Oryza sativa*. *Plant Physiol. Biochem.* **2001**, *39*, 173–177. [CrossRef]

10

AdRAP2.3, a Novel Ethylene Response Factor VII from *Actinidia deliciosa*, Enhances Waterlogging Resistance in Transgenic Tobacco through Improving Expression Levels of *PDC* and *ADH* Genes

De-Lin Pan †, Gang Wang †, Tao Wang, Zhan-Hui Jia, Zhong-Ren Guo and Ji-Yu Zhang *

Institute of Botany, Jiangsu Province and Chinese Academy of Sciences, Nanjing 210014, China; PPxsperfect@163.com (D.-L.P.); wg20092011@163.com (G.W.); immmorer@163.com (T.W.); 13915954315@163.com (Z.-H.J.); zhongrenguo@cnbg.net (Z.-R.G.)

* Correspondence: maxzhangjy@163.com

† These authors contributed equally to this work.

Abstract: APETALA2/ethylene-responsive factor superfamily (AP2/ERF) is a transcription factor involved in abiotic stresses, for instance, cold, drought, and low oxygen. In this study, a novel ethylene-responsive transcription factor named *AdRAP2.3* was isolated from *Actinidia deliciosa* 'Jinkui'. *AdRAP2.3* transcription levels in other reproductive organs except for the pistil were higher than those in the vegetative organs (root, stem, and leaf) in kiwi fruit. Plant hormones (Salicylic acid (SA), Methyl-jasmonate acid (MeJA), 1-Aminocyclopropanecarboxylic Acid (ACC), Abscisic acid (ABA)), abiotic stresses (waterlogging, heat, 4 °C and NaCl) and biotic stress (*Pseudomonas Syringae* pv. *Actinidiae*, *Psa*) could induce the expression of *AdRAP2.3* gene in kiwi fruit. Overexpression of the *AdRAP2.3* gene conferred waterlogging stress tolerance in transgenic tobacco plants. When completely submerged, the survival rate, fresh weight, and dry weight of transgenic tobacco lines were significantly higher than those of wile type (WT). Upon the roots being submerged, transgenic tobacco lines grew aerial roots earlier. Overexpression of *AdRAP2.3* in transgenic tobacco improved the pyruvate decarboxylase (PDC) and alcohol dehydrogenase (ADH) enzyme activities, and improved the expression levels of waterlogging mark genes *NtPDC*, *NtADH*, *NtHB1*, *NtHB2*, *NtPCO1*, and *NtPCO2* in roots under waterlogging treatment. Overall, these results demonstrated that *AdRAP2.3* might play an important role in resistance to waterlogging through regulation of *PDC* and *ADH* genes in kiwi fruit.

Keywords: ethylene-responsive factor; *Actinidia deliciosa*; *AdRAP2.3*; gene expression; waterlogging stress; regulation

---

1. Introduction

The plant APETALA2/ethylene-responsive factor (AP2/ERF) superfamily is one of the largest plant transcription factor families participating in plant development and resistance to biotic and abiotic stresses [1–4]. Transcription factors are known as regulating plant stress responses through binding to *cis*-acting elements in the promoters of stress-related genes or interacting with other transcription factors [5]. AP2/ERF superfamily members contain 60-70 highly conserved amino acids [6]. The AP2/ERF superfamily has been characterized based on either one or two AP2 domains: the AP2 subfamily has two AP2/ERF domains, the RAV subfamily has one AP2 domain and a B3 domain, C-repeat binding transcription factor/dehydrate responsive element binding factor (CBF/DREB) and the ERF subfamily have one AP2 domain [6]. ERF subfamily has conserved 14th alanine acid (A) and 19th aspartic acids (D) in the AP2/ERF domain. ERF subfamily transcription factors have been reported to be involved in biotic and abiotic stress [5]. The ERF subfamily was further divided into six small subgroups (V, VI, VII, VIII, IX, and X) based on the similarity of the amino acid sequences of their DNA-binding domain [1,7].

Members of the ERF subgroup VII have been identified to be involved in the hypoxic stress in *Arabidopsis thaliana*, including *RAP2.12*, *RAP2.2*, *RAP2.3*, *HRE1*, and *HRE2* [8,9]. The ERF VIIs were regulated by proteasome-mediated proteolysis via the oxygen-dependent branch of the N-end rule pathway [10,11]. Previous research showed that *RAP2.12* regulates central metabolic processes, for example, respiration, tricarboxylic acid (TCA) cycle, and amino acid metabolism [12]. *RAP2.2* has been proven to be induced by darkness, and overexpression of *RAP2.2* resulted in improving plant survival rate, increasing *ADH1* and *PDC1* expression levels, and increasing ADH and PDC enzyme activities [13]. In addition to *A. thaliana*, members of ERF subgroup VII have been found in other plants, for example, *Oryza sativa* [14] and *Petunia hybrida* [15].

The kiwi fruit is one of the most recently domesticated fruit crops. However, the majority of current kiwi fruit cultivars are susceptible to waterlogging stress. In many regions of China, especially in Eastern China, kiwi fruit plants suffer from excess rainfall during the summer rainy season. Waterlogging affects fruit yield severely and trees even die, which restricts the development of the kiwi fruit industry [16]. Kiwi fruit 'Jinkui' is tolerant to waterlogging, and understanding its waterlogging stress responses will be important for improving the tolerance of other kiwi fruit varieties. The transcriptome sequencing analysis in 'Jinkui' under waterlogging stress showed that the family of AP2/ERF transcription factors contained 14 upregulated and 14 downregulated members in the treatment samples. The transcription comp67160_0_seq1, which was a ERF VIIs member named *AdRAP2.3*, was strongly upregulated in roots of *Actinidia deliciosa* during the first 96 h after waterlogging treatment [17]. In order to analyze the function of the *AdRAP2.3* gene, the complete coding sequence (CDS) was cloned from the roots of 'Jinkui' under waterlogging stress according to the sequence comp67160_0_seq1 in this study. The expression patterns of *AdRAP2.3* in response to adverse stresses were investigated. The function of *AdRAP2.3* was further investigated by overexpression of *AdRAP2.3* in transgenic tobacco.

2. Results

2.1. Cloning and Sequence Analysis of AdRAP2.3

The *AdRAP2.3* gene was isolated from *A. deliciosa* roots according to the sequence comp67160_0_seq1 [17]. The gene *AdRAP2.3* contains an 837 bp complete open reading frame (ORF), encoding 278 amino acids (Figure S1) with a predicted molecular weight (MW) of 31.27 kDa, a theoretical isoelectric point (pI) of 5.32, an instability index of 37.29, and an average hydrophilic coefficient of −0.831. The protein is stable and hydrophilic. The predicted *AdRAP2.3* has a typical conserved DNA-binding domain (AP2/ERF domain) of 58 amino acids. Moreover, AdRAP2.3 has conserved alanine (A) and aspartic acid (D) residues at the 14th and 19th positions in the AP2/ERF domain, as is characteristic of the ERF subfamily (Figure S1).

The analysis of the secondary structure of the amino acid sequence showed that the amino acid sequence encoded by *AdRAP2.3* contains the AP2 domain, which is position from 60 aa to 120 aa (Figure S2). The position 60 to 90 is DNA binding sites according to Figure S3A. The three-dimensional structure of the protein encoded by the *AdRAP2.3* gene consists of one α-helix and three β-sheets according to Figure S3.

The 278 amino acid sequence predicted by the *AdRAP2.3* sequence was aligned with the amino acid sequences of other ERF VII subgroup proteins from *Arabidopsis*, *Oryza*, *Capsicum*, *Lycopersicon*, and *Actinidia*. The phylogenetic trees showed that RAP2.2 and RAP2.12 are closely related; SUB1C, SUB1B, and SUB1A-1 are closely related. PSR94738.1 is an ERF RAP2-3 like from *A. chinensis*, which is closely related to *AdRAP2.3* (Figure S4).

2.2. Expression Patterns of AdRAP2.3 in A. deliciosa

QRT-PCR assays were performed to investigate the expression pattern of *AdRAP2.3* in different organs and tissues of *A. deliciosa* (Figure 1). The expression level of *AdRAP2.3* is the highest in young fruit. *AdRAP2.3* transcription levels in other reproductive organs (except for pistils) were higher than those in the vegetative organs (root, stem, and leaf).

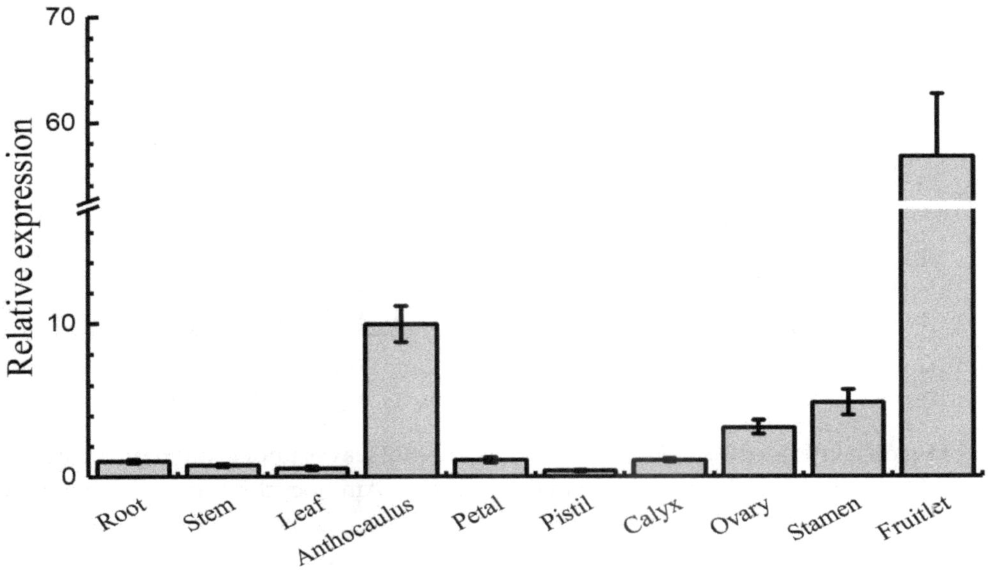

Figure 1. Expression analysis of gene *AdRAP2.3* in different kiwi fruit organs. *AdActin* transcription levels were used to normalize the samples. The mean and standard deviation were obtained from three independent experiments.

The biotic stresses were considered to be correlated with the hormone signals [18,19]. Therefore, the expression patterns of *AdRAP2.3* under the treatment of SA, MeJA, ACC and ABA were analyzed by real-time PCR. As shown in Figure 2A, after SA treatment, *AdRAP2.3* mRNA accumulation increases with time and reaches a maximum at 12 h. Upon treatment with MeJA, the transcription level is induced obviously and reaches a peak at 4 h, then reduces at 48 h (Figure 2B). The transcription levels were induced, with an induction peak at 4 h, and then declines over time until 48 h in kiwi fruit after treatment with ACC (Figure 3C). *AdRAP2.3* mRNA accumulation increases with time and reaches a maximum at 12 h and then decreases in kiwi fruit after ABA treatment (Figure 2D). These results indicated that plant hormones (SA, MeJA, ACC, and ABA) could induce the expression of *AdRAP2.3* gene in kiwi fruit.

Changes of *AdRAP2.3* transcription levels in *A. deliciosa* in response to various abiotic stresses were also analyzed by real-time PCR (Figure 3). As shown in Figure 3A, after 48 °C heat treatment, *AdRAP2.3* transcription level increases obviously at 2 h, reaching more than 30 times that in the control (0 h). The transcription level then decreases at 4 h (0 h). After 6 h recovery, the transcription level increases obviously. As shown in Figure 3B,C, after 4 °C and 0.2 M NaCl treatment, *AdRAP2.3* transcription level keeps going up during the first 48 h. *AdRAP2.3* mRNA accumulation increases about 400 times at 24 h and 600 times at 96 h after treatment with waterlogging (Figure 3D). Drought does not induce the expression of *AdRAP2.3* (Figure 3E). These results indicated that abiotic stresses including waterlogging, heat, low temperature (4 °C), and NaCl could induce the expression of *AdRAP2.3* gene in kiwi fruit. Compared with other environmental stresses, *AdRAP2.3* is strongly induced by waterlogging.

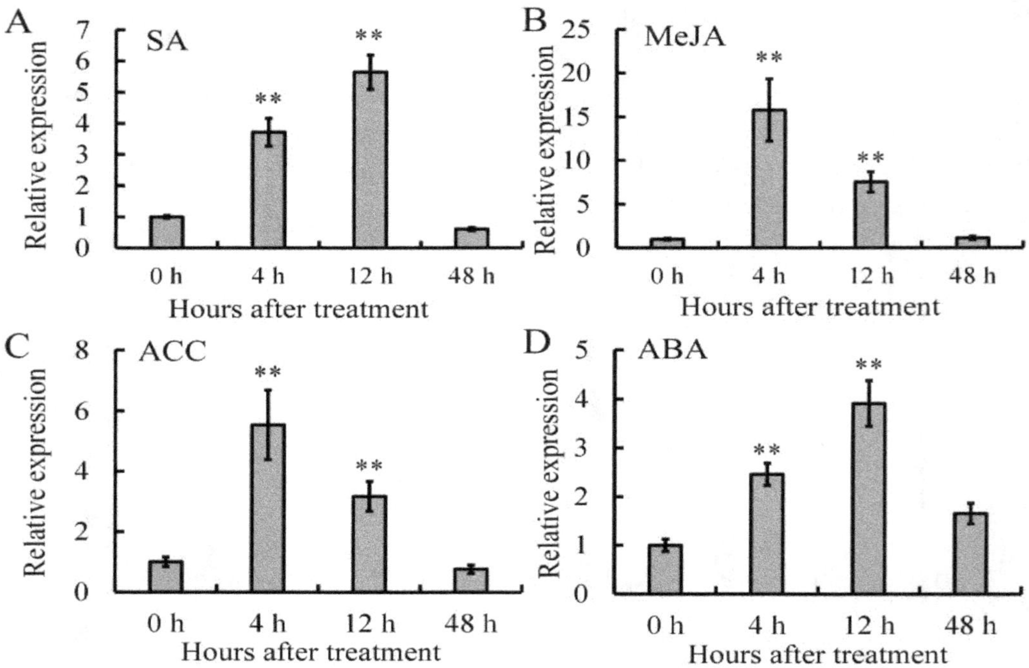

Figure 2. Expression analysis of *AdRAP2.3* in *A. deliciosa* 'Jinkui' leaves under different hormone treatments. (**A**) Salicylic acid (SA); (**B**) Methyl-jasmonate acid (MeJA); (**C**) 1-Aminocyclopropanecarboxylic Acid (ACC); (**D**) Abscisic acid (ABA). *AdActin* transcription levels were used to normalize the samples. The mean value and standard deviation were obtained from three independent experiments. The data represent averages ±Standard error (SE) of three biological repeats with three measurements per sample. ** indicates significant differences in comparison with the control at $p < 0.01$.

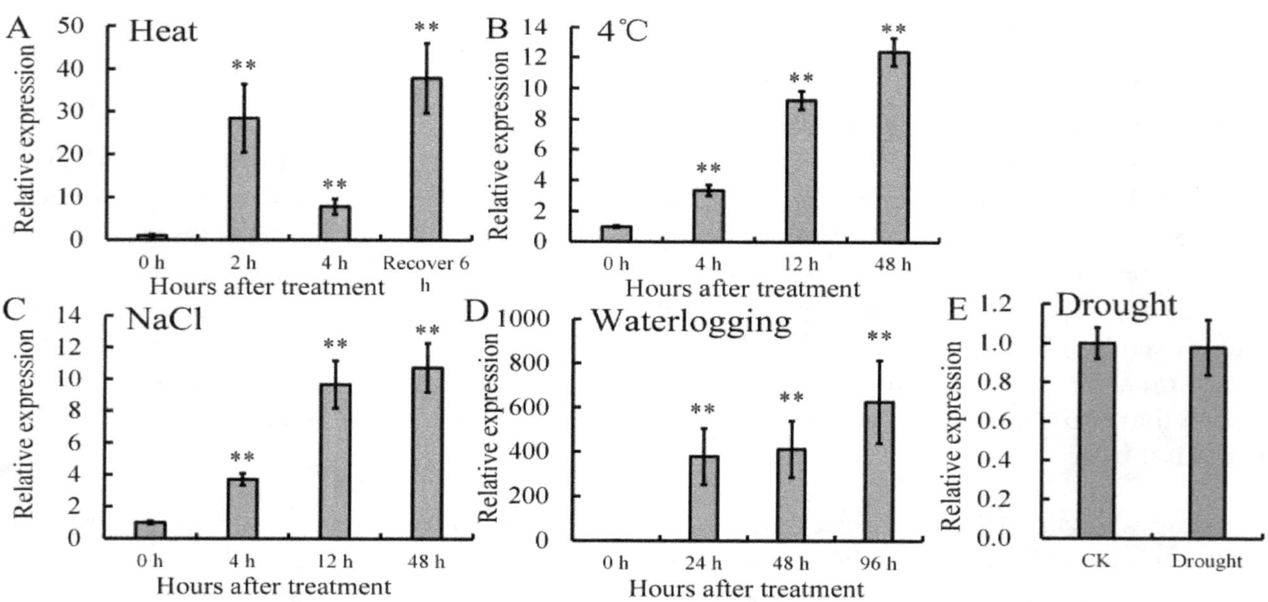

Figure 3. Expression analysis of *AdRAP2.3* gene under different abiotic stresses in kiwifruit. (**A**) Heat; (**B**) 4 °C; (**C**) NaCl; (**D**) waterlogging; (**E**) drought. *AdActin* transcription levels were used to normalize the samples. The mean value and standard deviation were obtained from three independent experiments. The data represent averages ± SE of three biological repeats with three measurements per sample. ** indicates significant differences in comparison with the control at $p < 0.01$.

Pseudomonas Syringae pv. *actinidiae* (*Psa*) is known as a serious disease to kiwi fruit; it causes cankers, cracks, and a reddish bacterial ooze on trunks [20]. So we investigated the *AdRAP2.3* expression in response to *Psa* infection, and the expression levels were analyzed in phloem treated with *Psa* infection (Figure 4). The *AdRAP2.3* transcription level increased significantly at 24 and 96 h, indicating that the *AdRAP2.3* gene is involved in responses to *Psa* stress.

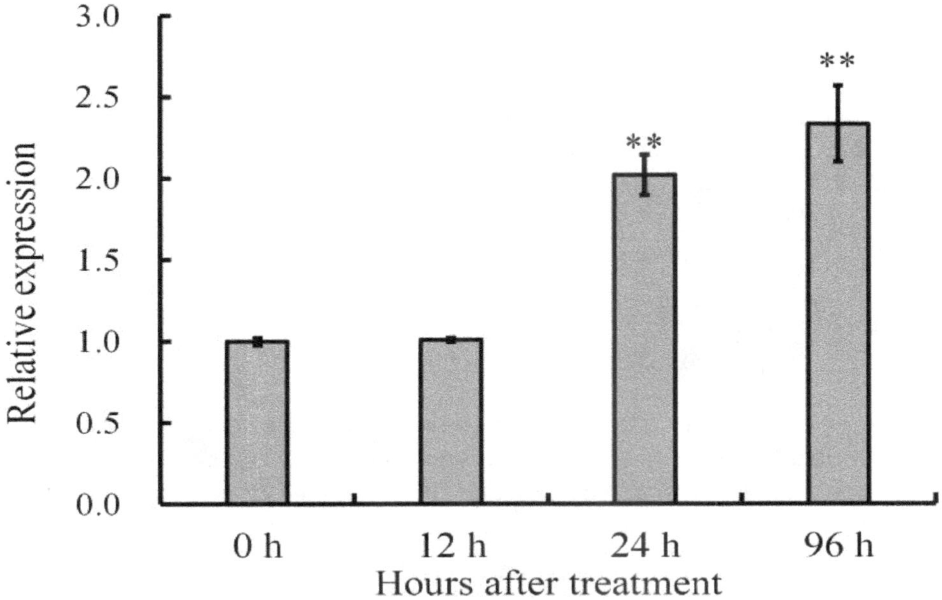

Figure 4. Expression analysis of *AdRAP2.3* gene under *Pseudomonas syringae* pv. *actinidiae* infection in kiwi fruit. *AdActin* transcription levels were used to normalize the samples. The mean value and standard deviation were obtained from three independent experiments. The data represent averages ± SE of three biological repeats with three measurements per sample. ** indicates significant differences in comparison with the control at $p < 0.01$.

2.3. Overexpression of AdRAP2.3 Enhanced Waterlogging Tolerance in Transgenic Plants

To investigate the function of kiwi fruit *AdRAP2.3* gene, transgenic tobacco plants overexpression of *AdRAP2.3* were generated. A total of 40 independent transgenic lines (T_0) were selected by hygromycin-resistance screening, and these transgenic lines were confirmed by β-glucuronidase (GUS) and PCR detection (Figure S5). The seeds from three representative *AdRAP2.3*-overexpressing lines (#1, #14 and #26) were selected for further functional analysis.

To investigated whether *AdRAP2.3* could increase waterlogging tolerance in transgenic tobacco plants. Seeds from three *AdRAP2.3*-overexpression lines and WT were planted on Murashige and Skoog (MS) medium for 10 days and then treated with completely submerged for 48 h (Figure 5). As shown in Figure 5A, the transgenic lines grow euphylla, the death rate of WT is significantly higher than those of transgenic lines (Figure 5B). Transgenic lines and WT were seeded on MS medium for 20 days and then treated with completely submerged for 7 d (Figure 6). The results showed that the transgenic lines grow better than WT after 5 d and 14 d recovery, and the fresh weight (Figure 6B) and dry weight (Figure 6C) of transgenic plants at recovery 14 d are significantly higher than those of WT respectively. To further verity that *AdRAP2.3* can confer the waterlogging tolerance of the overexpression lines, two-month-old transgenic plants and WT planted in soil were treated with root submerged (Figure 7). As shown in Figure 7B, after 7 d of treatment, transgenic lines grow aerial roots, but WT do not. These results showed that overexpression of *AdRAP2.3* could enhance the waterlogging tolerance in transgenic tobacco.

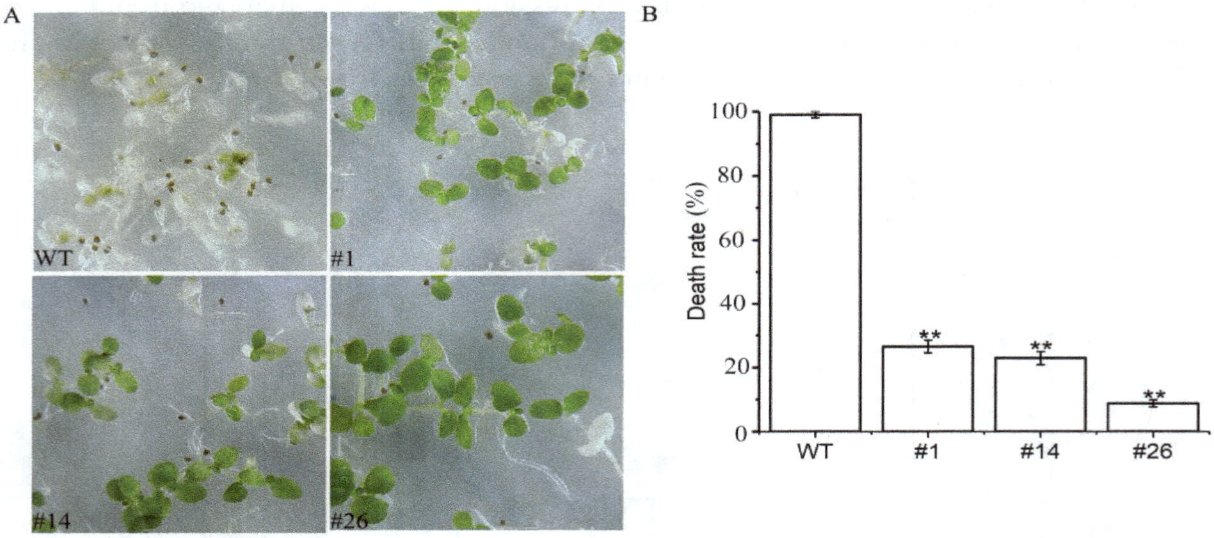

Figure 5. Phenotype (**A**) and death rate (**B**) in 10-day-old WT and transgenic tobacco line seedlings 48 h after they were completely submerged. WT: Wild type; #1,#14,#26: Transgenic tobacco lines. The data represent averages ± SE of three biological repeats with three measurements per sample. ** indicates significant differences in comparison with the WT at $p < 0.01$.

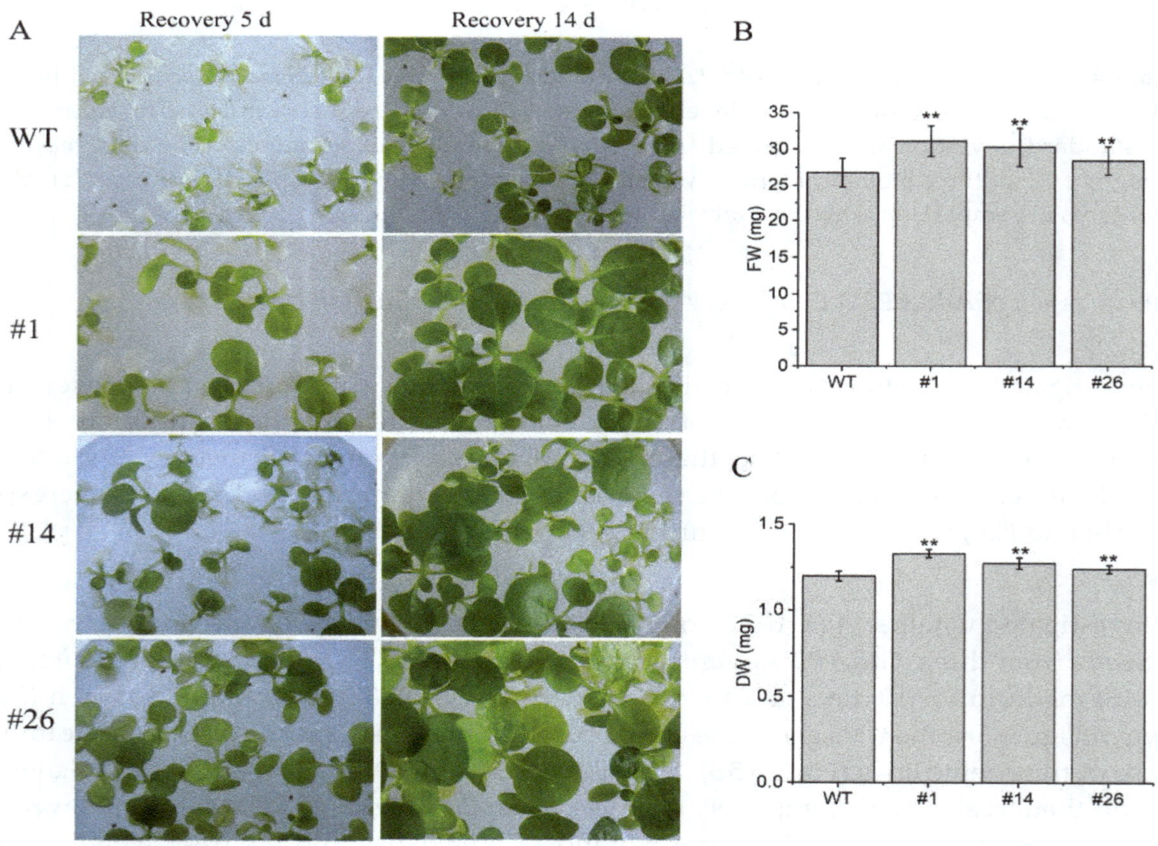

Figure 6. Phenotype, fresh weight (FW), and dry weight (DW) 7 d after completely submerged treatment for 20-day-old WT and transgenic tobacco lines seedlings. (**A**) Phenotype analysis; (**B**) fresh weight; (**C**) dry weight. WT: Wild type; #1, #14, #26: Transgenic tobacco lines. The data represent averages ± SE of three biological repeats with three measurements per sample. ** indicates significant differences in comparison with the WT at $p < 0.01$.

Figure 7. Morphological adaptations of two-month-old transgenic lines and WT plants were observed subjected to root submerged. (**A**) No waterlogging damage; (**B**) waterlogging for seven days. WT: Wild type; #1, #14, #26: Transgenic tobacco lines. Blue arrows indicated the aerial roots

2.4. Physiological Changes in Transgenic Plants under Waterlogging Stress

Alcoholic fermentation through the coupled activity of PDC and ADH enzyme is of great importance in a plant's ability to tolerate waterlogging [21–23]. All three transgenic lines showed improved resistance to waterlogging stress, and had similar results. The result of transgenic line #14 to improve waterlogging was a medium for all three transgenic lines. Transgenic line #14 was selected to explore the physiological mechanism of the waterlogging stress tolerance conferred by *AdRAP2.3* gene overexpression. PDC and ADH enzyme activities in root of WT and transgenic plants were measured in the control and in roots submerged for 24 d treatment (Figure 8). In the control, the PDC and ADH enzyme activities in the roots of transgenic lines are significantly higher than those of the WT. In terms of waterlogging stress, PDC and ADH enzyme activities in roots of transgenic lines are also significantly higher than those of WT. It can be concluded that overexpression of the *AdRAP2.3* gene can improve PDC and ADH enzyme activities.

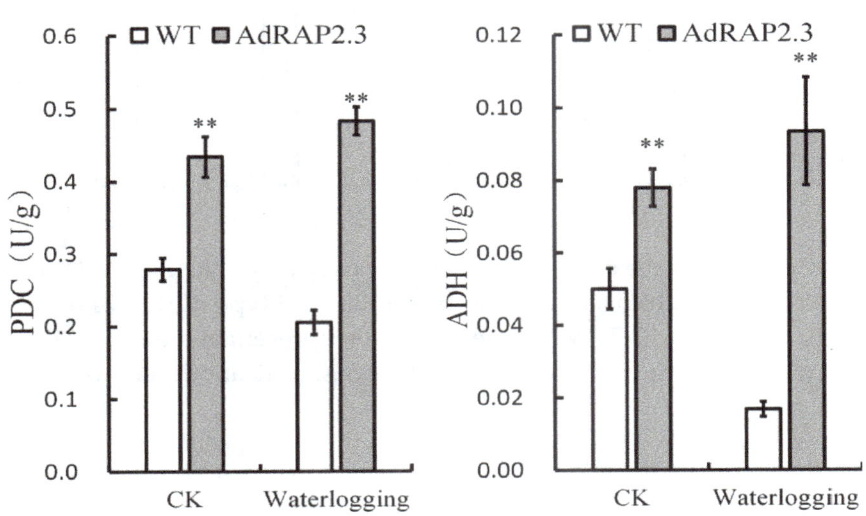

Figure 8. PDC and ADH enzyme activity measurements in roots of wild-type and AdRAP2.3 overexpressing lines #14. CK: normal growth conditions; WT: Wild type. The data represent averages ± SE of three biological repeats with three measurements per sample. ** indicates significant differences in comparison with the WT at $p < 0.01$.

2.5. Waterlogging-Related Genes Changes in Transgenic Plants under Waterlogging Stress

ADH, PDC, HB1, HB2, PCO1, and *PCO2* have been proved as marker genes in response to low oxygen stress [10,12]. There are no significantly difference between WT and transgenic lines on transcription levels of *NtPDC, NtADH, NtHB1, NtHB2,* and *NtPCO1* in roots under control condition (Figure 9). However, under waterlogging stress, the expression levels of *NtPDC, NtADH, NtHB1, NtHB2,* and *NtPCO1* in root of transgenic lines are significantly higher than those of WT (Figure 9). *NtPCO2* expression levels in roots are higher than those of WT under normal condition and waterlogging stress. Taken together, these results indicated overexpression *AdRAP2.3* can upregulate the expression levels of *NtPDC, NtADH, NtHB1, NtHB2, NtPCO1,* and *NtPCO2* under waterlogging stress.

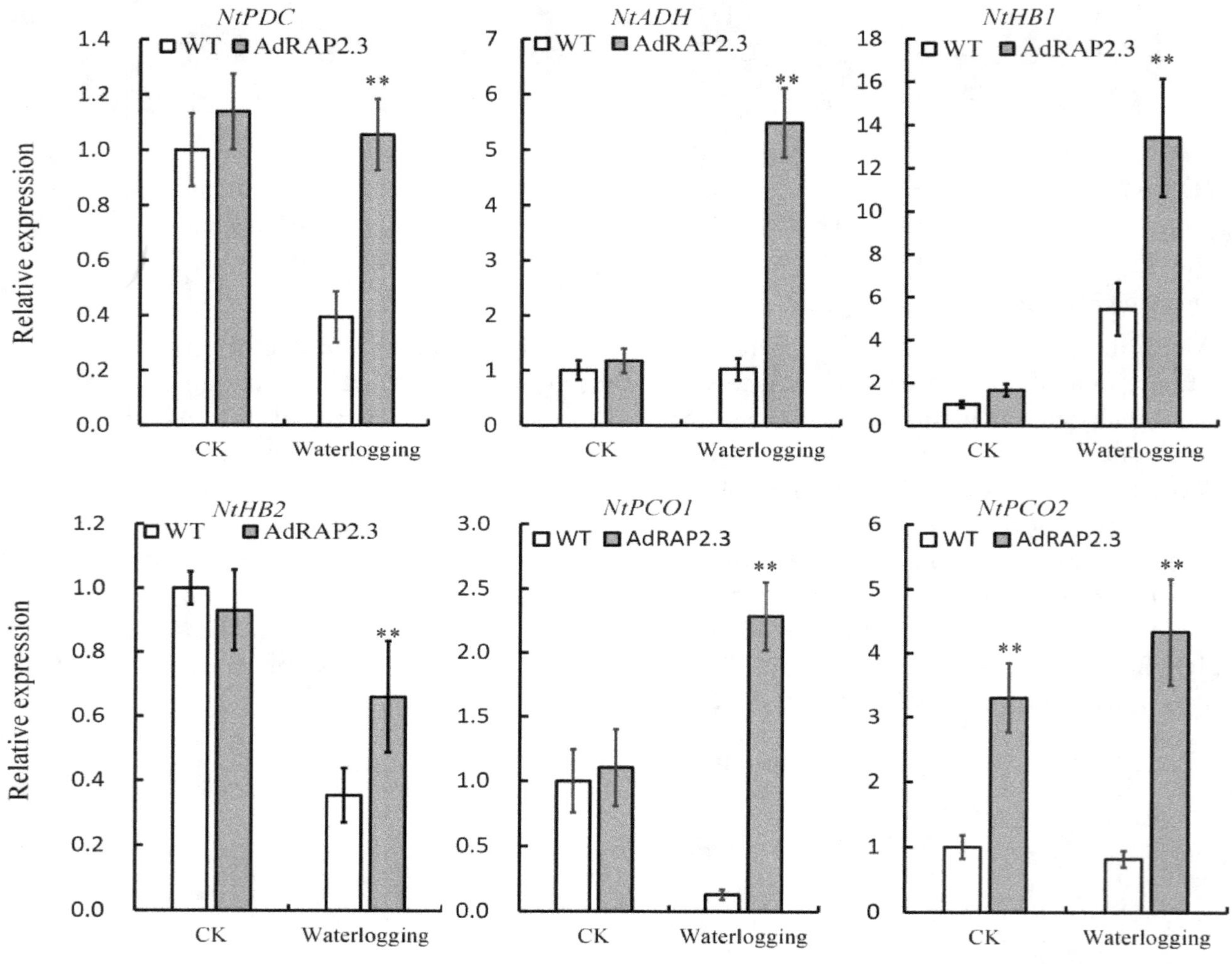

Figure 9. QRT-PCR analysis of the expression of the genes *NtPDC, NtADH, NtHB1, NtHB2, NtPCO1,* and *NtPCO2* in the root line of transgenic tobacco and the wild type during 24 d waterlogging stress. CK: Normal growth conditions; WT: Wild type. The data represent averages ± SE of three biological repeats with three measurements per sample. ** indicates significant differences in comparison with the WT at $p < 0.01$.

3. Discussion

3.1. Kiwi Fruit AdRAP2.3 Plays a Key Role in Resistance to Waterlogging Stress

Waterlogging or submergence caused O_2 deprivation in the soil [24,25]. ERF members are important regulators of low oxygen tolerance extensively studied in many plants [8,10,11]. The ERF family plays a crucial role in the determination of survival of *Arabidopsis* and rice, which could reduce oxygen availability [11,26]. Previous studies showed that there are five ERF VIIs genes in *Arabidopsis*; two ERF VIIs genes (*HYPOXIA RESPONSIVE ERF1/2*) were greatly enhanced at the transcriptional and translational levels by O_2 deprivation at multiple developmental stages, and the other three ERF VIIs genes were constitutively expressed (*RAP2.12*, *RAP2.2*, and *RAP2.3*) and further upregulated by darkness or ethylene in *A. thaliana* [11,27,28]. Transgenic *Arabidopsis* plants overexpressing *HRE1* showed an improved tolerance of anoxia [11]. The transcriptome sequencing analysis showed that there are 28 AP2/ERF transcription factors regulated by waterlogging in kiwi fruit. In this study, *A. deliciosa AdRAP2.3* gene was induced significantly by waterlogging, and overexpression of the *AdRAP2.3* gene conferred waterlogging stress tolerance in transgenic tobacco plants. When completely submerged, transgenic tobacco lines had a significantly higher survival rate, fresh weight, and dry weight compared to the WT. These results suggested that kiwi fruit *AdRAP2.3* plays a key role in resistance to waterlogging.

3.2. *AdRAP2.3 Could Enhance Resistance to Waterlogging through Promoting Pneumatophore Production*

Waterlogging may trigger different molecular and physiological disorders in plants. These include significant deterioration of plant water statues [29], decreases in leaf gas exchange variables, photoinhibition of photosystems (PSI and PSII) [30], and decreases in root system biomass. Using a light microscope, we see that the root anatomical structures of plants exhibit changes in root diameter, stele diameter, epidermal thickness, and xylem thickness under waterlogging stress [31]. In addition, the formation of aerenchym tissue is found after treated with waterlogging stress, including an increasing number of aerenchym cells and increasing length of aerenchym cells [31,32]. Aerenchyma formation under waterlogging stress is one of the most effective mechanisms to provide an adequate oxygen supply and overcome the stress-induced hypoxia imposed on plants [33]. The numerous pneumatophores contribute to morphological adaptations under waterlogging stress. Regardless of pneumatophores, stressed oil palm seedlings were able to adjust their leaf water status and gas exchange to cope with waterlogging [34]. Waterlogging subjects plant roots to an anoxic environment, limiting mitochondrial aerobic respiration and causing energy loss. In response to hypoxia stress, plants can temporarily compensate with anaerobic respiration [35]. In this study, the overexpression of *AdRAP2.3* induces pneumatophores under waterlogging in transgenic tobacco, suggesting that kiwi fruit *AdRAP2.3* could enhance resistance to waterlogging through promoting pneumatophore production.

3.3. *Kiwi Fruit AdRAP2.3 Enhances Waterlogging Resistance in Transgenic Tobacco through Improving Expression Levels of PDC and ADH Genes*

Three enzymes involved in anaerobic metabolic pathways, PDC, ADH, and lactate dehydrogenase (LDH), produce acetaldehyde, ethanol, and lactic acid, respectively. Although acetaldehyde and ethanol are thought to be harmful to plant cells, lactic acid is a major cause of root death through its induction of cytoplasmic acidification and pH reduction [36–38]. In this study, transgenic lines can improve PDC and ADH enzyme activities in the control. In response to long-term waterlogging stress, the PDC and ADH enzyme activities of WT decrease, whereas those of transgenic lines increase to sustain substrate-level adenosine triphosphate (ATP) production and promote hypoxia acclimation. These results indicate that *AdRAP2.3* promotes PDC and ADH enzyme activities.

Previously, a large number of high-throughput sequence (Tag-seq) analyses based on the Solexa Genome Analyzer platform were performed to analyze the gene expression profiling of plants under waterlogging stress. Differentially expressed genes (DEGs) are obtained, mainly linked to carbon metabolism, photosynthesis, reactive oxygen species generation/scavenging, and hormone synthesis/signaling [17,39–42]. Some waterlogging-responsive genes were isolated and identified, such as *RAP2.12*, *RAP2.2* [13], *HRE1*, *HRE2*, *AdPDC1* [43], *AdPDC2* [44], *AdADH1* [45], and *AdADH2* [45]. *RAP2.12* mRNA was described as sufficient for activating the anaerobic response in *Arabidopsis* [11]. The closest RAP2.12 homologue, RAP2.2, has been suggested to be functionally redundant in the induction of the anaerobic gene expression [13,46]. *HRE1* and *HRE2* are expressed at low levels under aerobic conditions and strongly upregulated by hypoxia [11]. Overexpression of *AdPDC1*, *AdPDC2*, *AdADH1*, and *AdADH2* in *Arabidopsis* enhanced waterlogging tolerance [43–45]. RAP2.12 was re-localized from the plasma membrane to the nucleus as O_2 concentrations declined, with increased accumulation of hypoxia-responsive mRNAs, including PDC1 and hypoxia-responsive ERF1/2 (HRE1/2) [10]. In this study, we further analyzed the expression of the hypoxia marker genes *NtADH*, *NtPDC*, *NtHB1*, *NtHB2*, *NtPCO1*, and *NtPCO2* in the roots of WT and a transgenic line under submerged root stress, and the results showed that *AdRAP2.3* regulates these hypoxia marker genes mRNA levels under waterlogging conditions. Thus, kiwi fruit *AdRAP2.3* enhances waterlogging resistance in transgenic tobacco by improving expression levels of *PDC* and *ADH* genes.

In summary, the results showed that the increase of *AdRAP2.3* expression during waterlogging stress was much higher than that during other environmental stresses and that the kiwi fruit *AdRAP2.3* gene is required during waterlogging stress. Upon roots being submerged, transgenic tobacco lines grew aerial roots earlier than the WT. Overexpression of the *AdRAP2.3* gene in transgenic tobacco improved the activities of the PDC and ADH enzymes, and the expression levels of waterlogging mark genes *NtPDC* and *NtADH* in roots under waterlogging treatment. Overall, these results suggested that *AdRAP2.3* might play an important role in resistance to waterlogging through regulation of *PDC* and *ADH* genes in kiwi fruit. In future studies we will concentrate on introducing the *AdRAP2.3* gene into kiwi fruit varieties with poor waterlogging resistance, such as 'Hongyang', to determine whether this can enhance resistance to waterlogging.

4. Materials and Methods

4.1. Plant Materials and Growth Conditions

The kiwi fruit cultivar 'Jinkui' was obtained from the Institute of Botany (Nanjing), Jiangsu Province and Chinese Academy of Sciences, China. The cutting seedlings from 'Jinkui' were grown in pots containing a 7:2:1 mixture of peat mold, vermiculite, and perlite in the greenhouse. Tissue culture seedlings of 'Jinkui' grew in a rooting medium (1/2 MS medium containing 0.6 mg/mL 1-naphthlcetic acid (NAA)) for one month. Shoots of 'Jinkui' were selected, surface sterilized, and grown in MS medium for one week. The conditions were: average temperature of 25 °C, relative air humidity of 60%, photoperiod of 16 h/8 h (light/dark), and quantum irradiance of 160 μmol m^{-2} s^{-1}.

4.2. Treatments

To analyze the tissue-specific gene expression, different organs and tissues of 'Jinkui', including the root, stem, leaf, anthocaulus, petal, pistil, calyx, ovary, stamen, and fruitlet (20 days after full blossom, DAFB), were collected. To analyze gene expression patterns in response to hormones, 0.1 mM SA, 0.05 mM MeJA, 0.01 mM ACC, and 0.01 mM ABA were sprayed on the surface of the culture tissue seedlings leaves, and the leaves were collected at 0, 4, 12, and 48 h to analyze the gene expression pattern in response to stresses including heat, cold, salt, waterlogging, and drought. For heat and cold stress, the tissue seedlings were cultured at 48 °C for 0, 2, and 4 h, had a 23 °C recovery for 6 h, and then were exposed to 4 °C for 0, 4, 12, and 48 h, after which the leaves were collected. For salt stress, the seedlings were cultured in 28 cm × 14 cm × 14 cm containers with 0.2 M NaCl, and the leaves were collected at 0, 4, 12, and 48 h. For waterlogging, the seedlings were waterlogged in a 28 cm × 14 cm × 14 cm container filled with tap water to 2.5 cm above the level of the soil surface, and roots were sampled at 0, 24, 48, and 96 h. For drought, the cutting seedlings were cultured without watering for 14 d (the control was irrigated normally) and the leaves were sampled [45]. To analyze the gene expression pattern in response to *Psa*, bacterial cells were suspended in distilled water and adjusted to an OD_{600} = 0.2, then injected into the 'Jinkui' shoots, which were carved with a knife for 0, 24, 48, or 96 h, and then phloem from the shoots was collected [47]. Different samples were snap frozen in liquid nitrogen and stored at −80 °C for later experiments.

4.3. RNA Extraction and cDNA Synthesis

Total RNA was extracted from samples according to a method reported previously [48]. The purity and content of total RNA were detected by a spectrophotometer (Bruker BioSpin GmbH, Rheinstetten, Germany) and 1.0% agarose electrophoresis. The cDNA was achieved with a PrimeScript^{TM} RT reagent kit with gDNA Eraser (Perfect Real Time, TaKaRa, Cat. #RR047Q, Dalian, China), which could eliminate the residual DNA. The cDNA samples were diluted 1:10 with sterile double distilled water and stored at −20 °C before being used.

4.4. Gene Clone and Sequence Analysis

The cDNA sample from the 'Jinkui' root treatment with waterlogging for four days was used to amplify the complete CDS of *AdRAP2.3*. Gene-specific primers F1 and F2 were designed according to the comp67160_0_seq1 sequence (Table S1) [17]. The open reading frame (ORF) was predicted by DNAstar 7.1.0(http://korwin-mikke.pl). Multiple alignments of the deduced amino acid sequence were performed using the the BioEdit software (v 7. 0. 5, Ibis Therapeutics, Carlsbad, CA, USA), and a phylogenetic tree was constructed by 1000 Bootstrap statistical tests with the Neighbor Joining (NJ) model by Mega 7.0 (https://www.megasoftware.net/). Molecular weight and isoelectric point were analyzed by Exspay (https://web.expasy.org/protparam/). The three-dimensional structure of *AdRAP2.3* was predicted by Swiss-plot (http://swissmodel.expasy.org/).

4.5. Gene Expression Analysis Using Quantitative Real-Time PCR

The qRT-PCR was carried out on an Applied Biosystems 7300 Real Time PCR System (Applied Biosystems, Waltham, MA, USA) using TaKaRa Company SYBR Premix Ex TaqTM II(Perfect Real Time, TaKaRa, code: DRR041A, Dalian, China). *AdActin* was used as internal reference gene to monitor cDNA abundance [49]. The quantitative PCR reaction system with specific primer contains 1 μL cDNA template, 10 μL 2 × SYBR Premix Ex Taq^{TM} II, 0.3μL (10 pm) of each primer (Table S1), and 8.4 μL ddH_2O. The primer sequences used are listed in Table S1. The reaction procedure is as follows: denaturation at 95 °C for 1 min, 95 °C for 20 s, 57 °C for 20 s, and 72 °C for 20 s; 45 cycles in total. Each sample set is repeated three times. After the reaction, the $2^{-\triangle\triangle Ct}$ method was used to analyze the gene expression level. The statistical significance was assessed using SPSS 17.0 (SPSS Corp., Chicago, IL, USA).

4.6. Construction Binary Vector and Transformation of Tobacco

The coding sequence of *AdRAP2.3* was amplified by PCR using a specific primer pair (F1 and F2) modified to include 5′BamH I and 3′Sac I restriction sites. The fragment was inserted into the binary vector pCAMBIA 1301 under the control of the *Cauliflower mosaic virus* (CaMV35S) promoter. Then, the plasmid was introduced into *Agrobacterium tumefaciens*. Agrobacterium-mediated transformation of tobacco was performed by the leaf disc method [50]. The transgenic plants were detected by GUS staining and PCR analysis. The seeds of transgenic lines were harvested at the same stage and stored for subsequent analysis.

4.7. Phenotype Analysis of Transgenic Tobacco under Waterlogging Resistance

Seeds from T_1 progeny transgenic lines (#1, #14, and #26) and WT were surface sterilized and sown on the MS medium for 10 days; the death rate was measured after they were completely submerged for 48 h. Seedlings of transgenic lines and WT grew on MS medium for 20 d, the waterlogging treatment was performed for 7 d, and then seedlings were returned to normal growth conditions for 14 d. After 14 d recovery, the fresh weight and dry weight were measured. *Nicotiana tabacum* seedlings (#1, #14, #26, and WT) were transplanted into pots from the medium and grown in a greenhouse for two months under a 16/8-h light/dark cycle at 22/25 °C and 60% relative humidity. The pots were flooded in 28 cm × 14 cm × 14 cm containers filled with tap water to 2.5 cm above the level of the soil surface. The seedlings' phenotypic changes during waterlogging stress assays were observed and photographed. The roots of transgenic lines and WT were collected after 24 d waterlogging and were later used for the measurement of enzyme activities and expression levels of downstream genes.

4.8. Measurement of Anaerobic Respiration and ADH and PDC Activities

The enzyme ADH and PDC activities in roots of transgenic lines #14 and WT after 24 d waterlogging were measured spectrophotometrically by monitoring the oxidation of Nicotinamide adenine dinucleotide (NADH) at 340 nm [21]. The PDC assay reaction was carried out for 60 s at 25 °C. The ADH assay reaction was carried out for 10 min at 37 °C. One unit of PDC or ADH was defined as the amount of enzyme required to decompose 1 μmol of NADH per minute per gram fresh weight; at least 10 independent plants were evaluated in each test, and all tests were repeated three times.

4.9. Expression Analysis of Downstream Genes

The expression levels of downstream genes in roots of transgenic line #14 and the WT after 24 d waterlogging were analyzed by qRT-PCR. The downstream genes *NtPDC, NtADH, NtHB1, NtHB2, NtPCO1*, and *NtPCO2* were acquired from NCBI (https://www.ncbi.nlm.nih.gov/). *NtTub* was used as an internal reference gene to monitor cDNA abundance [50]. The primers are listed in Table S1. Statistically significant differences were calculated with SPSS 17.0 (SPSS Corp., Chicago, IL, USA).

Author Contributions: J.-Y.Z. designed and initiated this study. G.W. and Z.-H.J. carried out the bioinformatics analyses. D.-L.P. performed the experiments. D.-L.P. and J.-Y.Z. wrote the manuscript. T.W. and Z.-R.G. helped in discussions of the manuscript. All authors read and approved the final manuscript.

Acknowledgments: This research is supported by the Natural Science Foundation of Jiangsu Province of China (General Program, BK20171328).

References

1. Nakano, T.; Suzuki, K.; Fujimura, T.; Shinshi, H. Genome-wide analysis of the ERF gene family in *Arabidopsis* and rice. *Plant Physiol.* **2006**, *140*, 411–432. [CrossRef] [PubMed]
2. Li, Z.J.; Tian, Y.S.; Xu, J.; Fu, X.Y.; Gao, J.J.; Wang, B.; Han, H.J.; Wang, L.J.; Peng, R.H.; Yao, Q.H. A tomato ERF transcription factor, SlERF84, confers enhanced tolerance to drought and salt stress but negatively regulates immunity against *Pseudomonas syringae* pv. tomato *DC3000*. *Plant Physiol. Biochem.* **2018**, *132*, 683–695. [CrossRef] [PubMed]
3. Riechmann, J.L.; Meyerowitz, E.M. The AP2/EREBP family of plant transcription factors. *Biol. Chem.* **1998**, *379*, 633–646. [PubMed]
4. Liang, Y.Q.; Li, X.S.; Zhang, D.Y.; Gao, B.; Yang, H.L.; Wang, Y.C.; Guan, K.Y.; Wood, A.J. *ScDREB8*, a novel A-5 type of DREB gene in the desert moss *Syntrichia caninervis*, confers salt tolerance to *Arabidopsis*. *Plant Physiol. Biochem.* **2017**, *120*, 242–251. [CrossRef] [PubMed]
5. Wu, D.Y.; Ji, J.; Wang, G.; Guan, C.F.; Jin, C. LchERF, a novel ethylene-responsive transcription factor from *Lycium chinense*, confers salt tolerance in transgenic tobacco. *Plant Cell Rep.* **2014**, *33*, 2033–2045. [CrossRef] [PubMed]
6. Li, X.P.; Zhu, X.Y.; Mao, J.; Zou, Y.; Fu, D.W.; Chen, W.X.; Lu, W.J. Isolation and characterization of ethylene response factor family genes during development, ethylene regulation and stress treatments in papaya fruit. *Plant Physiol. Biochem.* **2013**, *70*, 81–92. [CrossRef] [PubMed]

7. Sakuma, Y.; Liu, Q.; Dubouzet, J.G.; Abe, H.; Shinozaki, K.; Yamaguchi-Shinozaki, K. DNA-Binding specificity of the ERF/AP2 domain of Arabidopsis DREBs, transcription factors involved in dehydration- and cold-inducible gene expression. *Biochem. Biophys. Res. Commun.* **2002**, *290*, 998–1009. [CrossRef]
8. Gibbs, D.J.; Lee, S.C.; Isa, N.M.; Gramuglia, S.; Fukao, T.; Bassel, G.W.; Correia, C.S.; Corbineau, F.; Theodoulou, F.L.; Bailey-Serres, J.; et al. Homeostatic response to hypoxia is regulated by the N-end rule pathway in plants. *Nature* **2011**, *479*, 415–418. [CrossRef]
9. Gasch, P.; Fundinger, M.; Müller, J.T.; Lee, T.; Bailey-serres, J.; Mustroph, A. Redundant ERF-VII transcription factors bind an evolutionarily-conserved cis-motif to regulate hypoxia-responsive gene expression in *Arabidopsis*. *Plant Cell* **2015**, *28*, 160–180. [CrossRef]
10. Voesenek, L.A.C.J.; Bailey-Serres, J. Flood adaptive traits and processes: An overview. *New Phytol.* **2015**, *206*, 57–73. [CrossRef]
11. Licausi, F.; Kosmacz, M.; Weits, D.A.; Giuntoli, B.; Giorgi, F.; Voesenek, L.A.C.J.; Perata, P.; Dongen, J.T. Oxygen sensing in plants is mediated by an N-end rule pathway for protein destabilization. *Nature* **2011**, *479*, 419–422. [CrossRef] [PubMed]
12. Paul, M.V.; Iyer, S.; Amerhauser, C.; Lehmann, M.; Dongen, J.T.; Geigenberger, P. Oxygen sensing via the ethylene response transcription factor RAP2.12 affects plant metabolism and performance under both normoxia and hypoxia. *Plant Physiol.* **2016**, *172*, 141–153. [CrossRef] [PubMed]
13. Hinz, M.; Wilson, I.W.; Yang, J.; Buerstenbinder, K.; Llewellyn, D.; Dennis, E.S.; Sauter, M.; Dolferus, R. *Arabidopsis* RAP2.2: An ethylene response transcription factor that is important for hypoxia survival. *Plant Physiol.* **2010**, *153*, 757–772.
14. Perata, P.; Voesenek, L.A.C.J. Submergence tolerance in rice requires *Sub1A*, an ethylene-response-factor-like gene. *Trends Plant Sci.* **2007**, *12*, 43–46. [CrossRef] [PubMed]
15. Liu, J.X.; Li, J.Y.; Wang, H.N.; Fu, Z.D.; Liu, J.; Yu, Y.X. Identification and expression analysis of ERF transcription factor genes in petunia during flower senescence and in response to hormone treatments. *J. Exp. Bot.* **2011**, *62*, 825–840. [CrossRef] [PubMed]
16. Zhang, J.Y.; Pan, D.L.; Wang, G.; Xuan, J.P.; Wang, T.; Guo, Z.R. Genome-wide analysis and expression pattern of the AP2/ERF gene family in kiwifruit under waterlogging stress treatment. *Int. J. Environ. Agric. Res.* **2017**, *3*, 83–90.
17. Zhang, J.Y.; Huang, S.N.; Mo, Z.H.; Xuan, J.P.; Jia, X.D.; Wang, G.; Guo, Z.R. De novo transcriptome sequencing and comparative analysis of differentially expressed genes in kiwifruit under waterlogging stress. *Mol. Breed.* **2015**, *35*, 208. [CrossRef]
18. Vidhyasekaran, P. Abscisic acid signaling system in plant innate immunity. In *Plant Hormone Signaling Systems in Plant Innate Immunity. Signaling and Communication in Plants*; Springer: Dordrecht, The Netherlands, 2015; Volume 2, pp. 245–309.
19. Pál, M.; Szalai, G.; Kovács, V.; Gondor, O.K.; Janada, T. Salicylic acid-mediated abiotic stress tolerance. In *Salicylic Acid*; Hayat, S., Ahmad, A., Alyemeni, M., Eds.; Springer: Dordrecht, The Netherlands, 2013; pp. 183–247.
20. Janse, J.D.; Scortichini, M. Characterization of *Pseudomonas Syringae* pv. *actinidiae*, the causal agent of bacterial canker of kiwifruit by whole cell protein electrophoresis and fatty acid analysis. In *Pseudomonas Syringae Pathovars and Related Pathogens. Developments in Plant Pathology*; Rudolph, K., Burr, T.J., Mansfield, J.W., Stead, D., Vivian, A., von Kietzell, J., Eds.; Springer: Dordrecht, The Netherlands, 1997; Volume 9, p. 499.
21. Yin, D.M.; Chen, S.M.; Chen, F.D.; Guan, Z.Y.; Fang, W.M. Morpho-anatomical and physiological responses of two Dendranthema species to waterlogging. *Environ. Exp. Bot.* **2010**, *68*, 122–130. [CrossRef]
22. Ricard, B.; Couee, I.; Raymond, P.; Saglio, P.H.; Saintges, V.; Pradet, A. Plant metabolism under hypoxia and anoxia. *Plant Physiol. Biochem.* **1994**, 32, 1–10.
23. Tadege, M.; Dupuis, I.; Kuhlemeier, C. Ethanolic fermentation: New functions for an old pathway. *Trends Plant Sci.* **1999**, *4*, 320–325. [CrossRef]
24. Bailey-Serres, J.; Voesenek, L.A. Flooding stress: Acclimations and genetic diversity. *Ann. Rev. Plant Boil.* **2008**, *59*, 313–339. [CrossRef] [PubMed]
25. Yamauchi, T.; Watanabe, K.; Fukazawa, A.; Mori, H.; Abe, F.; Kawaguchi, K.; Oyanagi, A.; Nakazono, M. Ethylene and reactive oxygen species are involved in root aerenchyma formation and adaptation of wheat seedlings to oxygen-deficient conditions. *J. Exp. Bot.* **2014**, *65*, 261–273. [CrossRef] [PubMed]

26. Licausi, F.; Van Dongen, J.T.; Giuntoli, B.; Novi, G.; Santaniello, A.; Geigenberger, P.; Perata, P. HRE1 and HRE2, two hypoxia-inducible ethylene response factors, affect anaerobic responses in *Arabidopsis thaliana*. *Plant J.* **2010**, *62*, 302–315. [CrossRef] [PubMed]
27. Gibbs, D.J.; Bacardit, J.; Bachmair, A.; Holdsworth, M.J. The eukaryotic N-end rule pathway: Conserved mechanisms and diverse functions. *Trends Cell Biol.* **2014**, *24*, 603. [CrossRef] [PubMed]
28. Licausi, F.; Pucciariello, C.; Perata, P. New role for an old rule: N-end rule-mediated degradation of ethylene responsive factor proteins governs low oxygen response in plants. *J. Integr. Plant Boil.* **2013**, *55*, 31–39. [CrossRef] [PubMed]
29. Rasheed-Depardieu, C.; Parelle, J.; Tatin-Froux, F.; Parent, C.; Capelli, N. Short-term response to waterlogging in Quercus petraea and Quercus robur: A study of the root hydraulic responses and the transcriptional pattern of aquaporins. *Plant Physiol. Biochem.* **2015**, *97*, 323–330. [CrossRef] [PubMed]
30. Yan, K.; Zhao, S.J.; Cui, M.X.; Han, G.X.; Wen, P. Vulnerability of photosynthesis and photosystem I in Jerusalem artichoke (*Helianthus tuberosus* L.) exposed to waterlogging. *Plant Physiol. Biochem.* **2018**, *125*, 239–246. [CrossRef] [PubMed]
31. Purnobasuki, H.; Nurhidayati, T.; Hariyanto, S.; Jadid, N. Data of root anatomical responses to periodic waterlogging stress of tobacco (*Nicotiana tabacum*) varieties. *Data Brief* **2018**, *20*, 2012–2016. [CrossRef] [PubMed]
32. Yin, D.M.; Chen, S.M.; Chen, F.D.; Jiang, J.F. Ethylene promotes induction of aerenchyma formation and ethanolic; fermentation in waterlogged roots of *Dendranthema* spp. *Mol. Biol. Rep.* **2013**, *40*, 4581–4590. [CrossRef]
33. Zhang, X.; Fan, Y.; Shabala, S.; Koutoulis, A.; Shabala, L.; Johnson, P.; Hu, H.L.; Zhou, M.X. A new major-effect QTL for waterlogging tolerance in wild barley (*H. spontaneum*). *Theor. Appl. Genet.* **2017**, *130*, 1559–1568. [CrossRef]
34. Ponte, N.H.T.; Santos, R.I.; Filho, W.R.; Cunha, R.L.; Magalhaes, M.M.; Pinheiro, H.A. Morphological assessments evidence that higher number of pneumatophores improves tolerance to long-term waterlogging in oil palm (*Elaeis guineensis*) seedlings. *Flora* **2019**, *250*, 52–58. [CrossRef]
35. Wang, W.Q.; Zhang, F.S.; Zheng, Y.Z.; Mei, H.X. Comparison of morphology, physiology and mineral element contents among genotypes of sesame with different tolerance to waterlogging under anaerobic condition. *Chin. J. Appl. Ecol.* **2002**, *13*, 421–424.
36. Zhou, C.P.; Bai, T.; Wang, Y.; Wu, T.; Zhang, X.Z.; Xu, X.F.; Han, Z.H. Morpholoical and enzymatic responses to waterlogging in three *Prunus* species. *Sci. Hortic.* **2017**, *221*, 62–67. [CrossRef]
37. An, Y.Y.; Qi, L.; Wang, L.J. ALA pretreatment improves waterlogging tolerance of fig plants. *PLoS ONE* **2016**, *11*, e0147202. [CrossRef] [PubMed]
38. Bahmaniar, M.A. The influence of continuous rice cultivation and different waterlogging periods on the morphology, clay mineralogy, Eh, pH and K in paddy soils. *Eurasian Soil Sci.* **2008**, *41*, 87–92. [CrossRef]
39. Qi, X.H.; Xu, X.W.; Lin, X.J.; Zhang, W.J.; Chen, X.H. Identification of differentially expressed genes in cucumber (*Cucumis sativus* L.) root under waterlogging stress by digital gene expression profile. *Genomics* **2012**, *99*, 160–168. [CrossRef] [PubMed]
40. Qi, B.Y.; Yang, Y.; Yin, Y.L.; Xu, M.; Li, H.G. De novo sequencing, assembly, and analysis of the Taxodium 'Zhongshansa' roots and shoots transcriptome in response to short-term waterlogging. *BMC Plant Biol.* **2014**, *14*, 201. [CrossRef] [PubMed]
41. Xu, X.W.; Chen, M.Y.; Ji, J.; Xu, J.; Qi, X.H.; Chen, X.H. Comparative RNA-seq based transcriptome profiling of waterlogging response in cucumber hypocotyls reveals novel insights into the de novo adventitious root primordia initiation. *BMC Plant Biol.* **2017**, *17*, 129. [CrossRef]
42. Zhang, P.; Lyu, D.G.; Jia, L.T.; He, J.L.; Qin, S.J. Physiological and de novo transcriptome analysis of the fermentation mechanism of *Cerasus sachalinensis* roots in response to short-term waterlogging. *BMC Genom.* **2017**, *18*, 649. [CrossRef]
43. Zhang, J.Y.; Huang, S.N.; Wang, G.; Xuan, J.P.; Guo, Z.R. Overexpression of *Actinidia deliciosa pyruvate decarboxylase* 1 gene enhances waterlogging stress in transgenic *Arabidopsis thaliana*. *Plant Physiol. Biochem.* **2016**, *106*, 244–252. [CrossRef]
44. Luo, H.T.; Zhang, J.Y.; Wang, G.; Jia, Z.H.; Huang, S.N.; Wang, T.; Guo, Z.R. Functional characterization of waterlogging and heat stresses tolerance gene *Pyruvate decarboxylase* 2 from *Actinidia deliciosa*. *Int. J. Mol. Sci.* **2017**, *18*, 2377. [CrossRef] [PubMed]

45. Zhang, J.Y.; Huang, S.N.; Chen, Y.H.; Wang, G.; Guo, Z.R. Identification and characterization of two waterlogging responsive alcohol dehydrogenase genes (*AdADH1* and *AdADH2*) in *Actinidia deliciosa*. *Mol. Breed.* **2017**, *37*, 52. [CrossRef]
46. Weits, D.A.; Giuntoli, B.; Kosmacz, M.; Parlanti, S.; Hubberten, H.M.; Riegler, H.; Hoefgen, R.; Perata, P.; Dongen, J.T.; Licausi, F. Plant cysteine oxidases control the oxygen-dependent branch of the N-end-rule pathway. *Nat. Commun.* **2014**, *5*, 3425. [CrossRef] [PubMed]
47. Wang, T.; Wang, G.; Jia, Z.H.; Pan, D.L.; Zhang, J.Y.; Guo, Z.R. Transcriptome analysis of kiwifruit in response to *Pseudomonas syringae* pv. actinidiae infection. *Int. J. Mol. Sci.* **2018**, *19*, 373. [CrossRef] [PubMed]
48. Cai, B.H.; Zhang, J.Y.; Gao, Z.H.; Qu, S.C.; Tong, Z.G.; Mi, L.; Qiao, Y.S.; Zhang, Z. An improved method for isolation of total RNA from the leaves of *Fragaria* spp. *Jiangsu J. Agric. Sci.* **2008**, *24*, 875–877.
49. Yin, X.R.; Allan, A.C.; Xu, Q.; Burdon, J.; Dejnoprat, S.; Chen, K.S.; Fergusom, L.B. Differential expression of kiwifruit genes in response to postharvest abiotic stress. *Postharvest Biol. Technol.* **2012**, *66*, 1–7. [CrossRef]
50. Horsch, R.B. A simple and general method for transferring genes into plants. *Science* **1985**, *227*, 1229–1231.

11

A Conserved Glycine is Identified to be Essential for Desaturase Activity of IpFAD2s by Analyzing Natural Variants from *Idesia polycarpa*

Pan Wu [1,2], Lingling Zhang [1], Tao Feng [1], Wenying Lu [1,2], Huayan Zhao [3], Jianzhong Li [4] and Shiyou Lü [1,5,*]

1 Key Laboratory of Plant Germplasm Enhancement and Specialty Agriculture, Wuhan Botanical Garden, Chinese Academy of Sciences, Wuhan 430074, China; wupanx@126.com (P.W.); zhanglingling@wbgcas.cn (L.Z.); fengtao@wbgcas.cn (T.F.); luwenying16@mails.ucas.ac.cn (W.L.)
2 University of Chinese Academy of Sciences, Beijing 100049, China
3 Applied Biotechnology Center, Wuhan Institute of Bioengineering, Wuhan 430415, China; huayan_zhao@163.com
4 Tianjin Garrison hangu farm, Tianjin 300480, China; lvsy1021@gmail.com
5 Sino-Africa Joint Research Center, Chinese Academy of Sciences, Wuhan 430074, China
* Correspondence: shiyoulu@wbgcas.cn

Abstract: High amounts of polyunsaturated fatty acids (PUFAs) in vegetable oil are not desirable for biodiesel or food oil due to their lower oxidative stability. The oil from *Idesia polycarpa* fruit contains 65–80% (mol%) linoleic acid (C18:2). Therefore, development of *Idesia polycarpa* cultivars with low PUFAs is highly desirable for *Idesia polycarpa* oil quality. Fatty acid desaturase 2 (FAD2) is the key enzyme converting oleic acid (C18:1) to C18:2. We isolated four FAD2 homologs from the fruit of *Idesia polycarpa*. Yeast transformed with *IpFAD2-1*, *IpFAD2-2* and *IpFAD2-3* can generate appreciable amounts of hexadecadienoic acid (C16:2) and C18:2, which are not present in wild-type yeast cells, revealing that the proteins encoded by these genes have Δ^{12} desaturase activity. Only trace amounts of C18:2 and little C16:2 were detected in yeast cells transformed with *IpFAD2-4*, suggesting IpFAD2-4 displays low activity. We also analyzed the activity of several FAD2 natural variants of *Idesia polycarpa* in yeast and found that a highly conserved Gly376 substitution caused the markedly reduced products catalyzed by IpFAD2-3. This glycine is also essential for the activity of IpFAD2-1 and IpFAD2-2, but its replacement in other plant FAD2 proteins displays different effects on the desaturase activity, suggesting its distinct roles across plant FAD2s proteins.

Keywords: *Idesia polycarpa var*; glycine; FAD2; linoleic acid; oleic acid

---

1. Introduction

Vegetable oils are not only essential resources for nutritional applications, but also for sustainable industrial feedstocks, which are commonly used in paints, lubricants, soaps, biodiesel, etc. [1,2]. The demand for vegetable oils is quickly increasing due to the fast growing population across the world. To meet this demand, many efforts have been made to improve the yields of oil crops or to domesticate wild oilseed plants [3]. *Idesia polycarpa*, a member of *Flacourtiaceae* family, is a local tree species in some Asian countries including Korea, Japan, and China [4]. It is receiving more attention due to the high amount of oil in its fruits, which can potentially be used in the biodiesel industry [4]. In addition, the oil from *Idesia polycarpa* fruit is healthy and edible since it contains 65–80% (mol%) linoleic acid (C18:2). C18:2 is one of the essential polyunsaturated fatty acids in humans and cannot be synthesized by the human body, it can only be obtained from food [5]. However, high amounts of C18:2 in its oil also make it more prone to rancidity and thus decreases its flavor [6]. Previous reports showed that oleic acid (C18:1) had higher oxidative stability than C18:2, and thus the edible oils with higher ratios of C18:1/C18:2 are more desirable [7]. Similarly, an ideal biodiesel composition should also contain more monounsaturated fatty acids and less polyunsaturated fatty acids since high percentages of polyunsaturated fatty acids in biodiesel negatively affect its oxidative stability and cause high rates of nitrogen oxide emission [8]. Hence it would be valuable to breed *Idesia polycarpa* cultivars which produce oils with low C18:2 and high C18:1 contents and it would also be helpful to uncover the desaturation mechanism in woody plants.

The Δ_{12} fatty acid desaturase 2 (FAD2) is the key microsomal enzyme that converts C18:1 to C18:2 [9]. Many efforts have been made to modify the plant oil quality via manipulation of expression levels of FAD2, or through screening natural varieties with altered FAD2 activity. For example, the mutations of GmFAD2-1A and GmFAD2-1B greatly increased the levels of C18:1 in soybean seeds [10]. The *Arachis hypogaea AhFAD2* mutant was used as an introgression line for breeding peanut cultivars with high C18:1 and low C18:2 [11]. Numerous safflower breeding lines with high levels of C18:1 (75–84%) were selected from the natural variations in *FAD2* genes [12]. The mutation of a candidate protein, fatty acid desaturase-2 (FAD2-1D) gene from pima cotton produced less linoleic acid [13,14]. Olive oil extracted from the olive fruit has mostly high level of C18:1 (about 75%) and less C18:2 (about 5.5%), which might be attributed to the suppression of *FAD2* genes by si-RNA [15]. Taken together, the activity of FAD2 is crucial for determining the C18:1/C18:2 ratios in seed or fruit storage lipids and FAD2 is an ideal candidate for improving oil quality of oil crops or trees.

Plant FAD2 proteins belong to a large family of ER localized membrane-bound desaturases [16]. The relationship between structure and function of FAD2 proteins has been extensively studied in the past two decades. FAD2 proteins contain three to six predicated transmembrane domains (PTMDs) and three highly conserved histidine-rich motifs, which are key characteristics of all membrane-bound desaturases. In the conserved histidine-rich motifs, the histidines are proved to be crucial for FAD2 desaturase activity [17]. In addition, four relevant amino acid residues within a distance of five residues from the His boxes of AtFAD2 are responsible for the conversion of this monofunctional desaturase (Δ^{12} desaturase activity) into a bifunctional desaturase/hydroxylase [18]. Besides these His boxes, McCartney et al. [19] found that the deletion of *Arabidopsis thaliana* AtFAD2 C-terminus containing an ER retrieval motif resulted in loss of ER localization and enzyme activity in yeast cells. In addition, Hoffmann et al. [20] showed that a small membrane-peripheral region close to the active center of a monofunctional Δ^{12} desaturase from *Aspergillus nidulans* determines substrate specificity and regioselectivity. Despite much progress in the relationship between FAD2s structure and function, which has been elucidated in the past two decades, some unidentified factors affecting their enzymatic activity remain to be investigated.

C18:2 was the major component of fatty acids in both seed and pericarp of *Idesia ploycarpa*, which respectively accounted for 83.92% and 62.08% of the total fatty acids in the two organs [5]. Our previous study showed that four *IpFAD*2 genes exist in *Idesia ploycarpa* [5]. To ascertain if the IpFAD2 proteins are capable of desaturating C18:1 into C18:2, we isolated four *IpFAD2* genes from *Idesia ploycarpa*, and identified their activity in yeast cells. We also assessed the activity of natural FAD2 variants and identified a highly conserved glycine at position 376 of IpFAD2-3, which is critical for normal function of IpFAD2s. To determine if the function of this glycine is conserved across plant FAD2 proteins, we also evaluated the impacts of its substitution on the activity of FAD2 proteins from other plants. Our study provides a clue for genetically modifying the oil quality of *Idesia polycarpa*.

2. Results

2.1. Isolation of FAD2 Orthologs from Idesia ploycarpa

Our previous study showed that four *FAD2* orthologs are present in the fruit of *Idesia polycarpa* [5], which were renamed *IpFAD2-1* (c63420_g2), *IpFAD2-2* (c56614_g1), *IpFAD2-3* (c63420_g1), and *IpFAD2-4* (c50543_g1), respectively, in this study. The entire coding sequence (CDS) of *IpFAD2-1, IpFAD2-2, IpFAD2-3,* and *IpFAD2-4* were cloned from *Idesia ploycarpa*. The length of the predicted polypeptides encoded by *IpFAD2-1, IpFAD2-2, IpFAD2-3,* and *IpFAD2-4* CDS are 385, 382, 385, and 380 amino acids respectively. IpFAD2-1, IpFAD2-2, IpFAD2-3 and IpFAD2-4 shared 76.74%, 70.03%, 76.49% and 67.70% identity with AtFAD2 (Figure 1, Supplementary table S1). Four IpFAD2 proteins contain three conserved histidine-rich motifs, which are commonly present in all membrane-bound fatty acid desaturases (Figure 1) [21,22] and also contain an ER-localized motif in C-terminus [19]. The cDNA sequences of *IpFAD2-1, IpFAD2-2, IpFAD2-3,* and *IpFAD2-4* were submitted to the NCBI Genbank, their accession numbers were MH394208, MH394209, MH394210, and MK105894, respectively.

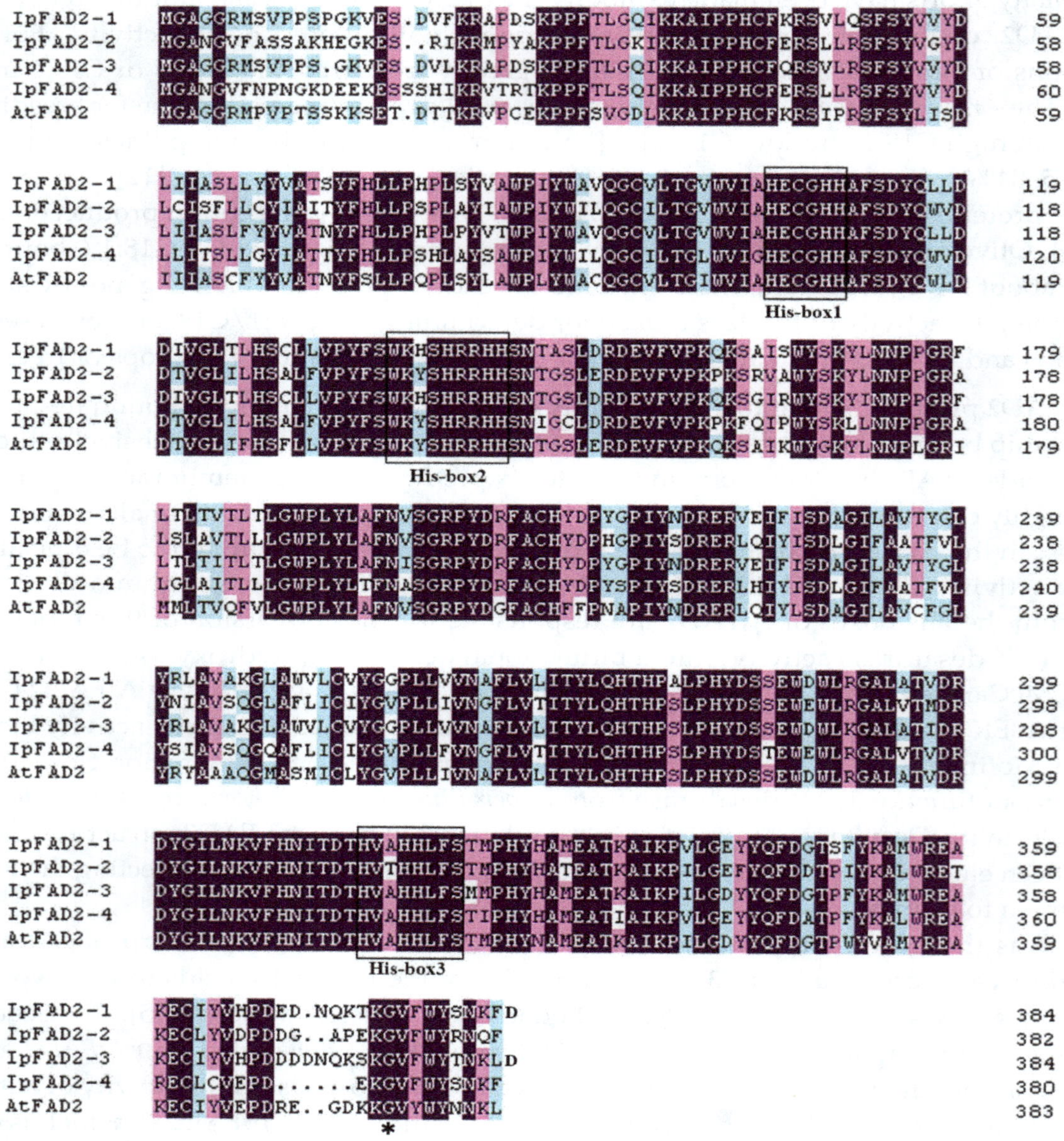

Figure 1. Alignments of predicted amino acid sequences encoded by FAD2 coding sequences from *Idesia polycarpa* and *Arabidopsis thaliana*. The three "histidine-rich motifs" are boxed. The GenBank accession numbers of *IpFAD2-1, IpFAD2-2, IpFAD2-3, IpFAD2-4, AtFAD2* are: MH394208, MH394209, MH394210, MK105894, and NP_187819.1 accordingly. The shading colors represent the identity level of amino acids. Black, magenta, and cyan indicate 100%, 80%, and 60% identity, respectively. The asterisk indicates the conserved glycine.

The phylogenetic relationship of the four IpFAD2s with other reported FAD2s was further elucidated. Similar to the previous report, the selected FAD2 proteins were grouped into two major clades, the house-keeping type and seed-type [16] (Figure 2). House-keeping type FAD2s are constitutively and abundantly expressed, while seed-type FAD2s are specifically or highly expressed in developing seeds [16]. As shown in Figure 2, IpFAD2-1 and IpFAD2-3 belong to the house-keeping group, whereas IpFAD2-2 and IpFAD2-4 fall into the seed-type group.

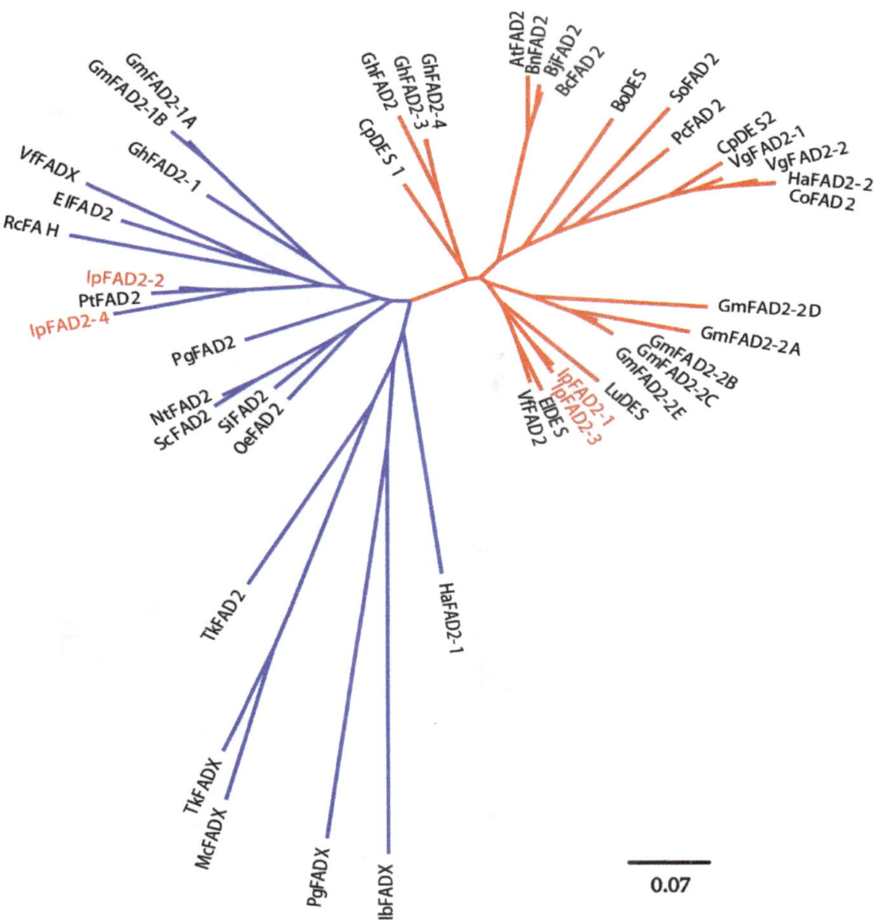

Figure 2. Cladogram of four IpFAD2s and other plant FAD2-related polypeptides. The house-keeping clade and the seed-type clade are labeled in red and blue accordingly. The four FAD2 homologs from *Idesia polycarpa* were designated IpFAD2-1, 2, 3, 4 and were labeled in red. The protein sequences used here were: GmFAD2-1A (Glyma.10G278000.1.p), GmFAD2-1B (Glyma.20G111000.1.p), GmFAD2-2A (Glyma.19G147300.1.p), GmFAD2-2B (Glyma.19G147400.1.p), GmFAD2-2C (Glyma.03G144500.1.p), GmFAD2-2D (Glyma.09G111900.1.p), and GmFAD2-2E (Glyma.15G195200.1.p) from *Glycine max*; CpDES (AAS19533) from *Cucurbita pepo*; IpFAD2-1 (MH394208), IpFAD2-2 (MH394209), IpFAD2-3 (MH394210), and IpFAD2-4 (MK105894) from *Idesia polycarpa*; HaFAD2-1 (AF251842) and HaFAD2-2 (AF251843) from *Helianthus annuus*; IbFADX (AF182520) from *Impatiens balsamina*; PgFADX (AY178446) from *Punica granatum*; McFADX (AF182521) from *Momordica charantia*; TkFADX (AY178444) and TkFAD2 (AY178445) from *Trichosanthes kirilowii*; SiFAD2 (AF192486) from *Sesamum indicum*; ScFAD2 (X92847) from *Solanum commersonii*; RcFAH (EU523112) from *Ricinus communis*; VfFADX (AF525535), and VfFAD2 (AF525534) from *Vernicia fordii*; GhFAD2-1 (X97016), GhFAD2-2 (Y10112), GhFAD2-3 (AF331163), and GhFAD2-4 (AY279315) from *Gossypium hirsutum*; PgFAD2 (AJ437139) from *Punica granatum*; CpDES (AAS19533, and CpDES2 (AAS19533) from *Cucurbita pepo*; AtFAD2 (L26296) from *Arabidopsis thaliana*; BnFAD2 (AF243045) from *Brassica napus*; BjFAD2 (X91139) from *Brassica juncea*; BcFAD2 (AF124360) from *Brassica carinata*; BoDES (AF074324) from *Borago officinalis*; SoFAD2 (AB094415) from *Spinacia oleracea*; PcFAD2 (U86072) from *Petroselinum crispum*; VgFAD2-1 (AF188263) and VgFAD2-2 (AF188264) from *Vernonia galamensis*; CoFAD2 (AF343065) from *Calendula officinalis*; LuDES (ACF49507) from *Linum usitatissimum*.

To identify which gene participated in the production of C18:2 content in fruit, we examined the expression patterns of these genes by RT-PCR analysis in seed and pericarp from fruit 80 days after pollination (DAP) (Figure 3). The expression levels of *IpFAD2-2* far exceeded those of the other *IpFAD2* genes, suggesting *IpFAD2-2* might be mainly responsible for producing high C18:2 content in fruit (Figure 3).

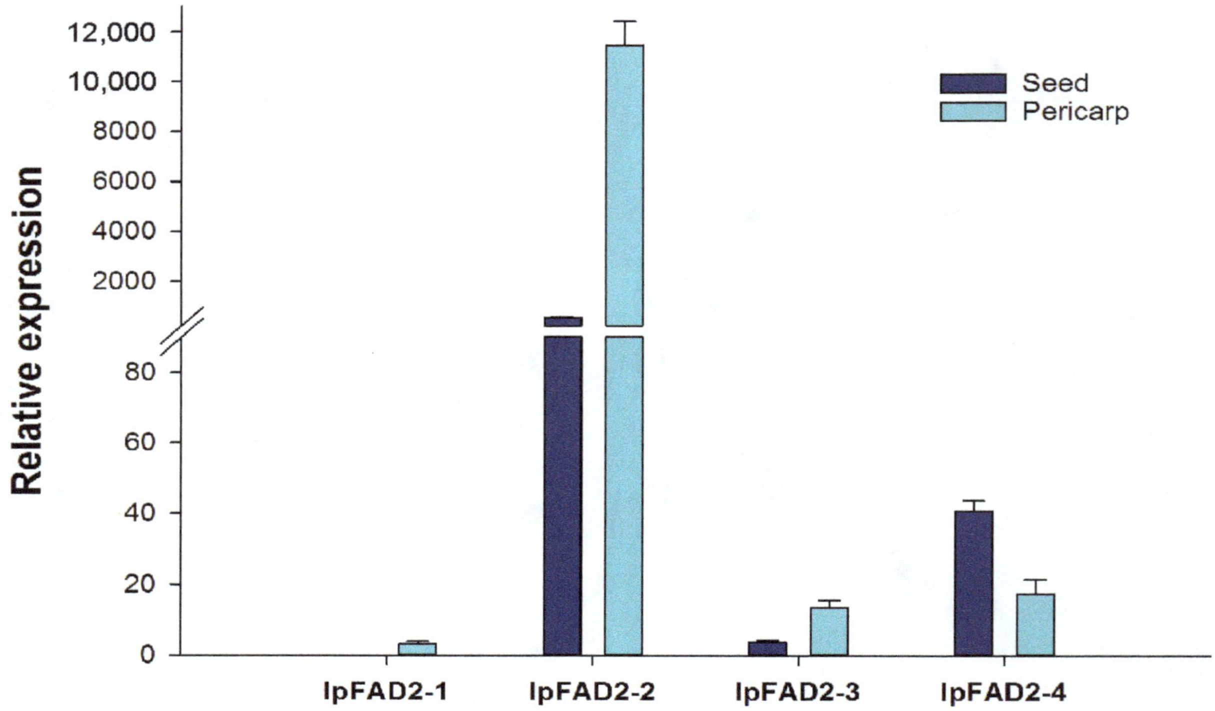

Figure 3. Relative expression levels of IpFAD2 genes in the pericarp and seed of I*desia* polycarpa at 80 days post pollination. Relative expression values were calculated using the 2−ΔΔCt method by using *EF1A* gene as an internal control.

2.2. Three IpFAD2s Possess Desaturase Activity

To examine if the proteins encoded by *IpFAD2-1*, *IpFAD2-2*, *IpFAD2-3*, and *IpFAD2-4* are involved in the desaturation process, four genes were transformed into the budding yeast *S. cerevisiae* INVSc1. Then we checked the expression levels of these transgenes by RT-PCR. As shown in Supplementary Figure S1, all transgenes are highly expressed. The fatty acid compositions in the transformed yeast cells were also analyzed by gas chromatography (GC) and the corresponding fatty acids were confirmed by gas chromatography mass spectrometry (GC-MS) (Supplementary Figure S2). The yeast cells harboring *IpFAD2-1*, *IpFAD2-2* and *IpFAD2-3* produced two novel fatty acids, C16:2 and C18:2, which were not generated in the yeast cells containing blank vector (Figure 4A–D, Supplementary Figure S3), suggesting that the three IpFAD2s have their own catalytic activity. Moreover, the proportion of C18:2 in yeast cells transformed with *IpFAD2-1*, I*pFAD2-2*, and *IpFAD2-3* is 9.97%, 8.96% and 11.43% of the total fatty acids, much higher than that of C16:2, which only accounts for 2.58%, 4.62%, and 4.45%, respectively. However, only trace amounts of C18:2 (0.66%) and little C16:2 were detected in the yeast cells containing *IpFAD2-4* (Figure 4E). These data indicate that IpFAD2-1, IpFAD2-2, and IpFAD2-3 possess high Δ^{12}-fatty acid desaturation activity using both C16:1 and C18:1 as substrates, furthermore C18:1 is a preferable substrate for all three IpFAD2 proteins, while IpFAD2-4 displays low activity (Figure 4A–E).

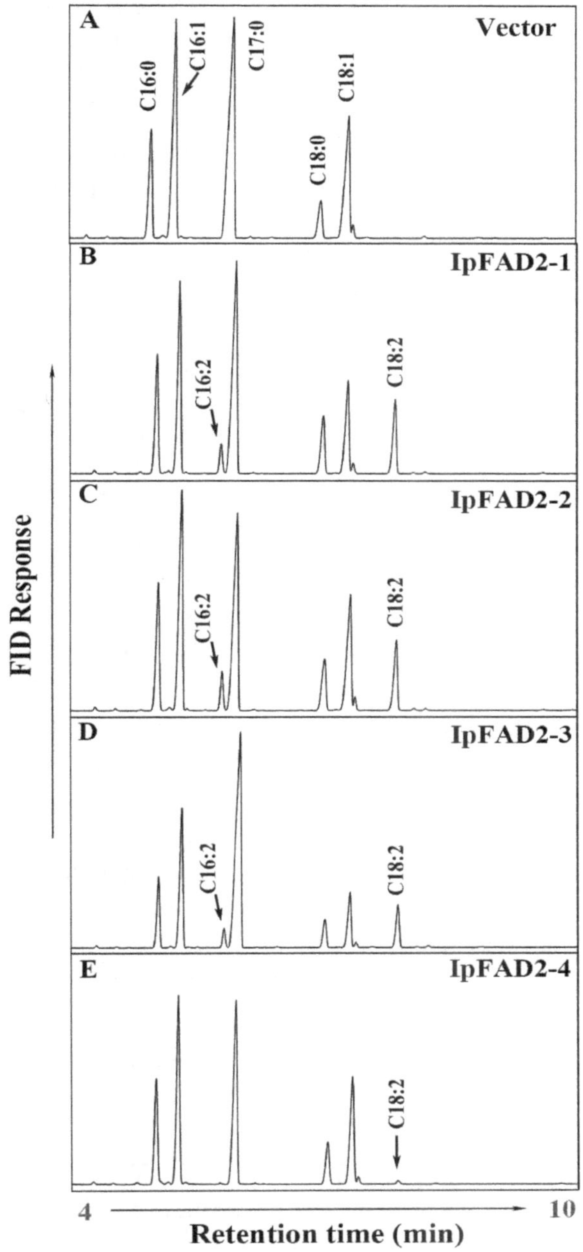

Figure 4. GC analysis of fatty acid methyl esters (FAMEs) isolated from yeast cells expressing IpFAD2s. The FAMEs of total lipid were extracted from yeast transformed with control empty control vector pESC-his (**A**), IpFAD2-1 (**B**), IpFAD2-2 (**C**), IpFAD2-3 (**D**), and IpFAD2-4 (**E**) under induction conditions and analyzed by gas chromatography/flame ionization detector (GC/FID). Major fatty acids peaks are labeled. The newly synthesized fatty acids corresponding to C16:2 Δ9,12 and C18:2 Δ9,12 are indicated by the arrows. Heptadecanoic acid methyl ester (C17:0) is used as the internal standard.

2.3. A Highly Conserved Glycine Residue Identified from the FAD2 Natural Variation is Required for FAD2 Desaturase Activity

Previous reports showed that some natural variations in FAD2 orthologs resulted in an elevated C18:1/C18:2 ratio in oil seed crops with C18:2 as the major fatty acid [10,11,14,23–25]. Thus it is feasible to find some natural FAD2 dysfunction variants in *Idesia ploycarpa*. A small population of five-year-old *Idesia ploycarpa* trees were examined. 32 Single nucleotide polymorphisms (SNPs) were found throughout the CDS region of IpFAD2s. The association between SNPs and the amino acid changes in IpFAD2s was summarized in Table 1. The SNPs caused synonymous mutations in the IpFAD2-1 CDS region, while the SNPs in IpFAD2-2 resulted in changes to five amino acids. The five IpFAD2-2 variants were named IpFAD2-2V1 (Y54S, V164I and V243M), IpFAD2-2V2 (Y54S), IpFAD2-2V3 (A69V), IpFAD2-2V4 (Q115/R and V164I), and IpFAD2-2V5 (V164I and V243M) (Supplementary Figure S4). Two of the five SNPs in IpFAD2-3 resulted in changes to two amino acids. The IpFAD2-3 variant was named as IpFAD2-3V1 (V253I and G376C) (Supplementary Figure S5). Seven of twelve SNPs in IpFAD2-4 resulted in changes to seven amino acids and the IpFAD2-4 variant was named a IpFAD2-4V1 (C151S, F164S, L172F, E289D and C365Y) and IpFAD2-4V2 (P31L, A71V, C151S and F164S) (Supplementary Figure S5).

Since SNPs caused the changes to amino acids in IpFAD2-2, IpFAD2-3, and IpFAD2-4, it is interesting to determine whether these amino acid variations affect desaturase activity of the three proteins. The constructs harboring wild type or FAD2 variants of IpFAD2-2, IpFAD2-3 and IpFAD2-4 were transformed into yeast *S. cerevisiae* INVSc1 and fatty acid composition was examined. The SNPs in IpFAD2-3s brought about the variations of two amino acids (V253I and G376C) in IpFAD2-3V1 (Table 1).

As shown in Figure 5A, the variations occurring in IpFAD2-2 and IpFAD2-4 have little effects on its desaturase activity as compared with their corresponding wild type form, whereas the two amino acid substitutions (V253I and G376C) severely affected the activity of IpFAD2-3V1 since the C18:2 content in IpFAD2-3V1 was only less than 10% of the wild type (Figure 5A). To ascertain which amino acid is responsible for this result, we performed single site mutation on IpFAD2-3 and obtained two *IpFAD2-3* mutants containing a single mutation with V253I or G376C. The G376C mutation caused the dramatic decreasing activity of IpFAD2-3, the percentage of C18:2 was greatly reduced to 8.3% of wild type (Figure 5A, Supplementary Figure S3). The activity of the variant containing V253I mutation was only slightly affected (Figure 5A, Supplementary Figure S3). These results suggested that Gly376 is essential for IpFAD2-3 activity. Here we noticed that Glycine was changed to Cysteine and thus deduced that the redox status might be concerned with the altered activity. To test this possibility, we replaced Gly376 with either alanine or serine. Our results showed that both replacements caused markedly decreased activity (Figure 5A). These results further illustrate the importance of G376 residues.

Table 1. Single nucleotide polymorphisms (SNPs) in the coding region of fatty acid desaturase 2 (*FAD2*) genes from *Idesia ploycarpa*.

FAD2 Gene	SNPSite	Amino Acid Position	SNP Mutation	Amino Acid Mutation	Mutation Type
IpFAD2-1 (1158 bp)	201	67	TAT→TAC	Tyr	S
	729	243	GCA→GCG	Ala	S
	765	255	TAT→TAC	Tyr	S
IpFAD2-2 (1149 bp)	161	54	TAT→TCT	Tyr→Ser	N
	206	69	GCC→GTC	Ala→Val	N
	344	115	CAG→CGG	Gln→Arg	N
	372	124	ATC→ATT	Ile	S
	399	133	TAC→TAT	Tyr	S
	490	164	AGT→AAT	Ser→Asn	N
	612	204	CGA→CGC	Arg	S
	624	208	CAC→CAT	His	S
	727	243	GTG→ATG	Val→Met	N
	1041	347	GAC→GAT	Asp	S
	1092	364	GTT→GTG	Val	S
	1113	371	CCA→CCC	Pro	S
IpFAD2-3 (1158 bp)	690	230	GGC→GGT	Gly	S
	696	232	CTC→CTT	Leu	S
	729	243	GTC→GTA	Val	S
	757	253	GTT→ATT	Val→Ile	N
	1126	376	GGC→TGC	Gly→Cys	N

Table 1. *Cont.*

FAD2 Gene	SNPSite	Amino Acid Position	SNP Mutation	Amino Acid Mutation	Mutation Type
IpFAD2-4 (1143 bp)	92	31	CCC→CTC	Pro→Leu	N
	212	71	GCC→GTC	Ala→Val	N
	279	93	CTA→CTC	Leu	S
	451	151	TGC→AGC	Cys→Ser	N
	480	160	CCA→CCG	Pro	S
	491	164	TTC→TCC	Phe→Ser	N
	514	172	CTC→TTC	Leu→Phe	N
	531	177	CCT→CCA	Pro	S
	867	289	GAA→GAT	Glu→Asp	N
	999	333	GCA→GCT	Ala	S
	1002	334	ACT→ACA	Thr	S
	1094	365	TGT→TAT	Cys→Tyr	N

S represents synonymous, N represents nonsynonymous.

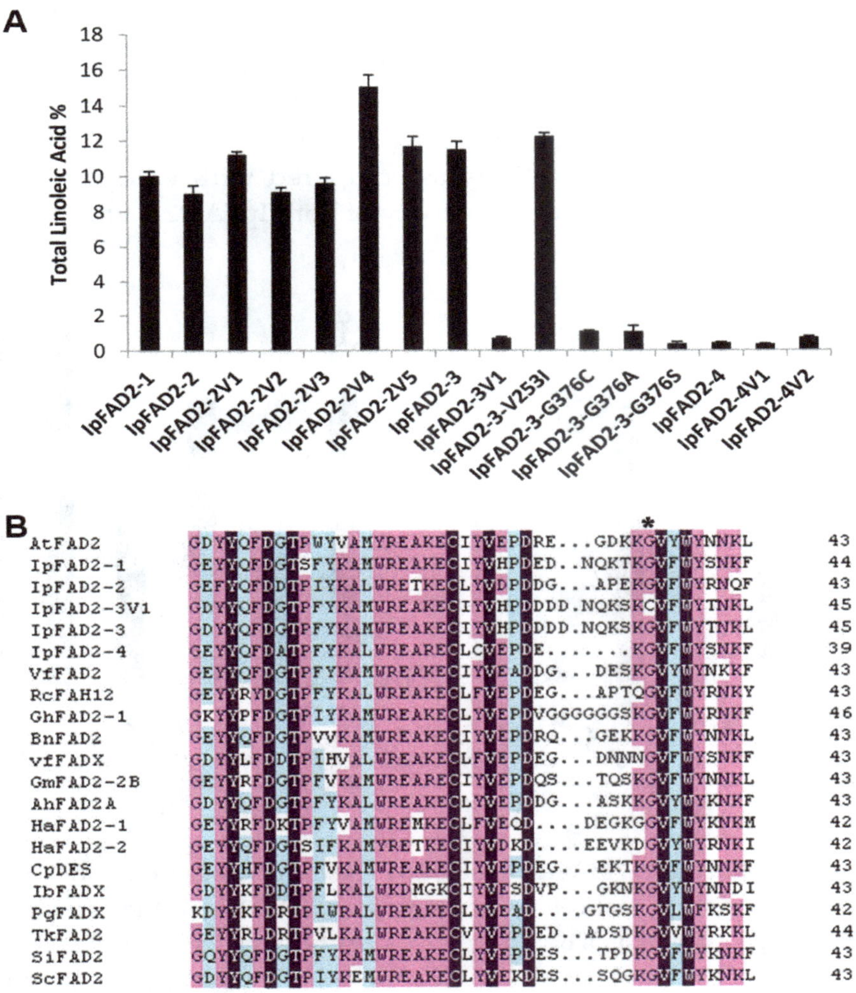

Figure 5. Total C18:2 accumulation in yeast transformed with different FAD2 alleles from *Idesia polycarpa*. (**A**) Total LA accumulation in transgenic yeast expressing natural FAD2 alleles and site-directed mutations from *Idesia polycarpa*. (**B**) Alignment of C-terminal amino acids of 21 FAD2 genes. The GenBank accession numbers of the proteins presented in this figure are shown in Figure 2. The asterisk indicates the conserved Glycine residue changed in IpFA2-3V1. The shading colors represent the identity level of amino acids. Black, magenta, and cyan indicate 100%, ~90%, and ~55% identity, respectively.

To find out the reasons why the substitution of the conserved glycine severely affects the activity of IpFAD2-3, we carefully examined plant FAD2 protein structure. This glycine residue is highly conserved across plant FAD2 proteins (Figure 5B). It is far away from the catalytic center consisting of three histidine-rich motifs [26], but is at -9 position relative to the C-terminus and adjacent to the mini ER retrieval sequence motif (Φ-X-X-K/R/D/E-Φ-COOH, YTNKL in the case of IpFAD2-3) [19]. Thus we hypothesized that the impacts of the glycine residue on IpFAD2-3 might act through disturbing the precise location of IpFAD2-3. To test this hypothesis, we made IpFAD2-3-GFP and IpFAD2-3-G376C-GFP constructs and co-transformed them with an ER membrane marker (CD3-959) into tobacco epidermal cells. Each fluorescent fusion protein was co-localized with the ER membrane marker CD3-959 (Supplementary Figure S6), consistent with the expression pattern of the wild type protein, indicating that G376C does not interrupt the localization of IpFAD2-3 and its impacts on enzyme activity could not act through mis-localizing the protein. The GFP fluorescent signals of tobacco epidermal cells containing either of IpFAD2-3-GFP or IpFAD2-3-G376C-GFP were similar to each other, suggesting that G376C does not affect IpFAD2-3 at protein level (Supplementary Figure S6).

2.4. Gly376 Has Different Effects on the Activity of FAD2 Proteins among Different Species

Since the highly conserved Gly376 is important for IpFAD2-3 activity, we wondered if it has a conserved function in all IpFAD2 proteins and in FAD2 proteins from other species. Firstly, we introduced the mutation into IpFAD2-1 and IpFAD2-2. The mIpFAD2-1 and mIpFAD2-2 variants display markedly reduced desaturase activity as compared with wild type, suggesting that the conserved glycine is also required for enzyme activity of both IpFAD2-1 and IpFAD2-2 (Figure 6).

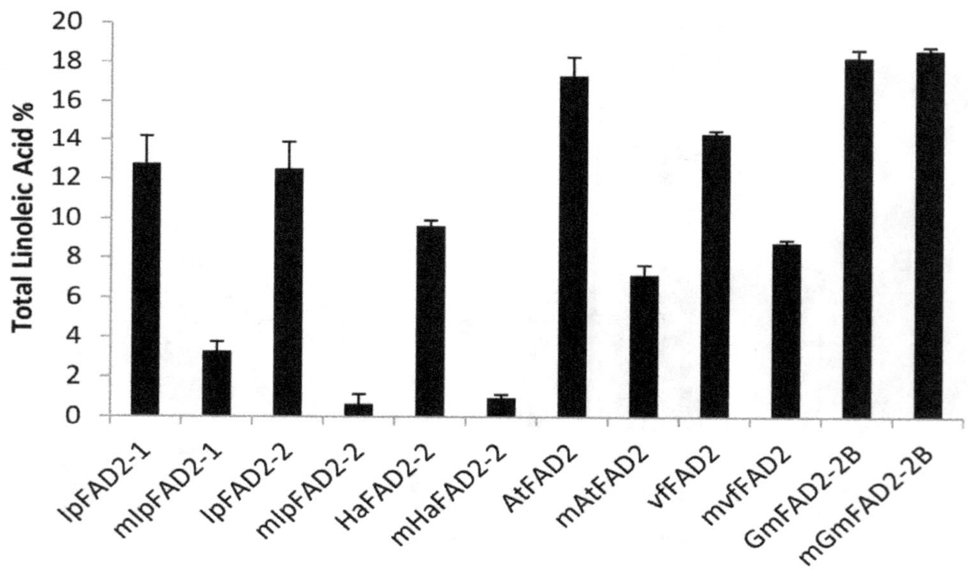

Figure 6. Total C18:2 accumulation in yeast containing FAD2 orthologs from different species. Levels of total C18:2 accumulation in yeast cells expressing plant FAD2s and the corresponding mutated form with the conserved glycine (G376, 375 or 374) substituted by cysteine. Ha, *Helianthus annuus*; At, *Arabidopsis thaliana*; Vf, *Vernicia fordii*; Gm, *Glycine max*. The corresponding mutated residues are: IpFAD2-1, G376C; IpFAD2-2, G374C; HaFAD2-2, G375C; AtFAD2, G375C; VfFAD2, G375C; GmFAD2-2B, G375C. "m" in front of each protein name represents the mutated form of the corresponding FAD2. Mean ± SD; $n = 3$.

To check if the conserved glycine also has similar roles across plant FAD2 proteins, we cloned *FAD2* genes from *Helianthus annuus*, *Arabidopsis thaliana*, *Vernicia fordii* and *Glycine max*, which are *HaFAD2-2*, *AtFAD2*, *VfFAD2* and *GmFAD2-2B* accordingly, and also replaced this glycine with cysteine in these FAD2 proteins. The wild type and mutated constructs were then transformed into yeast and FA composition was detected by GC. As shown in Figure 6, the yeast cells containing each wild-type

FAD2s produced a certain amount of C18:2, ranging from 9% to 18% of total fatty acids, indicating that all these selected FAD2s have functional Δ^{12} oleate desaturase activity in heterologous yeast cells. We then checked the C18:2 levels of the yeast cells containing the mutated constructs and found that the effects of the mutation (G376/C) on the activity of FAD2 proteins varied among different proteins. The levels of C18:2 in yeast cells containing the mutated HaFAD2-2 were markedly decreased to less than 10% of wild type (Figure 6). The activity of mutated AtFAD2 and VfFAD2 only showed moderate reduction since the mutants showed about 58% and 38% reductions. The mutated glycine in GmFAD2-2B has no effects on the production of C18:2 (Figure 6). Taken together, these results suggested that the effects of the highly conserved glycine at 376 position of IpFAD2-3 on plant FAD2 activity probably act in a species/protein-specific manner.

3. Discussion

The *FAD2* gene was first identified in Arabidopsis [27], since then many *FAD2* genes have been cloned from different plants [26,28–33]. To date, none of the microsomal Δ^{12} fatty acid desaturases associated with C18:2 have been identified in *Idesia ploycarpa*. Here we isolated four *IpFAD2* genes. Deduced amino acid sequences alignment showed that the four IpFAD2s shared 69.5–78.5% identity with AtFAD2, The *Idesia polycarpa* fruit, consisting of pericarp and seed, produces large amounts of fatty acids, among which about 83.92% of C18:2 is present in pericarp oils, that far exceeds seed oils [5]. To gain insight into the oil accumulation mechanism in pericarp and seed, we studied the expression patterns of four *IpFAD2* paralogues in these two organs by real time-PCR. Each gene shows distinct expression patterns (Figure 3). *IpFAD2-2* is highly expressed in both organs, its transcripts far exceed those of other *IpFAD2* genes, suggesting that *IpFAD2-2* might be a major gene responsible for C18:2 production in the fruit of *Idesia polycarpa*. The expression levels of *IpFAD2-2* varied between two parts, i.e., higher in pericarps and lower in seeds, suggesting its different roles in the two organs. Taken together, *IpFAD2-2* might be a candidate for genetically modifying the ratio of C18:1/C18:2 in *Idesia ploycarpa* fruits in the future. As compared with *IpFAD2-2*, the expression levels of the other three *IpFAD2* genes are very low in fruit (Figure 3). They also exhibit the distinct expression patterns found in pericarps and seeds. The expression of *IpFAD2-3* displays little differences between the two organs while *IpFAD2-1* was preferably expressed in the pericarp. The different expression patterns of these *IpFAD2* genes might cause the different ratios of C18:1/C18:2 in pericarps and seeds.

To identify the activity of IpFAD2s, we cloned these four genes and introduced them into yeast cells. With the exception of IpFAD2-4, the three other IpFAD2 proteins efficiently converted C18:1 to C18:2 in the yeast system (Figure 4). To investigate the reason why IpFAD2-4 displayed low activity, we carefully compared the amino acid sequences of IpFAD2-4 with that of the other proteins and found that IpFAD2-4 shared 85.1% with *IpFAD2-2* (Figures 1 and 2), suggesting that both genes might be derived from duplication events. RT-PCR results showed that it displayed similar patterns to *IpFAD2* though its expression levels are far lower than the latter. But the two proteins showed distinct activity in yeast cells. To precisely compare the activity of these proteins, we need to transform it into plants for functional identification in the near future.

With the rapid advances in biotechnology, genetic engineering has been widely used in identifying gene function or modifying plant quality in the lab due to its simplicity and easy-of-use. But its extensive application in nature is largely restricted since the impacts of genetically modified plants on nature are unpredictable. Natural variants have endured long-term natural selection and natural mutation is thus more reliable, stable, less toxic, and desirable for variety breeding. Natural variations in FAD2 coding region correlated with the C18:1 content have been identified from different plants. D150N and H101D from peanut, S117N and P137R from soybean have been shown to decrease the activity of FAD2 [10,34]. Most substitutions were in or near the His-box, which makes up the catalytic center. Here we checked 32 SNP sites present in FAD2 coding region among 30 *Idesia ploycarpa* natural variants. These SNP sites caused fourteen amino acid substitutions and twelve IpFAD2 alleles. Our data revealed that only the substitution of the highly conserved Gly376 severely affects the IpFAD2-3 activity (Figure 5A), suggesting its important role for IpFAD2-3. Its replacement also severely disrupted the activity of IpFAD2-1 and IpFAD2-2 (Figure 5A). We also expand its mutation to other plant FAD2 proteins and found that this glycine more or less affects the function of most of the tested plant FAD2 proteins except for GmFAD2-2B (Figure 6). Till now it is unknown how the glycine mutation causes the reduced products catalyzed by IpFAD2-3 proteins. Its mutation does not to affect the protein levels since the mutated protein can be normally expressed as shown in Supplementary Figure S6. The effects of this conserved glycine on enzyme activity might be associated with other unknown factors such as protein structure formation, phosphorylation and etc. All in all, the identified G376 in IpFAD2-3 could be a potential site for the manipulation of the desaturase activity of IpFAD2 by genome editing in the near future, and it will be applicable for genetically improving crop quality.

4. Materials and Methods

4.1. Plant Materials

The fresh leaves of 26 five-year-old *Idesia ploycarpa* female trees, which are growing at Huanggang, Hubei province, China, were collected, mixed and quickly frozen in liquid nitrogen for further RNA extraction. Col-0 *Arabidopsis thaliana* are growing in greenhouse condition. Sunflower (*Helianthus annuus*), tung tree (*Vernicia fordi*), and soybean (*Glycine max*) were collected from Wuhan Botanical Garden, Wuhan, China. The 80 days after pollination (DAP) fruits of *Idesia ploycarpa* cultivar 76A were quickly frozen in liquid nitrogen and stored at −80 °C until use.

4.2. Total RNA Extraction and Complementary DNA Synthesis

Total RNA was isolated from 100 mg of frozen leaves and seeds and pericarps of 80 DAP *Idesia ploycarpa* fruit with Trizol reagent (Life Technologies Corporation, Carlsbad, CA, USA) according to the manufacturer's protocol. RNA concentration was determined by NanoDropTM spectrophotometer ND2000 (Thermo Fisher Scientific, Wilmington, DE, USA). Total RNA was then treated with DNase I (Thermo Fisher Scientific, Wilmington, DE, USA) to eliminate residue DNA. About 500 ng DNA-free RNA was used as a template for first-strand complementary DNA (cDNA) synthesis. Reverse-transcription was performed with the M-MLV Reverse Transcriptase (Promega, Madison, WI, USA) and oligo(dT)$_{20}$ primer (Tsingke, Wuhan, China).

4.3. Gene Cloning and Sequence Analysis

Using leaf cDNA as template, we cloned the coding sequences of different *IpFAD2* with corresponding primers listed in Supplementary Table S2. The primers were designed according to the sequence submitted by Li et al. [5]. Due to lack of upstream sequence information, a degenerate primer was designed as the forward primer according to sequence homology used for *IpFAD2-4* cloning. The FAD2 fragments obtained were cloned into the pESC-his vector (Alilent Technologies, Santa Clara, CA, USA) directly and then sequenced. Multiple sequence alignments were performed using DNAman software. For phylogenetic relationship analysis, the protein sequences of IpFAD2 protein sequence and a number of plant FAD2 homologs were aligned with MAFFT v7.154b [35]. Maximum-likelihood (ML) tree was generated using FastTree v2.1.7 [36] and was visualized using FigTree (Available online: http://tree.bio.ed.ac.uk/software/figtree/).

4.4. Real-Time Quantitative PCR

Gene expression analysis was performed by RT-PCR using Applied Biosystems 7500 Fast Real-Time PCR System (Thermo Fisher Scientific, Wilmington, DE, USA). Primers with Tm (melting temperature) 60 °C and 18–20 bp in length were designed by Primer 3 (Supplementary Table S2). IpEF1A was selected as the internal reference gene. PCR reaction mix (20 μL per well) contained 10 μL TB Green Premix Ex Taq II (TliRNaseH Plus) (2X) (Takara, Tokyo, Japan), 0.8 μL forward and reverse primers (10 μM) (Tsingke, Wuhan, China), 0.4 μL ROX Reference Dye II (Takara, Tokyo, Japan), 50 ng cDNA and RNA-free water. The two-step thermal cycling conditions were 95 °C for 30 s, followed by 40 cycles of 95 °C for 5 s, 60 °C for 34 s. Corresponding gene expression level was analyzed with the 2-ΔΔCt method. Elongation factor 1-alpha was used as the internal control to normalize the relative amount of mRNAs for all samples.

4.5. Site-Directed Mutagenesis

Mutagenesis was done according to the fast mutagenesis system (Transgen, Beijing, China). In brief, the mutated plasmids were amplified with two primers containing the mutations (Supplementary Table S2) using the TransStart FastPfu DNA polymerase (Transgen, Beijing, China). The PCR conditions were as follows: initial denaturation at 94 °C for 5 min, followed by 25 cycles of 94 °C for 20 s, 55 °C for 20 s, and 72 °C for 3 min, final extension at 72 °C for 10 min. The amplicons were subsequently digested with DMT (Transgen, Beijing, China) enzyme for eliminating the methylated parental plasmid and then purified from agrose gels. The purified products were transformed into DMT competent cells (Transgen, Beijing, China). The mutated clones selected on plates containing antibiotics were verified by sequencing.

4.6. Yeast Transformation and Heterologous Expression of IpFAD2 Variants

Constructs containing the *IpFAD2* genes were transformed into *Saccharomyces cerevisiae* INVSc1 (Invitrogen, Carlsbad, CA, USA) by the LiAc/SS carrier DNA/PEG method. Transformants were incubated in yeast nitrogen base (YNB) liquid medium at 28 °C for 36 h with rotary shaking at 180 rpm and then spread on synthetic defined medium without histidine (SD-his, Clonetech, Mountain View, CA, USA) solidified medium supplemented with glucose. The colonies growing on SD-his medium were then cultured in SD-his liquid medium for another two days and then centrifuged. The pellets were washed with distilled water twice and sub-cultured in SD-his liquid medium containing galactose (2%, *w*/*v*) for 48 h. Yeast cells were collected for fatty acid analysis.

4.7. Analysis of Fatty Acid Composition in Yeast

The induced yeast cells transformed with *FAD2* cDNA fragments were pelleted and washed with distilled water twice, and total lipids were extracted with hexane and methylated with 5M KOH-methanol. The heptadecanoic acid methyl ester (C17:0) was used as the internal standard. Fatty acid methyl esters were measured by gas chromatography with an Agilent 7820A (Alilent Technologies, Santa Clara, CA, USA). The samples were separated on an Agilent DB-23 capillary column (Alilent Technologies, Santa Clara, CA, USA). The column temperature was programmed with an initial temperature of 180 °C for 1 min, ramping at 3 °C/min to 240 °C, and then holding for 39 min.

4.8. Subcellular Localization Assay

The coding sequences of IpFAD2-3 and IpFAD2-3V1 were amplified with specific primers (Supplemental Table S2). The amplified fragments were cloned into the PMDC83 vector, which generated Pro35S::IpFAD2-3::GFP and Pro35S::IpFAD2-3V1::GFP fusion constructs. The obtained plasmids were transferred into *Agrobacterium tumefaciens* (GV3101) using the freeze-thaw method, and subsequently transformed into leaves of *Nicotiana benthamiana* by infiltration. To precisely localize which compartments the FAD2 proteins reside in, CD3-959 (35S-ER-mCherry), an ER marker [37] was co-transformed with the FAD proteins. The fluorescent signals generated by GFP and mCherry fusion proteins were observed by confocal microscopy.

5. Conclusions

We identified four *FAD2* homologs from fruits of *Idesia polycarpa*. Heterologous expression in yeast showed that three IpFAD2s have strong Δ^{12} fatty acid desaturase activity. Natural variation together with site-directed mutagenesis analysis reveals one natural variation (G376C in IpFAD2-3) that strongly hinders the catalytic activity of IpFAD2-3. Even though this amino acid is highly conserved among plant FAD2 proteins, the effects of its mutation on the Δ^{12} oleate desaturase activity of tested FAD2 proteins are different. Our findings will be helpful to advance understanding the roles of FAD2 proteins in woody plants and also provide a new potential site of IpFAD2s for modifying the ratio of C18:1/C18:2 of *Idesia polycarpa* fruit in the future through genetic engineering. Further characterization of the mechanisms of the effects of G376C substitutions in different FAD2s, either at the enzyme activity level or on other modulatory molecules, is currently under way in our group.

Author Contributions: Conceptualization, P.W. and L.Z.; methodology, T.F. and W.L.; investigation, P.W.; resources, J.L.; writing—original draft preparation, P.W.; writing—review and editing, S.L. and H.Z.; funding acquisition, S.L.

References

1. Dyer, J.M.; Stymne, S.; Green, A.G.; Carlsson, A.S. High-value oils from plants. *Plant J.* **2008**, *54*, 640–655. [CrossRef] [PubMed]
2. Lummiss, J.A.M.; Oliveira, K.C.; Pranckevicius, A.M.T.; Santos, A.G.; dos Santos, E.N.; Fogg, D.E. Chemical Plants: High-Value Molecules from Essential Oils. *J. Am. Chem. Soc.* **2012**, *134*, 18889–18891. [CrossRef] [PubMed]
3. Eriksson, D.; Merker, A. Cloning and functional characterization of genes involved in fatty acid biosynthesis in the novel oilseed crop *Lepidium campestre* L. *Plant Breed.* **2011**, *130*, 407–409. [CrossRef]
4. Yang, F.X.; Su, Y.Q.; Li, X.H.; Zhang, Q.; Sun, R.C. Preparation of biodiesel from *Idesia polycarpa var.* vestita fruit oil. *Ind. Crop Prod.* **2009**, *29*, 622–628. [CrossRef]
5. Li, R.J.; Gao, X.; Li, L.M.; Liu, X.L.; Wang, Z.Y.; Lu, S.Y. De novo assembly and characterization of the fruit transcriptome of *idesia polycarpa* reveals candidate genes for lipid biosynthesis. *Front. Plant Sci.* **2016**, *7*, 801. [CrossRef] [PubMed]
6. Pandey, M.K.; Wang, M.L.; Qiao, L.X.; Feng, S.P.; Khera, P.; Wang, H.; Tonnis, B.; Barkley, N.A.; Wang, J.P.; Holbrook, C.C.; et al. Identification of QTLs associated with oil content and mapping *FAD2* genes and their relative contribution to oil quality in peanut (*Arachis hypogaea* L.). *BMC Genet.* **2014**, *15*, 133. [CrossRef] [PubMed]
7. Ge, Y.; Chang, Y.; Xu, W.L.; Cui, C.S.; Qu, S.P. Sequence variations in the *FAD2* gene in seeded pumpkins. *Genet. Mol. Res.* **2015**, *14*, 17482–17488. [CrossRef] [PubMed]
8. Qu, J.; Mao, H.Z.; Chen, W.; Gao, S.Q.; Bai, Y.N.; Sun, Y.W.; Geng, Y.F.; Ye, J. Development of marker-free transgenic *Jatropha* plants with increased levels of seed oleic acid. *Biotechnol. Biofuels* **2012**, *5*, 10. [CrossRef] [PubMed]
9. Miquel, M.; Browse, J. Arabidopsis mutants deficient in polyunsaturated fatty acid synthesis. Biochemical and genetic characterization of a plant oleoyl-phosphatidylcholine desaturase. *J. Biol. Chem.* **1992**, *267*, 1502–1509. [PubMed]
10. Pham, A.T.; Lee, J.D.; Shannon, J.G.; Bilyeu, K.D. Mutant alleles of *FAD2-1A* and *FAD2-1B* combine to produce soybeans with the high oleic acid seed oil trait. *BMC Plant Biol.* **2010**, *10*, 195. [CrossRef]
11. Janila, P.; Pandey, M.K.; Shasidhar, Y.; Variath, M.T.; Sriswathi, M.; Khera, P.; Manohar, S.S.; Nagesh, P.; Vishwakarma, M.K.; Mishra, G.P.; et al. Molecular breeding for introgression of fatty acid desaturase mutant alleles (*ahFAD2A* and *ahFAD2B*) enhances oil quality in high and low oil containing peanut genotypes. *Plant Sci.* **2016**, *242*, 203–213. [CrossRef] [PubMed]
12. Cao, S.; Zhou, X.R.; Wood, C.C.; Green, A.G.; Singh, S.P.; Liu, L.; Liu, Q. A large and functionally diverse family of *Fad2* genes in safflower (*Carthamus tinctorius* L.). *BMC Plant Biol.* **2013**, *13*, 5. [CrossRef] [PubMed]
13. Shockey, J.; Dowd, M.; Mack, B.; Gilbert, M.; Scheffler, B.; Ballard, L.; Frelichowski, J.; Mason, C. Naturally occurring high oleic acid cottonseed oil: Identification and functional analysis of a mutant allele of *Gossypium barbadense fatty acid desaturase-2*. *Planta* **2017**, *245*, 611–622. [CrossRef] [PubMed]
14. Sturtevant, D.; Horn, P.; Kennedy, C.; Hinze, L.; Percy, R.; Chapman, K. Lipid metabolites in seeds of diverse *Gossypium* accessions: Molecular identification of a high oleic mutant allele. *Planta* **2017**, *245*, 595–610. [CrossRef] [PubMed]

15. Unver, T.; Wu, Z.; Sterck, L.; Turktas, M.; Lohaus, R.; Li, Z.; Yang, M.; He, L.; Deng, T.; Escalante, F.J.; et al. Genome of wild olive and the evolution of oil biosynthesis. *Proc. Natl. Acad. Sci. USA* **2017**, *114*, E9413–E9422. [CrossRef] [PubMed]
16. Dar, A.A.; Choudhury, A.R.; Kancharla, P.K.; Arumugam, N. The *FAD2* gene in plants: Occurrence, regulation, and role. *Front. Plant Sci.* **2017**, *8*, 1789. [CrossRef] [PubMed]
17. Avelange-Macherel, M.H.; Macherel, D.; Wada, H.; Murata, N. Site-directed mutagenesis of histidine residues in the delta 12 acyl-lipid desaturase of Synechocystis. *FEBS Lett.* **1995**, *361*, 111–114. [CrossRef]
18. Broun, P.; Shanklin, J.; Whittle, E.; Somerville, C. Catalytic plasticity of fatty acid modification enzymes underlying chemical diversity of plant lipids. *Science* **1998**, *282*, 1315–1317. [CrossRef] [PubMed]
19. McCartney, A.W.; Dyer, J.M.; Dhanoa, P.K.; Kim, P.K.; Andrews, D.W.; McNew, J.A.; Mullen, R.T. Membrane-bound fatty acid desaturases are inserted co-translationally into the ER and contain different ER retrieval motifs at their carboxy termini. *Plant J.* **2004**, *37*, 156–173. [CrossRef]
20. Hoffmann, M.; Hornung, E.; Busch, S.; Kassner, N.; Ternes, P.; Braus, G.H.; Feussner, I. A small membrane-peripheral region close to the active center determines regioselectivity of membrane-bound fatty acid desaturases from *Aspergillus nidulans*. *J. Biol. Chem.* **2007**, *282*, 26666–26674. [CrossRef]
21. Los, D.A.; Murata, N. Structure and expression of fatty acid desaturases. *BBA-Lipids Lipid Met.* **1998**, *1394*, 3–15. [CrossRef]
22. Shanklin, J.; Cahoon, E.B. Desaturation and related modifications of fatty acids. *Annu. Rev. Plant Phys.* **1998**, *49*, 611–641. [CrossRef] [PubMed]
23. Jung, S.; Swift, D.; Sengoku, E.; Patel, M.; Teule, F.; Powell, G.; Moore, K.; Abbott, A. The high oleate trait in the cultivated peanut [*Arachis hypogaea L.*]. I. Isolation and characterization of two genes encoding microsomal oleoyl-PC desaturases. *Mol. Gen. Genet.* **2000**, *263*, 796–805. [CrossRef] [PubMed]
24. Thapa, R.; Carrero-Colon, M.; Crowe, M.; Gaskin, E.; Hudson, K. Novel *FAD2-1A* alleles confer an elevated oleic acid phenotype in soybean seeds. *Crop Sci.* **2016**, *56*, 226–231. [CrossRef]
25. Ustun, R.; Uzun, B. Breeding for introgression of *FAD2-1A* and *FAD2-1B* genes to local soybean cultivars of Turkey. *J. Biotechnol.* **2017**, *256*, S103. [CrossRef]
26. Dyer, J.M.; Chapital, D.C.; Kuan, J.C.W.; Mullen, R.T.; Turner, C.; McKeon, T.A.; Pepperman, A.B. Molecular analysis of a bifunctional fatty acid conjugase/desaturase from tung. Implications for the evolution of plant fatty acid diversity. *Plant Physiol.* **2002**, *130*, 2027–2038. [CrossRef] [PubMed]
27. Okuley, J.; Lightner, J.; Feldmann, K.; Yadav, N.; Lark, E.; Browse, J. Arabidopsis *Fad2* gene encodes the enzyme that is essential for polyunsaturated lipid-synthesis. *Plant Cell* **1994**, *6*, 147–158. [CrossRef] [PubMed]
28. Kishore, K.; Sinha, S.K.; Kumar, R.; Gupta, N.C.; Dubey, N.; Sachdev, A. Isolation and characterization of microsomal omega-6-desaturase gene (*fad2-1*) from soybean. *Indian J. Exp. Biol.* **2007**, *45*, 390–397.
29. Li, L.Y.; Wang, X.L.; Gai, J.Y.; Yu, D.Y. Molecular cloning and characterization of a novel microsomal oleate desaturase gene from soybean. *J. Plant Physiol.* **2007**, *164*, 1516–1526. [CrossRef]
30. Pirtle, I.L.; Kongcharoensuntorn, W.; Nampaisansuk, M.; Knesek, J.E.; Chapman, K.D.; Pirtle, R.M. Molecular cloning and functional expression of the gene for a cotton Delta-12 fatty acid desaturase (*FAD-2*). *BBA-Gene Struct. Expr.* **2001**, *1522*, 122–129. [CrossRef]
31. Sakai, H.; Kajiwara, S. Cloning and functional characterization of a Delta 12 fatty acid desaturase gene from the basidiomycete *Lentinula edodes*. *Mol. Genet. Genomics* **2005**, *273*, 336–341. [CrossRef] [PubMed]
32. Schierholt, A.; Becker, H.C.; Ecke, W. Mapping a high oleic acid mutation in winter oilseed rape (*Brassica napus* L.). *Theor. Appl. Genet.* **2000**, *101*, 897–901. [CrossRef]
33. Tang, G.Q.; Novitzky, W.P.; Griffin, H.C.; Huber, S.C.; Dewey, R.E. Oleate desaturase enzymes of soybean: Evidence of regulation through differential stability and phosphorylation. *Plant J.* **2005**, *44*, 433–446. [CrossRef] [PubMed]
34. Bruner, A.C.; Jung, S.; Abbott, A.G.; Powell, G.L. The naturally occurring high oleate oil character in some peanut varieties results from reduced oleoyl-PC desaturase activity from mutation of aspartate 150 to asparagine. *Crop Sci.* **2001**, *41*, 522–526. [CrossRef]
35. Katoh, K.; Standley, D.M. MAFFT multiple sequence alignment software version 7: Improvements in performance and usability. *Mol. Biol. Evol.* **2013**, *30*, 772–780. [CrossRef]
36. Price, M.N.; Dehal, P.S.; Arkin, A.P. FastTree 2-approximately maximum-likelihood trees for large alignments. *PLoS ONE* **2010**, *5*, e9490.
37. Nelson, B.K.; Cai, X.; Nebenfuhr, A. A multicolored set of in vivo organelle markers for co-localization studies in Arabidopsis and other plants. *Plant J.* **2007**, *51*, 1126–1136. [CrossRef]

12

Identification of Two Novel R2R3-MYB Transcription factors, *PsMYB114L* and *PsMYB12L*, Related to Anthocyanin Biosynthesis in *Paeonia suffruticosa*

Xinpeng Zhang [1,†], Zongda Xu [2,†], Xiaoyan Yu [2,*], Lanyong Zhao [2,*], Mingyuan Zhao [2], Xu Han [2] and Shuai Qi [2]

1 State Key Laboratory of Crop Biology, College of Horticulture Science and Engineering, Shandong Agricultural University, Taian 271018, China; zhxpnd123@163.com

2 College of Forestry, Shandong Agricultural University, Taian 271018, China; xuzoda@163.com (Z.X.); zhaomingy9@163.com (M.Z.); hanxusdau@163.com (X.H.); shuaiqi@sdau.edu.cn (S.Q.)

* Correspondence: yxyxst20040214@163.com (X.Y.); sdzly369@163.com (L.Z.)

† These authors contributed equally to this work.

Abstract: Flower color is a charming phenotype with very important ornamental and commercial values. Anthocyanins play a critical role in determining flower color pattern formation, and their biosynthesis is typically regulated by R2R3-MYB transcription factors (TFs). *Paeonia suffruticosa* is a famous ornamental plant with colorful flowers. However, little is known about the R2R3-MYB TFs that regulate anthocyanin accumulation in *P. suffruticosa*. In the present study, two R2R3-MYB TFs, namely, *PsMYB114L* and *PsMYB12L*, were isolated from the petals of *P. suffruticosa* 'Shima Nishiki' and functionally characterized. Sequence analysis suggested that *PsMYB114L* contained a bHLH-interaction motif, whereas *PsMYB12L* contained two flavonol-specific motifs (SG7 and SG7-2). Subsequently, the in vivo function of *PsMYB114L* and *PsMYB12L* was investigated by their heterologous expression in *Arabidopsis thaliana* and apple calli. In transgenic *Arabidopsis* plants, overexpression of *PsMYB114L* and of *PsMYB12L* caused a significantly higher accumulation of anthocyanins, resulting in purple-red leaves. Transgenic apple calli overexpressing *PsMYB114L* and *PsMYB12L* also significantly enhanced the anthocyanins content and resulted in a change in the callus color to red. Meanwhile, gene expression analysis in *A. thaliana* and apple calli suggested that the expression levels of the flavonol synthase (*MdFLS*) and anthocyanidin reductase (*MdANR*) genes were significantly downregulated and the dihydroflavonol 4-reductase (*AtDFR*) and anthocyanin synthase (*AtANS*) genes were significantly upregulated in transgenic lines of *PsMYB114L*. Moreover, the expression level of the *FLS* gene (*MdFLS*) was significantly downregulated and the *DFR* (*AtDFR/MdDFR*) and *ANS* (*AtANS/MdANS*) genes were all significantly upregulated in transgenic lines plants of *PsMYB12L*. These results indicate that *PsMYB114L* and *PsMYB12L* both enhance anthocyanin accumulation by specifically regulating the expression of some anthocyanin biosynthesis-related genes in different plant species. Together, these results provide a valuable resource with which to further study the regulatory mechanism of anthocyanin biosynthesis in *P. suffruticosa* and for the breeding of tree peony cultivars with novel and charming flower colors.

Keywords: *P. suffruticosa*; R2R3-MYB; overexpression; anthocyanin; transcriptional regulation

---

1. Introduction

Paeonia suffruticosa is a very popular ornamental flowering plant that was first cultivated more than 1600 years ago in China and is currently distributed worldwide. This species is in the Paeoniaceae family and has been named 'the king of flowers' for its showy and colorful flowers [1]. Approximately 1500 cultivars of *P. suffruticosa* with a variety of flower colors have been produced by breeders worldwide [2]. Among the many flower colors of this species, most fit into two clusters: monochrome color (red, pink, white, purple, black, blue, green, and yellow) and double color. Cultivars with a double-color phenotype are rarer and more sought after, and thus have great ornamental and commercial value [3]. Among them, *P. suffruticosa* 'Shima Nishiki', a well-known chimeric cultivar, was selected from the bud mutation of *P. suffruticosa* 'Taiyoh'. 'Shima Nishiki' usually has red and pink petals on the same flower, and this aesthetically pleasing double-color phenotype can be stably inherited [4]. Therefore, the 'Shima Nishiki' cultivar is regarded as an important experimental material with which to study the molecular regulatory mechanism of flower color and in the breeding of new cultivars [5].

Anthocyanins are important soluble flavonoid compounds that are widely distributed in the leaves, flowers, fruits, seeds and other tissues of many plants [6]. Anthocyanin composition and concentration are usually closely related to flower color intensity [7,8]. The anthocyanin biosynthetic pathway is well known to be highly conserved in many ornamental plants [9–14]. Anthocyanin biosynthesis and accumulation are usually regulated by a series of structural genes and regulatory genes [15,16]. The structural genes encode enzymes associated with anthocyanin biosynthesis, including chalcone synthase (CHS), chalcone isomerase (CHI), flavanone 3-hydroxylase (F3H), flavonoid 3′-hydroxylase (F3′H), dihydroflavonol 4-reductase (DFR) anthocyanin synthase (ANS), Flavonol synthase (FLS), and anthocyanidin reductase (ANR) [17–19] (Figure 1). Among them, FLS is a dedicated enzyme involved in flavonol biosynthesis, and ANR is a key enzyme for proanthocyanidin biosynthesis. The regulatory genes can be divided into three families R2R3-MYB, bHLH, and WD40 [20–22] and they usually form a regulatory complex to activate the expression of anthocyanin biosynthetic genes [23–25].

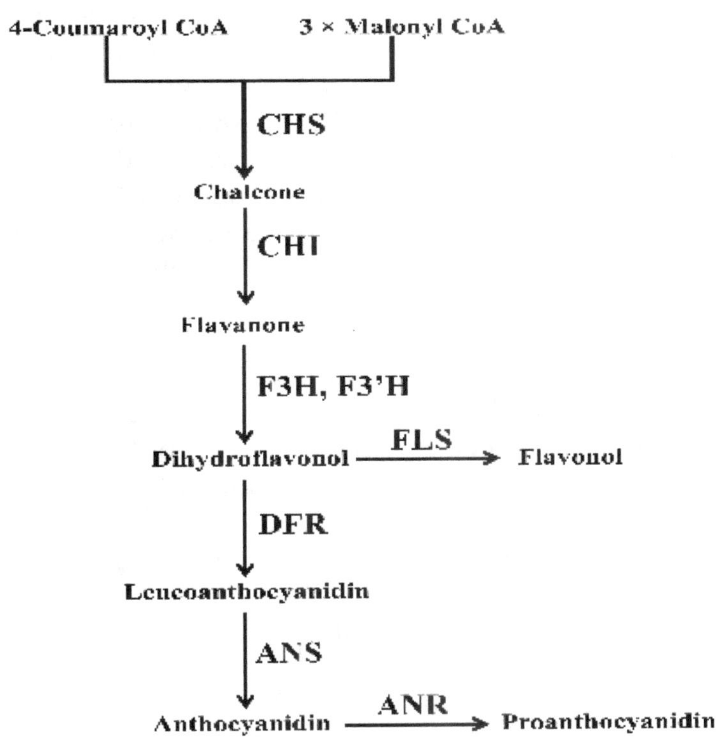

Figure 1. A general schematic diagram of the metabolic pathway related to anthocyanin biosynthesis. CHS, chalcone synthase; CHI, chalcone isomerase; F3H, flavanone 3-hydroxylase; F3′H, flavonoid 3′-hydroxylase; DFR, dihydroflavonol 4-reductase; ANS, anthocyanidin synthase; FLS, flavonol synthase; ANR, anthocyanidin reductase.

Many structural genes have been characterized and cloned in *P. suffruticosa* [12,26–28]. In the MYB-bHLH-WDR (MBW) complex, R2R3-MYB transcription factors (TFs) usually play critical roles in anthocyanin biosynthesis and accumulation [29,30]. Many R2R3-MYB TFs involved in anthocyanin biosynthesis have been isolated and characterized from various plants, including *Arabidopsis thaliana* [24], *Zea mays* [31], *Vitis vinifera* [32], *Malus* crabapple [33], *Petunia hybrida* [34], *Antirrhinum majus* [29], *Dendranthema morifolium* [35] and *Phalaenopsis aphrodite* [14]. In *P. suffruticosa*, most previous studies were focused primarily on the preliminary investigation of R2R3-MYB TFs based on transcriptome sequencing and qRT-PCR analyses [12,36–38], whereas whether and how R2R3-MYB TFs control anthocyanin biosynthesis and accumulation in *P. suffruticosa* are almost unknown.

In the present study, two novel R2R3-MYB TFs, namely, *PsMYB114L* and *PsMYB12L*, were cloned in *P. suffruticosa*. Subsequently, the expression patterns of *PsMYB114L* and *PsMYB12L* were determined at five developmental stages in *P. suffruticosa* 'Shima Nishiki'. Furthermore, the function of these two TFs was further verified by heterologous expression in *Arabidopsis* and apple calli. These results will provide valuable insights into understanding the putative roles of *PsMYB114L* and *PsMYB12L* in regulating anthocyanin biosynthesis in *P. suffruticosa*.

2. Results

2.1. Cloning and Analysis of the PsMYB114L and PsMYB12L Genes

Based on the functional annotation and gene expression analysis of transcriptome sequencing data in *P. suffruticosa* 'Shima Nishiki' [39], we filtered two MYB unigenes exhibiting relatively high expression differences between the red and pink petals as the targeted genes of this study.

The full-length cDNA sequences of the two novel MYB genes were obtained with PCR amplification. By conducting GenBank BLAST searches of the amino acid sequences of these two genes, we found that these genes have the highest homology with transcription factor *MYB114*-like [*Quercus suber*] and transcription factor *MYB12*-like [*Juglans regia*], respectively. Therefore, we named these genes *PsMYB114L* and *PsMYB12L*. Sequencing results revealed that *PsMYB114L* (Figure S1A,B) and *PsMYB12L* (Figure S1C,D) contained an open reading frame (ORF) of 600 and 1140 bp encoding 199 and 379 amino acids and that their predicted proteins had a molecular mass of 22.81 and 42.61 kDa and a theoretical isoelectric point (pI) of 8.53 and 4.86, respectively.

Multiple sequence alignment of amino acids revealed that *PsMYB114L* and *PsMYB12L*, belonging to the SANT superfamily (which typically consists of tandem repeats of three alpha-helices arranged in a helix-turn-helix motif, with each alpha helix containing a bulky aromatic residue), and other known R2R3-MYB TFs related to anthocyanin biosynthesis contained a highly conserved R2R3 DNA-binding domain. The presence of this conserved domain means that *PsMYB114L* and *PsMYB12L* are also R2R3-MYB TFs and may perform similar functions in regulating anthocyanin biosynthesis. Furthermore, *PsMYB114L* had a bHLH-interaction motif ([D/E]Lx2[R/K]x3Lx6Lx3R) in the R3 domain at the N terminus and did not have any conserved motifs at the C terminus (Figure 2A). Moreover, *PsMYB12L* did not have any bHLH-interaction motifs at the N terminus, whereas it contained two flavonol-specific motifs [40], namely, SG7 ([K/R][R/x][R/K]xGRT[S/x][R/G]xx[M/x]K) and SG7-2 ([W/x][L/x]LS), at the C terminus (Figure 2B).

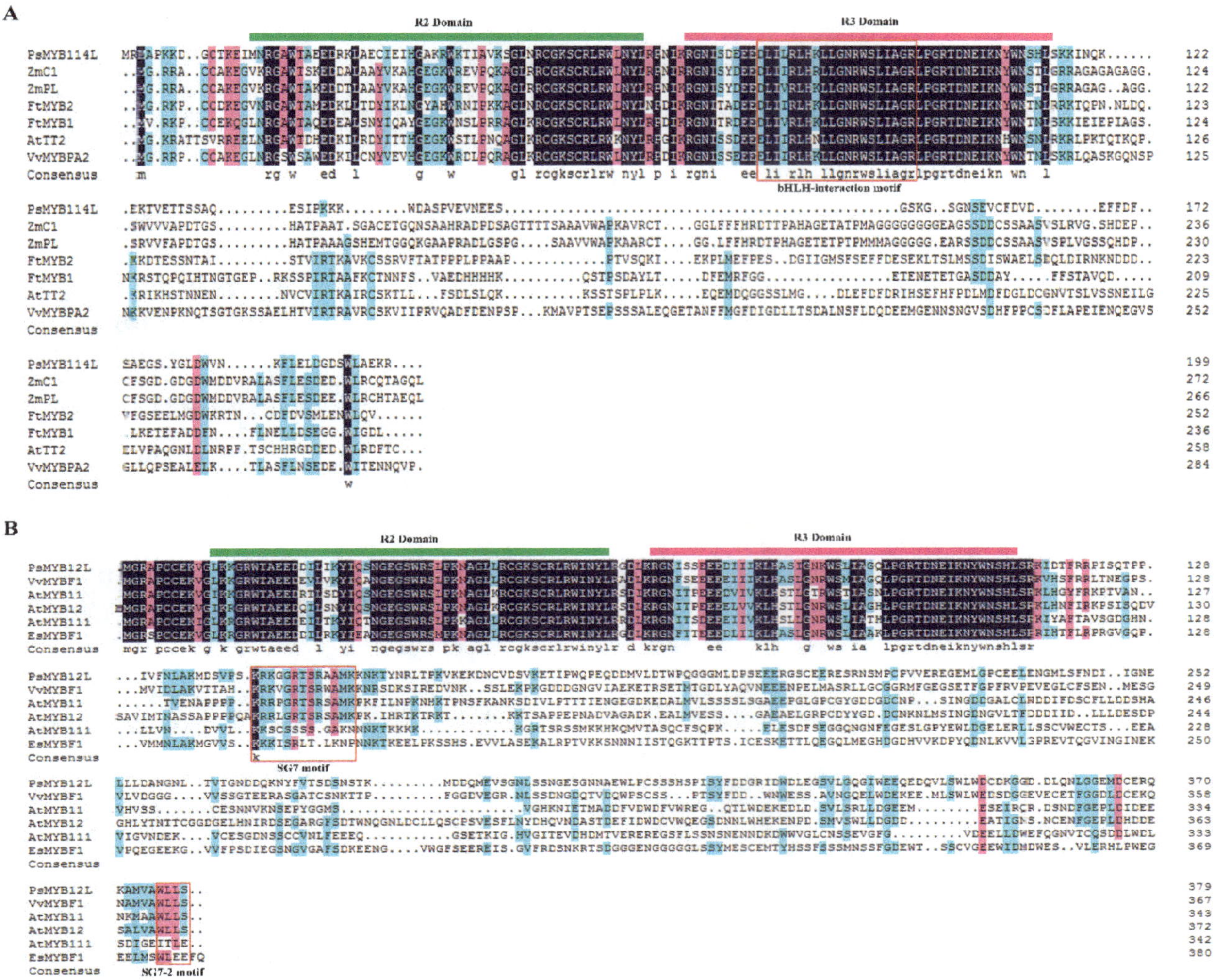

Figure 2. Amino acid sequence alignment analysis of the *PsMYB114L* (**A**) and *PsMYB12L* (**B**) genes with other known R2R3-MYB TFs. The green and pink long lines indicate the R2 and R3 domain, respectively. The red boxes show the conserved bHLH-interaction, SG7 and SG7-2 motifs. The NCBI GenBank accession numbers of these sequences are as follows: *ZmC1, Zea mays,* AF320613.3; *ZmPL, Zea mays,* NM_001112415.1; *FtMYB2, Fagopyrum tataricum,* JF313346.1; *FtMYB1, Fagopyrum tataricum,* JF313344.1; *AtTT2, Arabidopsis thaliana,* NM_122946.3; *VvMYBPA2, Vitis vinifera,* NM_001281024.1; *VvMYBF1, Vitis vinifera,* FJ948477.2; *AtMYB11, Arabidopsis thaliana,* NM_116126.3; *AtMYB12, Arabidopsis thaliana,* NM_130314.4; *AtMYB111, Arabidopsis thaliana,* NM_124310.3; *EsMYBF1, Epimedium sagittatum,* KU365320.1.

To better evaluate the phylogenetic relationships of *PsMYB114L*, *PsMYB12L* and 16 other known MYB TFs related to the regulation of anthocyanin biosynthesis, a phylogenetic tree was constructed based on the amino acid sequences of these 18 MYB TFs from different species using the neighbor-joining method. The phylogenetic analysis indicated that these 18 MYB TFs were classified into four groups (Flavonol, Anthocyanin, Anthocyanin/Proanthocyanidin and Proanthocyanidin) based on their specific roles in the flavonoid biosynthesis pathway. Among them, *PsMYB114L* had the closest phylogenetic relationship with *ZmC1* and *ZmPL*, which are involved in regulating anthocyanin biosynthesis, whereas *PsMYB12L* belongs to a subgroup of MYB proteins that includes VvMYBF1, EsMYBF1, AtMYB11, AtMYB12 and AtMYB111, which regulate flavonol synthesis and had the closest phylogenetic relationship with *VvMYBF1* (Figure 3).

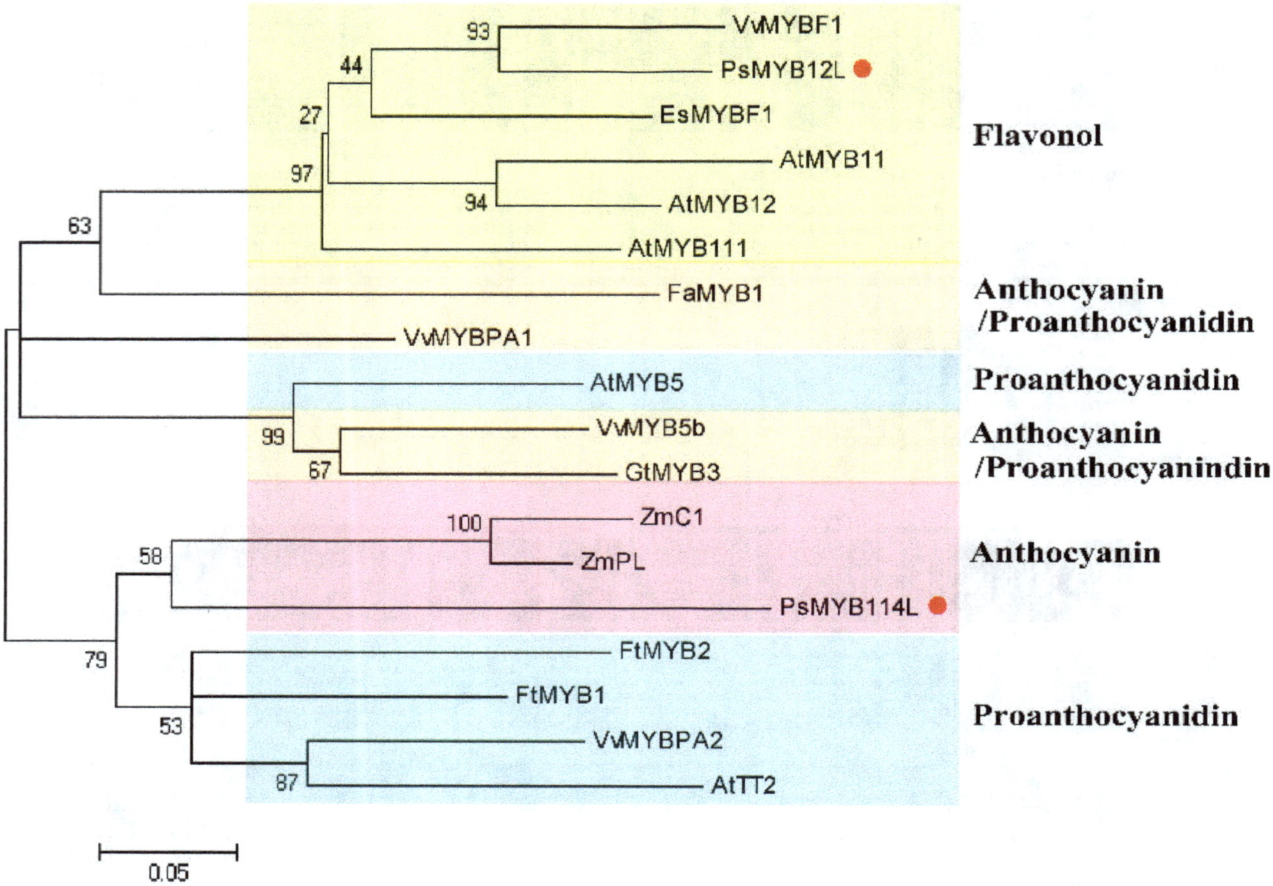

Figure 3. Phylogenetic analysis of the *PsMYB114L* and *PsMYB12L* genes with R2R3-MYB TFs from other species. The NCBI GenBank accession numbers of these sequences are as follows: *FaMYB1, Fragaria x ananassa,* AF401220.1; *VvMYBPA1, Vitis vinifera,* NM_001281231.1; *VvMYB5b, Vitis vinifera,* NM_001280925.1; *AtMYB5, Arabidopsis thaliana,* AF401220.1; *GtMYB3, Gentiana triflora,* AB289445.1; *ZmC1, Zea mays,* AF320613.3; *ZmPL, Zea mays,* NM_001112415.1; *FtMYB2, Fagopyrum tataricum,* JF313346.1; *FtMYB1, Fagopyrum tataricum,* JF313344.1; *AtTT2, Arabidopsis thaliana,* NM_122946.3; *VvMYBPA2, Vitis vinifera,* NM_001281024.1; *VvMYBF1, Vitis vinifera,* FJ948477.2; *AtMYB11, Arabidopsis thaliana,* NM_116126.3; *AtMYB12, Arabidopsis thaliana,* NM_130314.4; *AtMYB111, Arabidopsis thaliana,* NM_124310.3; *EsMYBF1, Epimedium sagittatum,* KU365320.1.

2.2. *Subcellular Localization of PsMYB114L and PsMYB12L*

To examine the subcellular localization of *PsMYB114L* and *PsMYB12L*, the recombinant vector (*PsMYB114L*-GFP/*PsMYB12L*-GFP) and the control vector (pCAMBIA1301-GFP) were introduced into the tobacco leaves. Our results were basically consistent with those of previous studies [41,42]. The green fluorescent protein (GFP) fluorescence of the control vector was clearly distributed throughout the entire cell (Figure 4A), and the *PsMYB114L*-GFP/*PsMYB12L*-GFP vector displayed a strong fluorescence signal in the nucleus and cytoplasm of tobacco cells (Figure 4B,C). Therefore, we speculated that the two R2R3-MYB TFs (*PsMYB114L*/*PsMYB12L*) were simultaneously localized and functioned in the nucleus and cytoplasm.

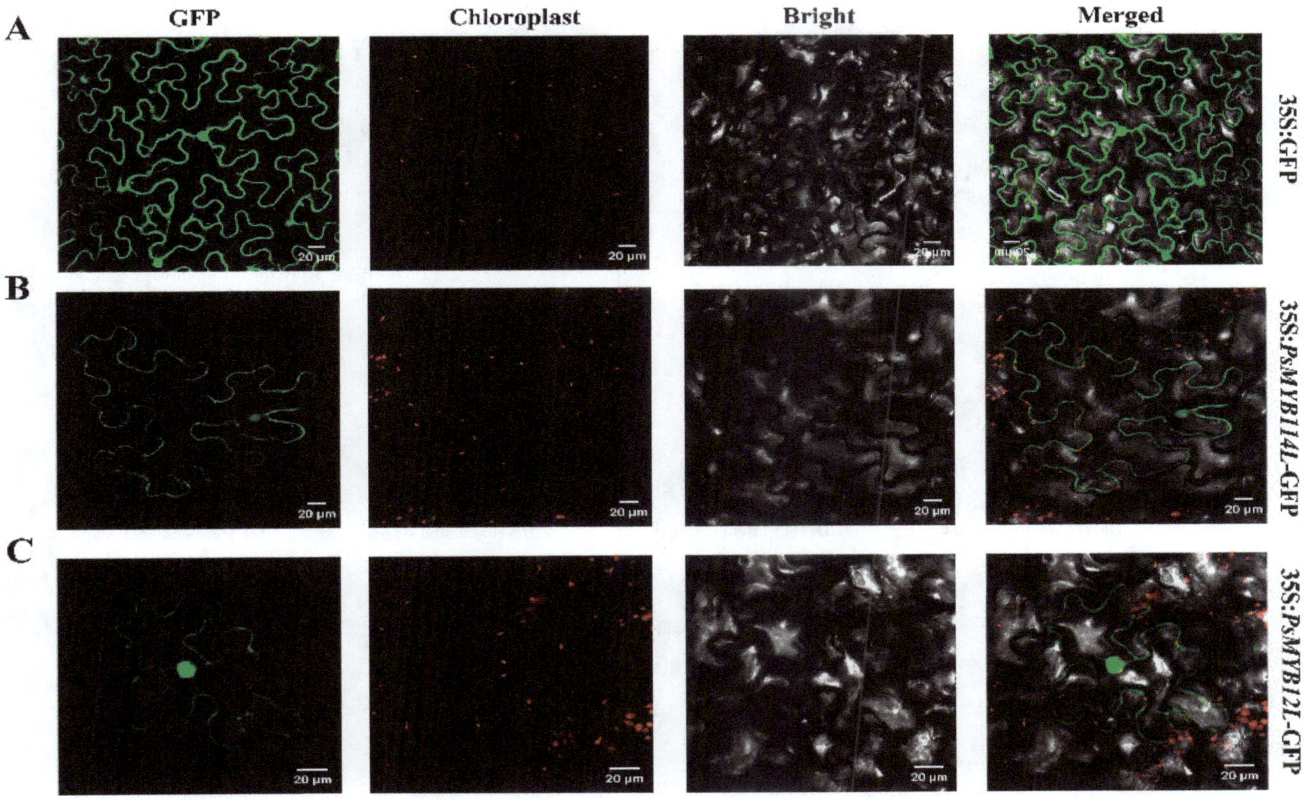

Figure 4. Subcellular localization analysis of the *PsMYB114L* and *PsMYB12L* genes. (**A**) Control vector (pCAMBIA1301-GFP) expressed in epidermal cells of tobacco leaves. (**B**) Recombinant vector (*PsMYB114L*-GFP) expressed in epidermal cells of tobacco leaves. (**C**) Recombinant vector (*PsMYB12L*-GFP) expressed in epidermal cells of tobacco leaves. White lines at the bottom right of the picture represent 20 μm in the respective pixel. GFP, GFP fluorescence; Chloroplast, Chloroplast fluorescence; Bright, Bright field; Merged, Superposition of bright field and fluorescence.

2.3. *Expression Patterns of PsMYB114L and PsMYB12L in P. suffruticosa 'Shima Nishiki'*

qRT-PCR analysis was conducted to survey the expression patterns of *PsMYB114L* and *PsMYB12L* in *P. suffruticosa* 'Shima Nishiki' (Figure 5). Petal samples of this cultivar were collected at five developmental stages (Figure S2). The expression levels of the *PsMYB114L* gene peaked at S3 and then decreased from S3 to S5, whereas the *PsMYB12L* gene exhibited the highest expression at S4. Furthermore, the expression levels of the eight anthocyanin biosynthesis-related genes (*PsCHS, PsCHI, PsF3H, PsF3′H, PsDFR, PsANS, PsFLS,* and *PsANR*) were analyzed. Among these genes, *PsF3′H, PsDFR,* and *PsANS* showed a trend similar to that of *PsMYB12L*, whereas *PsFLS* and *PsANR* showed a trend similar to that of *PsMYB114L*.

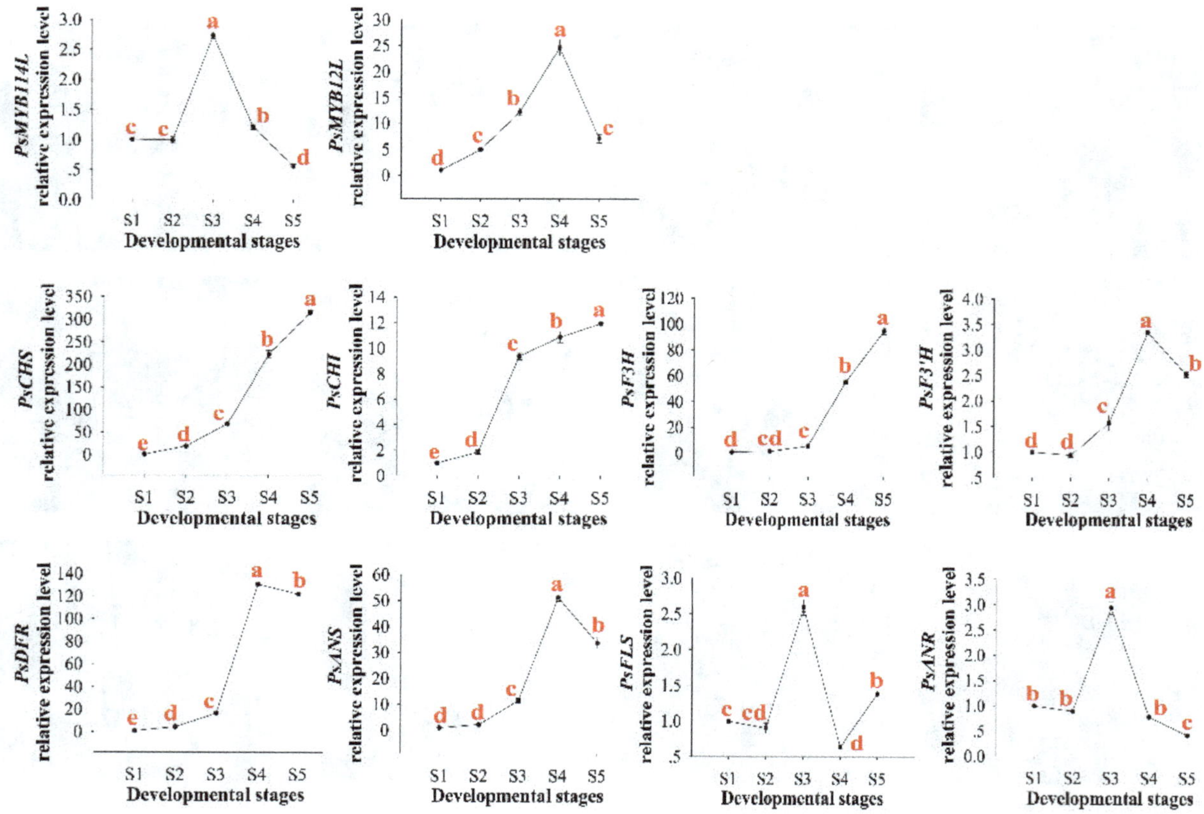

Figure 5. The expression patterns of the *PsMYB114L* gene, *PsMYB12L* gene and anthocyanin biosynthesis-related structural genes in *P. suffruticosa* 'Shima Nishiki'. S1, flower bud emerging stage; S2 small bell-like flower-bud stage; S3, large bell-like flower-bud stage; S4, bell-like flower-bud extending stage; S5, color exposing stage. Different lowercase letters indicate significant differences at $p < 0.05$.

2.4. Overexpression of PsMYB114L and PsMYB12L in Arabidopsis

To characterize the functions of *PsMYB114L* and *PsMYB12L*, these two genes under the expression of the 35S promoter were genetically transformed into *Arabidopsis*. Phenotypic investigations of the transgenic lines of *PsMYB114L* and *PsMYB12L* revealed that their leaves were much deeper in color than those of Col-0 and showed a purple-red color (Figure 6A). Meanwhile, these transgenic lines of the two genes were confirmed by PCR analysis (Figure 6B). Furthermore, the total anthocyanin content results indicated that the transgenic lines of *PsMYB114L* and *PsMYB12L* produced much more anthocyanin than Col-0 (Figure 6C,D).

Additionally, the expression levels of anthocyanin biosynthesis-related genes (*AtCHS*, *AtCHI*, *AtF3H*, *AtF3′H*, *AtDFR*, *AtANS*, *AtFLS*, and *AtANR*) in the Col-0 and the transgenic *Arabidopsis* plants of *PsMYB114L* and *PsMYB12L* were analyzed with qRT-PCR experiments. Compared with the Col-0, overexpression of *PsMYB114L* upregulated the expression of most of the genes (*AtCHS*, *AtCHI*, *AtF3H*, *AtF3′H*, *AtDFR*, and *AtANS*) in transgenic *PsMYB114L* plants; among them, both of the *AtDFR*/*AtANS* genes showed a relatively high difference between the Col-0 and transgenic plants, whereas *AtFLS* and *AtANR* were downregulated in transgenic *PsMYB114L* plants (Figure 6E).

For *PsMYB12L* overexpression in *Arabidopsis*, the expression levels of all eight genes were upregulated in transgenic *PsMYB12L* plants. Among them, the four genes (*AtCHS*, *AtCHI*, *AtDFR*, and *AtANS*) all showed a relatively high difference between the Col-0 and transgenic plants (Figure 6F).

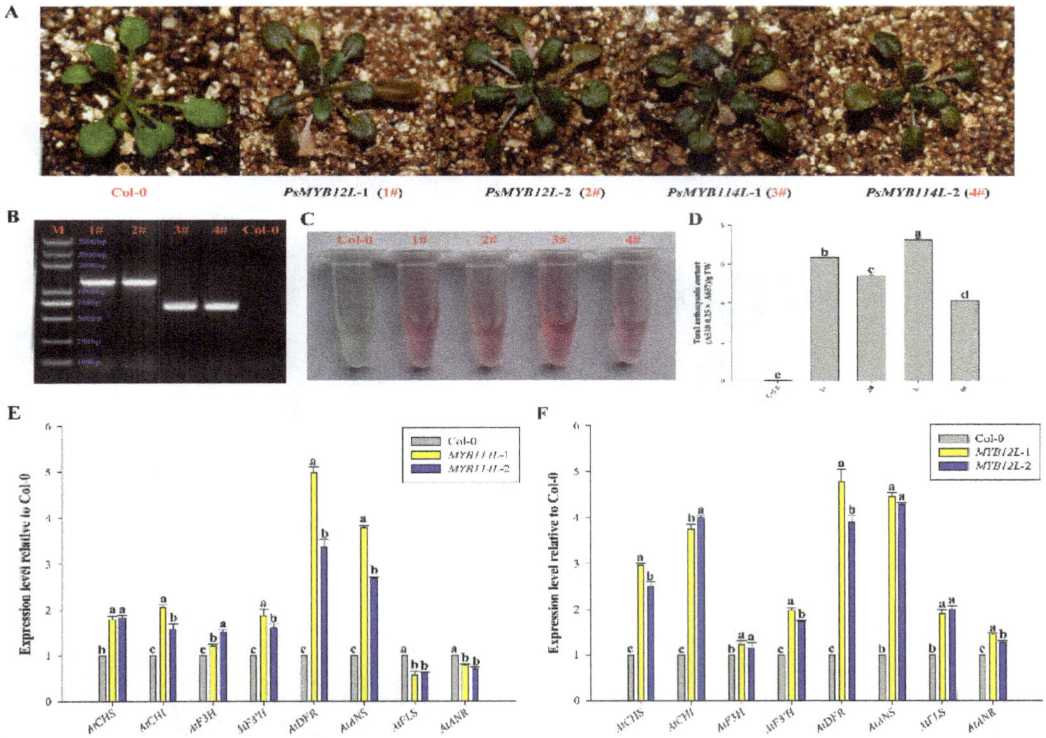

Figure 6. Overexpression analysis of the *PsMYB114L* and *PsMYB12L* genes in *Arabidopsis*. (**A**) Comparison of leaf colors in transgenic *Arabidopsis* plants and Col-0. (**B**) Results of positive PCR detection in transgenic *Arabidopsis* plants. (**C**) Anthocyanin extraction solutions for transgenic *Arabidopsis* plants and Col-0. (**D**) Total anthocyanin content in transgenic *Arabidopsis* plants and Col-0. (**E**) Expression analysis of anthocyanin biosynthesis-related genes in transgenic *Arabidopsis* plants of *PsMYB114L* and Col-0. (**F**) Expression analysis of anthocyanin biosynthesis-related genes in transgenic *Arabidopsis* plants of *PsMYB12L* and Col-0. Col-0, *Arabidopsis thaliana* ecotype Columbia; 1# and 2#, two transgenic lines of the *PsMYB12L* gene; 3# and 4#, two transgenic lines of the *PsMYB114L* gene. Different lowercase letters indicate significant differences at $p < 0.05$.

2.5. Overexpression of PsMYB114L and PsMYB12L in Apple Calli

For further functional validation, the two genes (*PsMYB114L* and *PsMYB12L*) were ectopically expressed in the calli of 'Orin' apple. Interestingly, after light and low-temperature treatments, the WT had almost no phenotypic changes, but an especially obvious color change was observed in the transgenic lines of *PsMYB114L* and *PsMYB12L* (Figures 7A and 8A). PCR amplification confirmed that these transgenic apple calli carry *PsMYB114L* and *PsMYB12L* (Figures 7B and 8B). With regard to the total anthocyanin content, the transgenic lines of *PsMYB114L* and *PsMYB12L* all accumulated markedly higher amounts of anthocyanins than did the WT (Figures 7C,D and 8C,D).

Additionally, the expression levels of anthocyanin biosynthesis-related genes (*MdCHS*, *MdCHI*, *MdF3H*, *MdF3′H*, *MdDFR*, *MdANS*, *MdFLS*, and *MdANR*) in the WT and the transgenic lines of *PsMYB114L* and *PsMYB12L* were analyzed by qRT-PCR. Compared with the WT, overexpression of *PsMYB114L* downregulated the expression of most of the genes, specifically, *MdCHS*, *MdCHI*, *MdF3H*, *MdF3′H*, *MdFLS*, and *MdANR*, and upregulated the expression of *MdDFR* and *MdANS* in transgenic *PsMYB114L* calli (Figure 7E).

For *PsMYB12L* overexpression, the expression levels of most genes, including *MdCHS*, *MdF3H*, *MdF3′H*, *MdDFR*, *MdANS*, and *MdANR*, were upregulated, but those of *MdCHI* and *MdFLS* were downregulated in transgenic *PsMYB12L* calli (Figure 8E).

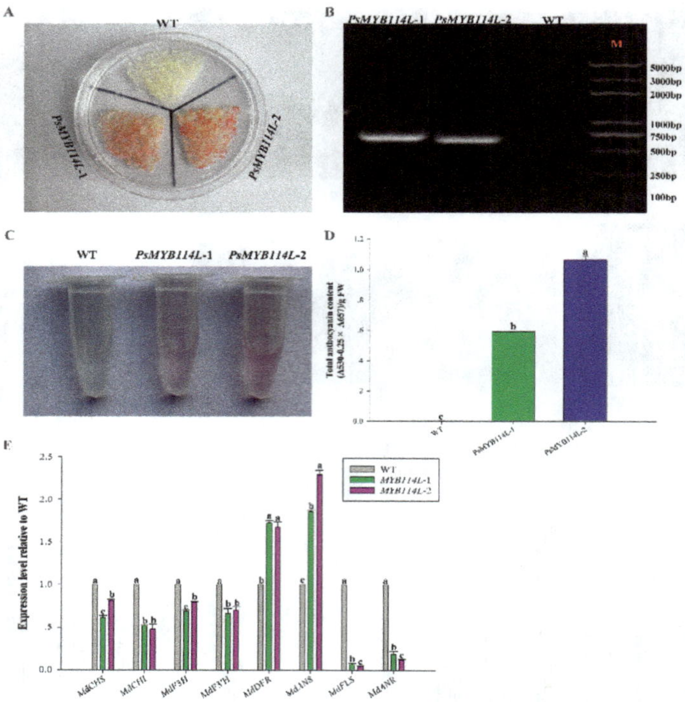

Figure 7. Overexpression analysis of the *PsMYB114L* gene in apple calli. (**A**) Colors observed in transgenic apple calli and the WT. (**B**) Results of positive PCR detection in transgenic apple calli. (**C**) Anthocyanin extraction solutions for transgenic apple calli and the WT. (**D**) Total anthocyanin content in transgenic apple calli and the WT. (**E**) Expression analysis of anthocyanin biosynthesis-related genes in transgenic apple calli and the WT. WT, Wild-type 'Orin' apple calli; *PsMYB114L*-1 and *PsMYB114L*-2, two transgenic lines of the *PsMYB114L* gene. Different lowercase letters indicate significant differences at $p < 0.05$.

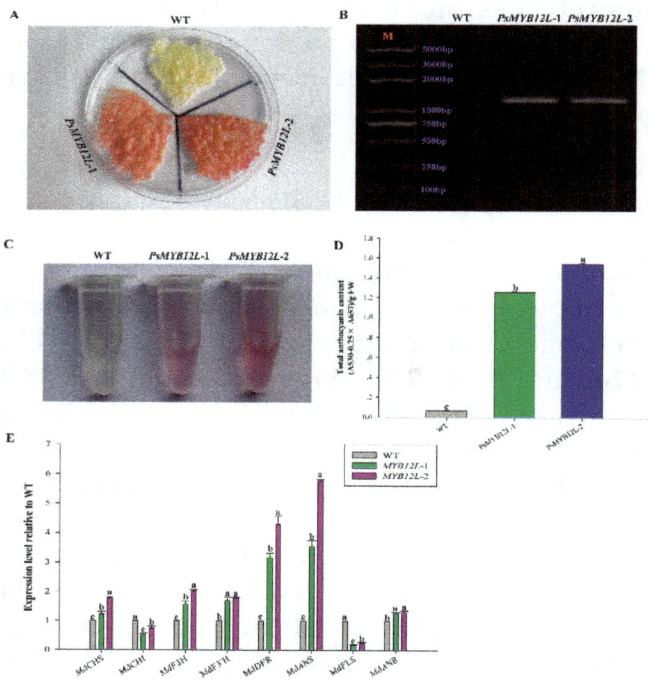

Figure 8. Overexpression analysis of the *PsMYB12L* gene in apple calli. (**A**) Colors observed in transgenic apple calli and the WT. (**B**) Results of positive PCR detection in transgenic apple calli. (**C**) Anthocyanin extraction solutions for transgenic apple calli and the WT. (**D**) Total anthocyanin content in transgenic apple calli and the WT. (**E**) Expression analysis of anthocyanin biosynthesis-related genes in transgenic apple calli and the WT. *PsMYB12L*-1 and *PsMYB12L*-2, two transgenic lines of the *PsMYB12L* gene. Different lowercase letters indicate significant differences at $p < 0.05$.

3. Discussion

Flower color is a very important trait in many ornamental plants and has a close association with their ornamental and commercial value. Many prior studies have shown that anthocyanins are a key factor influencing flower color [43–45]. R2R3-MYB TFs comprise one of the largest gene families in plants and play key roles in regulating anthocyanin accumulation by activating the expression of structural genes involved in the anthocyanin biosynthetic pathway [46,47]. However, the role of R2R3-MYB TFs in regulating flower color in *P. suffruticosa* has seldom been functionally verified. Therefore, determining how certain R2R3-MYB TFs regulate anthocyanin production in *P. suffruticosa* would aid in breeding improved cultivars with desirable flower colors.

In the present study, two novel R2R3-MYB TFs (*PsMYB114L* and *PsMYB12L*) possibly involved in anthocyanin biosynthesis were successfully cloned and characterized from the petals of *P. suffruticosa* 'Shima Nishiki' and found to contain full-length cDNA of 600 and 1140 bp encoding 199 and 379 amino acids, respectively. The amino acid sequence alignment between *PsMYB114L*/*PsMYB12L* and other known R2R3-MYB TFs involved in anthocyanin regulation indicated that the R2R3 domain distributions of these R2R3-MYB TFs were highly similar, but a bHLH-interaction motif ([D/E]Lx2[R/K]x3Lx6Lx3R) existed in the R3 domain of *PsMYB114L*, whereas *PsMYB12L* did not contain this motif for interaction with bHLH proteins. In *Arabidopsis*, based on a similar function, 125 TFs of R2R3-MYB gene-family members were classified into more than 25 subgroups [48]. Furthermore, many previous studies demonstrated that subgroup 7 [49,50], characterized by both the SG7 ([K/R][R/x][R/K]xGRT[S/x][R/G]xx[M/x]K) and SG7-2 ([W/x][L/x]LS) motifs, specifically regulated flavonol biosynthesis. *PsMYB12L* contained these two motifs (SG7 and SG7-2) at the C terminus of the protein, but *PsMYB114L* lacked these two motifs.

Phylogenetic analysis indicated that *PsMYB12L* and 5 flavonol-regulating R2R3-MYB TFs (*VvMYBF1*, *EsMYBF1*, and *AtMYB11/12/111*) belonging to subgroup 7 [30,51] were clustered together, and *PsMYB114L* and certain R2R3-MYB TFs belonging to subgroup 5 (*AtTT2*, *ZmC1*, *VvMYBPA2*, etc.) [52–55] had relatively higher homology. Based on the motif analysis of amino acid sequences and phylogenetic analysis, *PsMYB114L* might regulate anthocyanin production by combinatorially interacting with a basic helix-loop-helix (bHLH) factor [25,56,57]. *PsMYB12L* might independently regulate the expression of anthocyanin biosynthesis-related genes without the MBW complex [58].

In addition, we conducted further ectopic transgenic studies by overexpressing *PsMYB114L*/*PsMYB12L* in *Arabidopsis* and apple calli. In contrast to the green-colored leaves of the Col-0 *A. thaliana* ecotype and the white-colored WT apple calli, the leaves of these transgenic *Arabidopsis* plants turned purple-red and the transgenic calli of *PsMYB114L* and *PsMYB12L* were red, which was in agreement with their remarkably higher anthocyanin content. The color and total anthocyanin content analyses of *Arabidopsis* and apple calli indicated that these two R2R3-MYB TFs contribute to anthocyanin accumulation in transgenic lines.

Subsequently, qRT-PCR analysis of seven anthocyanin biosynthesis-related genes (*MdCHS*, *MdCHI*, *MdF3H*, *MdF3′H*, *MdDFR*, *MdANS*, *MdFLS*, and *MdANR*) was further performed in *Arabidopsis* and apple calli. In terms of *PsMYB114L*, the qRT-PCR results in *Arabidopsis* showed that the expression levels of *AtDFR* and *AtANS* were significantly upregulated, whereas *AtFLS* and *AtANR* were downregulated to a certain extent compared with the levels in the Col-0. Furthermore, the qRT-PCR results in apple calli showed that the expression levels of *MdDFR* and *MdANS* were upregulated to a certain extent, whereas *MdFLS* and *MdANR* (especially *MdFLS*) were significantly downregulated compared with the levels in the WT. Meanwhile, based on the results of expression patterns of *PsMYB114L* in *P. suffruticosa* 'Shima Nishiki', we have known that *PsMYB114L* have a positive correlation with *PsFLS* and *PsANR*. By comparing these three qRT-PCR results in *Arabidopsis*, apple calli, and *P. suffruticosa*, we found differences in the expression patterns of some anthocyanin biosynthesis-related genes. Previous studies have showed that many R2R3-MYB TFs usually regulate flavonoid biosynthesis by interacting with the promoter of the targeted structural genes [55,58]. For promoter region, in general, the sequence of the same structural gene in different plant species also differs greatly. Therefore, it is possible that the same MYB TFs performed different regulatory mechanisms of flavonoid biosynthesis in different species [59]. Dihydroflavonol is the direct substrate for two key genes (*FLS* and *DFR*) in the flavonoid biosynthetic pathways, and these two genes usually show a competitive interaction in producing colored anthocyanidin and colorless flavonols [60]. In this study, the strong upregulation of *AtDFR* and *AtANS* may have played key roles in activating the branch of the anthocyanin biosynthesis, resulting in purple-red leaves in the transgenic *Arabidopsis* plants of *PsMYB114L*, whereas the strong downregulation of *MdFLS* would inhibit the branch of the flavonol biosynthesis, resulting in the production of anthocyanins and a red-colored phenotype in the transgenic calli of *PsMYB114L*. Furthermore, because *PsMYB114L* has a bHLH-interaction motif, it may form an MBW complex and contribute to anthocyanin accumulation by regulating the expression of these key genes (*AtDFR*, *AtANS*, *MdFLS*, and *MdANR*) in *Arabidopsis* and apple calli.

With regard to *PsMYB12L*, the qRT-PCR results showed that the expression levels of the four genes (*AtCHS*, *AtCHI*, *AtDFR* and *AtANS*) were all significantly upregulated in the transgenic *Arabidopsis* plants. Moreover, the expression levels of *MdDFR* and *MdANS* were significantly upregulated in the transgenic calli of *PsMYB12L*, but *MdFLS* was significantly downregulated. Meanwhile, based on the results of expression patterns of *PsMYB12L* in *P. suffruticosa* 'Shima Nishiki', we have known that *PsMYB12L* have a positive correlation with *PsDFR*, *PsANS*, and *PsF3′H*. By comparing these three qRT-PCR results in *Arabidopsis*, apple calli, and *P. suffruticosa*, we can found that the two key anthocyanin biosynthesis-related genes (*DFR* and *ANS*) showed a very similar expression pattern. We considered that *PsMYB12L* should be a specific transcriptional regulator on *DFR* and *ANS* genes in these three species. Furthermore, we also found differences in the expression patterns of the *FLS* gene in *Arabidopsis* and apple calli, and considered that the expression difference of the *FLS* gene is likely caused by the promoter sequence specificity of this gene in these two species [61]. Based on the motif analysis of *PsMYB12L*, we speculated that the TF may be a flavonol-specific MYB regulator. Many flavonol-specific MYB TFs have been isolated and functionally verified in various plants, such as *A. thaliana*, *Vitis vinifera*, and *Epimedium sagittatum* [50,62,63]. Furthermore, many flavonol-specific MYB TFs negatively regulate anthocyanin accumulation by inducing higher expression of the *FLS* gene. By overexpressing *AtMYB12* in tobacco, the expression of *NtCHS*, *NtCHI*, and *NtFLS* was specifically activated; moreover, the flowers of the transgenic plants were paler in color than their wild-type counterparts [64]. Ectopic expression analysis of *EsMYBF1* in transgenic tobacco indicated that *NtCHS*, *NtCHI*, *NtF3H*, and *NtFLS* were upregulated but *NtDFR* and *NtANS* were significantly downregulated, and the accumulation of anthocyanins in transgenic tobacco flowers was also remarkably decreased [63]. A study on the overexpression of *PpMYB15* in tobacco showed that it can significantly activate the expression of *NtCHS*, *NtCHI*, *NtF3H*, and *NtFLS*, while it had no effects on the expression of *NtDFR* and *NtANS*, resulting in pale-pink or pure white flowers in transgenic tobacco plants [40]. Compared with the expression of anthocyanin biosynthesis-related genes documented in the abovementioned studies, in this study *AtCHS/MdCHS*, *AtCHI*, *AtF3H/MdF3H*, and *AtFLS* had a somewhat similar expression pattern and *MdFLS*, *AtDFR/MdDFR* and *AtANS/MdANS* exhibited the opposite pattern. However, the lower expression of the *MdFLS* gene and the higher expression of *AtDFR/MdDFR* and *AtANS/MdANS* were consistent with the significantly higher anthocyanin accumulation in transgenic lines of *PsMYB12L*. Beacuse *PsMYB12L* has the flavonol-specific motif and lacks the bHLH-interaction motif, it alone enhances anthocyanin production by regulating the expression of these key genes (*AtDFR / MdDFR*, *AtANS/MdANS*, and *MdFLS*) independently of bHLH cofactors in *Arabidopsis* and apple calli.

4. Materials and Methods

4.1. Plant Materials

The tree peony cultivar *P. suffruticosa* 'Shima Nishiki' was grown in the experimental nursery of Forestry College, Shandong Agricultural University, Tai'an, Shandong, China. Flower samples were collected at five early flower-bud developmental stages (flower bud emerging stage (S1), small bell-like flower-bud stage (S2), large bell-like flower-bud stage (S3), bell-like flower-bud extending stage (S4), and color exposing stage (S5)) (Figure S2) [65]. All these samples were immediately frozen in liquid nitrogen and then stored at –80 °C for further experiments.

The *A. thaliana* ecotype Columbia (Col-0) was used for genetic transformation and phenotypic analysis in the present study. The plants were grown under a 16 h light/ 8 h dark photoperiod at 23 °C/21 °C

Furthermore, calli of the wild type (WT) of 'Orin' apple were subcultured on Murashige and Skoog (MS) medium with 1.5 mg L^{-1} 6-benzyl adenine (6-BA) and 0.5 mg L^{-1} 2,4-dichlorophenoxyacetic acid (2,4-D) at room temperature (24 °C) in a continuous dark environment at 15-day intervals [66]. Subsequently, the calli were used for genetic transformation and phenotypic analysis.

4.2. Total RNA Extraction and cDNA Synthesis

Total RNA was extracted from all samples according to instructions of the EASY Spin Plant RNA Rapid Extraction Kit (Aidlab Biotech, Beijing, China). The purity and concentration of all RNA samples were assessed using a Nanodrop 2000C spectrophotometer (Thermo Fisher Scientific, Wilmington, Delaware, DE, USA), and RNA quality was detected using 1 % agarose gels. Furthermore, cDNA was synthesized with 1 μg of total RNA using 5× All-In-One RT MasterMix (with an AccuRT Genomic DNA Removal Kit) (ABM, Vancouver, BC, Canada).

4.3. Cloning of the PsMYB114L and PsMYB12L Genes in P. suffruticosa

In this study, based on the transcriptome sequencing data of *P. suffruticosa* 'Shima Nishiki' in our laboratory, two R2R3-MYB transcription factors were filtered by analyzing the functional annotations of MYB unigenes and performing gene expression analysis.

The cDNA of the 'Shima Nishiki' cultivar's petals was used as the template. The full-length coding sequence (CDS) of the *PsMYB114L* (MK518073) and *PsMYB12L* (MK518074) genes was amplified using PCR. The complete 5′ CDS of the *PsMYB114L* and *PsMYB12L* genes was identified from the transcriptome sequencing data of *P. suffruticosa* 'Shima Nishiki'. The cDNA 3′ end sequence of these candidate genes was obtained using nested PCR technology using *PsMYB114L*-1-F/*PsMYB114L*-2-F and *PsMYB12L*-1-F/*PsMYB12L*-2-F as forward primers (Table S1), respectively, and B26 was used as the common reverse primer. The full-length cDNA of the *PsMYB114L* and *PsMYB12L* genes was amplified with the forward primers *PsMYB114L*-F1/*PsMYB12L*-F1 and the reverse primers *PsMYB114L*-R1/*PsMYB12L*-R1 (Table S1). The PCR program of gene amplification was as follows: initial denaturation at 95 °C for 1 min, followed by 30 cycles of 98 °C for 10 s, 60 °C for 15 s and 68 °C for 60 s. The PCR products were purified and cloned into the pTOPO-Blunt Simple vector for sequencing.

4.4. Subcellular Localization

The full-length cDNA without the termination codon of *PsMYB114L*/*PsMYB12L* was amplified with special primers (*PsMYB114L*-GFPF/*PsMYB12L*-GFPF and *PsMYB114L*-GFPR/*PsMYB12L*-GFPR) (Table S1) with restriction sites (*Xba* I and *Kpn* I) and subcloned into the pCAMBIA1301-GFP vector between the *Xba* I and *Kpn* I sites to create the *PsMYB114L*-GFP/*PsMYB12L*-GFP fusion construct. The recombinant vectors (*PsMYB114L*-GFP/*PsMYB12L*-GFP) and control vector (pCAMBIA1301-GFP) were then introduced into tobacco leaves by agroinfiltration. These infiltrated plants were grown for over 72 h in a growth chamber, and the GFP fluorescence of samples was observed under a Nikon C2-ER confocal laser scanning microscope (Nikon, Tokyo, Japan) [67].

4.5. Overexpression Vector Construction

The full-length cDNA of the *PsMYB114L* and *PsMYB12L* genes from the petals of *P. suffruticosa* 'Shima Nishiki' was amplified using recombinant primers (*PsMYB114L*-F2/*PsMYB12L*-F2 and

PsMYB114L-R2/*PsMYB12L*-R2) (Table S1) with restriction sites (*Spe* I and *BstE* II). Based on the predesigned vector construction procedure, the pCAMBIA1304 empty vector and the pTOPO-Blunt Simple vector containing the target genes (*PsMYB114L* and *PsMYB12L*) with restriction sites were double digested separately between the *Spe* I and *BstE* II sites and then recombined (Figure S3A–C). Subsequently, the two recombinant vectors pCAMBIA1304-*PsMYB114L* (Figure S3D) and pCAMBIA1304-*PsMYB12L* (Figure S3E) were verified successfully by PCR and sequencing with the forward vector validation primer 1304Ve-F and the reverse primers *PsMYB114L*-R2/*PsMYB12L*-R2 (Table S1). These two overexpression constructs were also introduced into *Agrobacterium tumefaciens* strain GV3101 using the freeze-thaw method.

4.6. Stable Transformation of Arabidopsis

The transformation of *Arabidopsis* was performed using the floral dip transformation method [68]. An *A. tumefaciens* infection solution (OD600 = 0.8–1.2) containing 5 % sucrose and 0.01 % Silwet L-77 was prepared to infect inflorescences, and the infection time per inflorescence was 15 s. Subsequently, these plants were transferred to a dark treatment for 24 h. These steps were repeated twice more according to the growth state of the plant. Mature T1 seeds were harvested, surface sterilized, and then sown on MS medium with 30 mg L^{-1} hygromycin B to screen for positive transformants. The resistant seedlings were transplanted into soil and then placed in a light incubator (16 h light/8 h dark at 23 °C/21 °C). When these transgenic *Arabidopsis* plants had grown to a certain size, they were further verified with gene-specific primers by PCR.

4.7. Stable Transformation of Apple Calli

To transform apple calli, 15-day-old WT apple calli were incubated with *A. tumefaciens* infection solution that carried pCAMBIA1304-PsMYB114L/pCAMBIA1304-PsMYB12L for 20 min, and the apple calli were then cocultured on MS medium supplemented with 0.5 mg L^{-1} 2,4-D and 1.5 mg L^{-1} 6-BA for 2 days at 24 °C in the dark. Subsequently, the apple calli were washed three times with sterile water and transferred to a selective medium that contained 15 mg L^{-1} hygromycin B for transgene selection. The transgenic apple calli were cocultured in the selective medium containing appropriate concentrations of an antibiotic and transferred to a light incubator with constant light (photon flux density of ~100 μmol s^{-1} m^{-2}) and low-temperature (15 °C) treatments for phenotypic observation [69,70].

4.8. Measurement of Total Anthocyanin Content

Total anthocyanin were extracted from the rosette leaves of 25-day-old *Arabidopsis* plants and apple calli cultured for 7 days. Anthocyanin extraction was performed using a methanol–HCl method [71]. Approximately 0.1 g of each sample was incubated in 5 mL of 0.1 % acidic methanol solution (CH_3OH:HCl:H_2O = 70:0.1:29.9, *v*/*v*/*v*) overnight in the dark at 4 °C. The absorbance of each extract was measured at 530 and 657 nm with a UV-1600 spectrophotometer (SHIMADZU, Kyoto, Japan). The total anthocyanin content was calculated using the following equation: $Q_{\text{Total Anthocyanin}} = (A530 - 0.25 \times A657) \times FM^{-1}$. There were three biological replicates for each sample.

4.9. Quantitative Real-Time PCR (qRT-PCR) Analysis

qRT-PCR was performed to analyze the expression levels of anthocyanin biosynthesis-related genes in all plant materials in this study. The qRT-PCR experiments were conducted using SYBR® Premix Ex Taq™ (Tli RNaseH Plus) (TaKaRa, Kyoto, Japan) on a Bio-Rad CFX96™ Real-Time system (Bio-Rad, Hercules, CA, USA) with three biological replicates according to the manufacturer's instructions. The PCR conditions were as follows: 95 °C for 30 s, 40 cycles of 95 °C for 5 s and 60 °C for 30 s and then a dissociation stage at 95 °C for 10 s, 65 °C for 5 s and 95 °C for 5 s. The *Psubiquitin* gene, *AtActin2* gene and *MdActin* gene were used as internal controls to normalize the expression levels in *P. suffruticosa*, *A. thaliana* and *Malus domestica*, respectively. All gene-specific primers used in this study are shown in Table S1 [39,66]. The relative expression levels of genes were calculated using the $2^{-\Delta\Delta Ct}$ method [72].

4.10. Sequence and Statistical Analysis

Multiple sequence alignment was performed using DNAMAN 8.0 software (Lynnon Biosoft, San Ramon, CA, USA). Homology search of sequences was carried out using the GenBank BLAST. Phylogenetic tree construction of sequences was performed using MEGA 5.0 software (Arizona State University, Tempe, AZ, USA) with the bootstrap values from 1000 replicates. Primers were designed using Primer Premier 5.0 software (PREMIER Biosoft International, Palo Alto, CA, USA). All experiments were repeated three times, and the data are expressed as the mean ± standard error. Variance analyses were performed using SPSS software ver. 17.0 (SPSS Inc., Chicago, IL, USA). p-values of < 0.05 were considered statistically significant.

5. Conclusions

In conclusion, two novel R2R3-MYB TFs, namely *PsMYB114L* and *PsMYB12L*, were successfully cloned from the petals of *P. suffruticosa* 'Shima Nishiki' and functionally characterized by heterologous expression in *Arabidopsis* and apple calli. Based on the above results, we preliminarily demonstrated the potential functional roles of *PsMYB114L* and *PsMYB12L* in regulating anthocyanin biosynthesis. These results provide a valuable resource for further understanding the molecular regulatory mechanisms of anthocyanin biosynthesis and accumulation in *P. suffruticosa* and breeding improved cultivars of *P. suffruticosa* with desirable flower colors in the future.

Author Contributions: X.Y. and L.Z. conceived and designed the research. X.Z. and Z.X. participated in the specific design of the study. X.Z. and Z.X. performed the experiments and the data analysis, and drafted the manuscript. M.Z., X.H., and S.Q. contributed analysis tools and helped analyze the data. All authors contributed to manuscript revision and approved the final version.

References

1. Zhang, J.J.; Shu, Q.Y.; Liu, Z.A.; Ren, H.X.; Wang, L.S.; De Keyser, E. Two EST-derived marker systems for cultivar identification in tree peony. *Plant Cell Rep.* **2012**, *31*, 299–310. [CrossRef] [PubMed]
2. Shi, Q.Q.; Zhou, L.; Wang, Y.; Li, K.; Zheng, B.Q.; Miao, K. Transcriptomic analysis of *Paeonia delavayi* wild population flowers to identify differentially expressed genes involved in purple-red and yellow petal pigmentation. *PLoS ONE* **2015**, *10*, e0135038. [CrossRef] [PubMed]
3. Zhao, D.Q.; Tao, J. Recent advances on the development and regulation of flower color in ornamental plants. *Front. Plant Sci.* **2015**, *6*, 261. [CrossRef] [PubMed]
4. Zhang, X.P.; Zhao, M.Y.; Guo, J.; Zhao, L.Y.; Xu, Z.D. Anatomical and biochemical analyses reveal the mechanism of double-color formation in *Paeonia suffruticosa* 'Shima Nishiki'. *3 Biotech* **2018**, *8*, 420. [CrossRef] [PubMed]
5. Noman, A.; Aqeel, M.; Deng, J.M.; Khalid, N.; Sanaullah, T.; He, S.H. Biotechnological advancements for improving loral attributes in ornamental plants. *Front. Plant Sci.* **2017**, *8*, 530. [CrossRef] [PubMed]
6. Li, C.H.; Qiu, J.; Yang, G.S.; Huang, S.R.; Yin, J.M. Isolation and characterization of a R2R3-MYB transcription factor gene related to anthocyanin biosynthesis in the spathes of *Anthurium andraeanum* (Hort.). *Plant Cell Rep.* **2016**, *35*, 2151–2165. [CrossRef] [PubMed]
7. Grotewold, E. The genetics and biochemistry of floral pigments. *Annu. Rev. Plant Biol.* **2006**, *57*, 761–780. [CrossRef] [PubMed]

8. Miyagawa, N.; Miyahara, T.; Okamoto, M.; Hirose, Y.; Sakaguchi, K.; Hatano, S.; Ozeki, Y. Dihydroflavonol 4-reductase activity is associated with the intensity of flower colors in delphinium. *Plant Biotechnol.* **2015**, *32*, 249–255. [CrossRef]
9. Zhang, J.L.; Pan, D.R.; Zhou, Y.F.; Wang, Z.C.; Hua, S.M.; Hou, L.L.; Sui, F.F. Cloning and expression of genes involved in anthocyanins synthesis in ornamental sunflower. *Acta Hortic. Sin.* **2009**, *36*, 73–80.
10. Chen, S.M.; Li, C.H.; Zhu, X.R.; Deng, Y.M.; Sun, W.; Wang, L.S.; Chen, F.D.; Zhang, Z. The identification of flavonoids and the expression of genes of anthocyanin biosynthesis in the chrysanthemum flowers. *Biol. Plant.* **2012**, *56*, 458–464. [CrossRef]
11. Zhao, D.Q.; Tao, J.; Han, C.X.; Ge, J.T. Flower color diversity revealed by differential expression of flavonoid biosynthetic genes and flavonoid accumulation in herbaceous peony (*Paeonia lactiflora* Pall.). *Mol. Biol. Rep.* **2012**, *39*, 11263–11275. [CrossRef] [PubMed]
12. Zhang, C.; Wang, W.N.; Wang, Y.J.; Gao, S.L.; Du, D.N.; Fu, J.X.; Dong, L. Anthocyanin biosynthesis and accumulation in developing flowers of tree peony (*Paeonia suffruticosa*) 'Luoyang Hong'. *Postharvest Biol. Technol.* **2014**, *97*, 11–22. [CrossRef]
13. Shi, S.G.; Yang, M.; Zhang, M.; Wang, P.; Kang, Y.X.; Liu, J.J. Genome-wide transcriptome analysis of genes involved in flavonoid biosynthesis between red and white strains of *Magnolia sprengeri* pamp. *BMC Genom.* **2014**, *15*, 706. [CrossRef] [PubMed]
14. Hsu, C.C.; Chen, Y.Y.; Tsai, W.C.; Chen, W.H.; Chen, H.H. Three R2R3-MYB transcription factors regulate distinct floral pigmentation patterning in *Phalaenopsis* spp. *Plant Physiol.* **2015**, *168*, 175–191. [CrossRef] [PubMed]
15. Nakatsuka, T.; Nishihara, M.; Mishiba, K.; Yamamura, S. Temporal expression of flavonoid biosynthesis-related genes regulates flower pigmentation in gentian plants. *Plant Sci.* **2005**, *168*, 1309–1318. [CrossRef]
16. Jaakola, L. New insights into the regulation of anthocyanin biosynthesis in fruits. *Trends Plant Sci.* **2013**, *18*, 477–483. [CrossRef] [PubMed]
17. Tanaka, Y.; Nakamura, N.; Togami, J. Altering flower color in transgenic plants by RNAi-mediated engineering of flavonoid biosynthetic pathway. *Methods Mol. Biol.* **2008**, *442*, 245–257. [PubMed]
18. Petroni, K.; Tonelli, C. Recent advances on the regulation of anthocyanin synthesis in reproductive organs. *Plant Sci.* **2011**, *181*, 219–229. [CrossRef] [PubMed]
19. Tian, J.; Chen, M.C.; Zhang, J.; Li, K.T.; Song, T.T.; Zhang, X.; Yao, Y.C. Characteristics of dihydrofavonol 4-reductase gene promoters from different leaf colored Malus crabapple cultivars. *Hortic. Res.* **2017**, *4*, 17070. [CrossRef] [PubMed]
20. Hichri, I.; Barrieu, F.; Bogs, J.; Kappel, C.; Delrot, S.; Lauvergeat, V. Recent advances in the transcriptional regulation of the flavonoid biosynthetic pathway. *J. Exp. Bot.* **2011**, *62*, 2465–2483. [CrossRef] [PubMed]
21. Huang, Y.J.; Song, S.; Allan, A.C.; Liu, X.F.; Yin, X.R.; Xu, C.J.; Chen, K.S. Differential activation of anthocyanin biosynthesis in *Arabidopsis* and tobacco over-expressing an R2R3 MYB from Chinese bayberry. *Plant Cell Tiss. Org. Cult.* **2013**, *1113*, 491–499. [CrossRef]
22. Schaart, J.G.; Dubos, C.; Romero De La Fuente, I.; van Houwelingen, A.M.M.L.; de Vos, R.C.H.; Jonker, H.H.; Xu, W.J.; Routaboul, J.M.; Lipinec, L.; Bovy, A.G. Identification and characterization of MYB-bHLH-WD40 regulatory complexes controlling proanthocyanidin biosynthesis in strawberry (*Fragaria* × *ananassa*) fruits. *New Phytol.* **2013**, *197*, 454–467. [CrossRef] [PubMed]
23. Koes, R.; Verweij, W.; Quattrocchio, F. Flavonoids: A colorful model for the regulation and evolution of biochemical pathways. *Trends Plant Sci.* **2005**, *10*, 236–242. [CrossRef] [PubMed]
24. Gonzalez, A.; Zhao, M.Z.; Leavitt, J.M.; Lloyd, A.M. Regulation of the anthocyanin biosynthetic pathway by the TTG1/bHLH/Myb transcriptional complex in *Arabidopsis* seedlings. *Plant J.* **2008**, *53*, 814–827. [CrossRef] [PubMed]
25. Albert, N.W.; Davies, K.M.; Lewis, D.H.; Zhang, H.B.; Montefiori, M.; Brendolise, C.; Boase, M.R.; Ngo, H.; Jameson, P.E.; Schwinn, K.E. A conserved network of transcriptional activators and repressors regulates anthocyanin pigmentation in eudicots. *Plant Cell* **2014**, *24*, 962–980. [CrossRef] [PubMed]
26. Zhou, L.; Wang, Y.; Peng, Z. Molecular characterization and expression analysis of chalcone synthase gene during flower development in tree peony (*Paeonia suffruticosa*). *Afr. J. Biotechnol.* **2011**, *10*, 1275–1284.
27. Zhou, L.; Wang, Y.; Peng, Z. Cloning and expression analysis of dihydroflavonol 4-reductase gene *PsDFR1* from tree peony (*Paeonia suffruticosa*). *Plant Physiol. J.* **2011**, *47*, 885–892.

28. Zhou, L.; Wang, Y.; Ren, L.; Shi, Q.Q.; Zheng, B.Q.; Miao, K.; Guo, X. Overexpression of *PsCHI1*, a homologue of the chalcone isomerase gene from tree peony (*Paeonia suffruticosa*), reduces the intensity of flower pigmentation in transgenic tobacco. *Plant Cell Tiss. Org. Cult.* **2014**, *116*, 285–295. [CrossRef]
29. Schwinn, K.; Venail, J.; Shang, Y.J.; Mackay, S.; Alm, V.; Butelli, E.; Oyama, R.; Bailey, P.; Davies, K.; Martin, C. A small family of *MYB*-regulatory genes controls foral pigmentation intensity and patterning in the Genus *Antirrhinum*. *Plant Cell* **2006**, *18*, 831–851. [CrossRef] [PubMed]
30. Dubos, C.; Stracke, R.; Grotewold, E.; Weisshaar, B.; Martin, C.; Lepiniec, L. MYB transcription factors in *Arabidopsis*. *Trends Plant Sci.* **2010**, *15*, 573–581. [CrossRef] [PubMed]
31. Carey, C.C.; Strahle, J.T.; Selinger, D.A.; Chandler, V.L. Mutations in the *pale aleurone color1* regulatory gene of the *Zea mays* anthocyanin pathway have distinct phenotypes relative to the functionally similar *TRANSPARENT TESTA GLABRA1* gene in *Arabidopsis thaliana*. *Plant Cell* **2004**, *16*, 450–464. [CrossRef] [PubMed]
32. Walker, A.R.; Lee, E.; Bogs, J.; McDavid, D.A.J.; Thomas, M.R.; Robinson, S.P. White grapes arose through the mutation of two similar and adjacent regulatory genes. *Plant J.* **2007**, *49*, 772–785. [CrossRef] [PubMed]
33. Tian, J.; Feng, Z.; Zhang, J.; Song, T.T.; Wan, H.H.; Zhang, M.L.; Yao, Y.C. McMYB10 regulates coloration via activating *McF3′H* and later structural genes in ever-red leaf crabapple. *Plant Biotechnol. J.* **2015**, *13*, 948–961. [CrossRef] [PubMed]
34. Albert, N.W.; Lewis, D.H.; Zhang, H.B.; Schwinn, K.E.; Jameson, P.E.; Davies, K.M. Members of an R2R3-MYB transcription factor family in *Petunia* are developmentally and environmentally regulated to control complex floral and vegetative pigmentation patterning. *Plant J.* **2011**, *65*, 771–784. [CrossRef] [PubMed]
35. Zhu, L.; Shan, H.; Chen, S.M.; Jiang, J.F.; Gu, C.S.; Zhou, G.Q.; Chen, Y.; Song, A.P.; Chen, F.D. The heterologous expression of the chrysanthemum R2R3-MYB transcription factor *CmMYB1* alters lignin composition and represses flavonoid synthesis in *Arabidopsis thaliana*. *PLoS ONE* **2013**, *8*, e65680. [CrossRef] [PubMed]
36. Zhang, Y.X.; Zhang, L.; Gai, S.P.; Liu, C.Y.; Lu, S. Cloning and expression analysis of the R2R3-*PsMYB1* gene associated with bud dormancy during chilling treatment in the tree peony (*Paeonia suffruticosa*). *Plant Growth Regul.* **2015**, *75*, 667–676. [CrossRef]
37. Gao, L.X.; Yang, H.X.; Liu, H.F.; Yang, J.; Hu, Y.H. Extensive transcriptome changes underlying the flower color intensity variation in *Paeonia ostii*. *Front. Plant Sci.* **2016**, *6*, 1205. [CrossRef] [PubMed]
38. Shi, Q.Q.; Li, L.; Zhang, X.X.; Luo, J.R.; Li, X.; Zhai, L.J.; He, L.X.; Zhang, Y.L. Biochemical and comparative transcriptomic analyses identify candidate genes related to variegation formation in *Paeonia rockii*. *Molecules* **2017**, *22*, 1364. [CrossRef] [PubMed]
39. Zhang, X.P.; Zhao, L.Y.; Xu, Z.D.; Yu, X.Y. Transcriptome sequencing of Paeonia suffruticosa 'Shima Nishiki'to identify differentially expressed genes mediating double-color formation. *Plant Physiol. Biochem.* **2018**, *123*, 114–124. [CrossRef] [PubMed]
40. Cao, Y.L.; Xie, L.F.; Ma, Y.Y.; Ren, C.H.; Xing, M.Y.; Fu, Z.S.; Wu, X.Y.; Yin, X.R.; Xu, C.J.; Li, X. *PpMYB15* and *PpMYBF1* Transcription factors are involved in regulating flavonol biosynthesis in peach fruit. *J. Agric. Food Chem.* **2019**, *67*, 644–652. [CrossRef] [PubMed]
41. Gong, B.H.; Yi, J.; Sui, J.J.; Wu, J.; Wu, Z.; Cheng, Y.H.; Wu, C.Y.; Liu, C.; Yi, M.F. Cloning and expression analysis of LlHsfA1 from *Lilium longiforum*. *Acta Hortic. Sin.* **2014**, *41*, 1400–1408.
42. Ding, K.; Ma, P.D.; Jia, Y.Y.; Pei, T.L.; Bai, Z.Q.; Liang, Z.S. Subcellular localization and transactivation analysis of three R2R3-MYB in *Salvia miltiorrhiza* Bunge. *Acta Agric. Boreali-Occidentalis Sin.* **2018**, *27*, 586–594.
43. Moreau, C.; Ambrose, M.J.; Turner, L.; Hill, L.; Noel Ellis, T.H.; Hofer, J.M.I. The *b* gene of pea encodes a defective flavonoid 3′,5′-hydroxylase and confers pink flower color. *Plant Physiol.* **2012**, *159*, 759–768. [CrossRef] [PubMed]
44. Davies, K.M.; Albert, N.W.; Schwinn, K.E. From landing lights to mimicry: The molecular regulation of flower colouration and mechanisms for pigmentation patterning. *Funct. Plant Biol.* **2012**, *39*, 619–638. [CrossRef]
45. Li, X.; Lu, M.; Tang, D.Q.; Shi, Y.M. Composition of carotenoids and flavonoids in narcissus cultivars and their relationship with flower color. *PLoS ONE* **2015**, *10*, e0142074. [CrossRef] [PubMed]
46. Zhang, Q.; Hao, R.J.; Xu, Z.D.; Yang, W.R.; Wang, J.; Cheng, T.R.; Pan, H.T.; Zhang, Q.X. Isolation and functional characterization of a R2R3-MYB regulator of *Prunus mume* anthocyanin biosynthetic pathway. *Plant Cell. Tiss. Org. Cult.* **2017**, *131*, 417–429. [CrossRef]

47. Feng, K.; Xu, Z.S.; Que, F.; Liu, J.X.; Wang, F.; Xiong, A.S. An R2R3-MYB transcription factor, OjMYB1, functions in anthocyanin biosynthesis in *Oenanthe javanica*. *Planta* **2018**, *247*, 301–315. [CrossRef] [PubMed]
48. Stracke, R.; Werber, M.; Weisshaar, B. The *R2R3-MYB* gene family in *Arabidopsis thaliana*. *Curr. Opin. Plant Biol.* **2001**, *4*, 447–456. [CrossRef]
49. Stracke, R.; Ishihara, H.; Huep, G.; Barsch, A.; Mehrtens, F.; Niehaus, K.; Weisshaar, B. Differential regulation of closely related R2R3-MYB transcription factors controls flavonol accumulation in different parts of the *Arabidopsis thaliana* seedling. *Plant J.* **2007**, *50*, 660–677. [CrossRef] [PubMed]
50. Czemmel, S.; Stracke, R.; Weisshaar, B.; Cordon, N.; Harris, N.N.; Walker, A.R.; Robinson, S.P.; Bogs, J. The grapevine R2R3-MYB transcription factor VvMYBF1 regulates flavonol synthesis in developing grape berries. *Plant Physiol.* **2009**, *151*, 1513–1530. [CrossRef] [PubMed]
51. Schwinn, K.E.; Ngo, H.; Kenel, F.; Brummell, D.A.; Albert, N.W.; McCallum, J.A.; Pither-Joyce, M.; Crowhurst, R.N.; Eady, C.; Davies, K.M. The onion (*Allium cepa* L.) *R2R3-MYB* gene *MYB1* regulates anthocyanin biosynthesis. *Front. Plant Sci.* **2016**, *7*, 1865. [CrossRef] [PubMed]
52. Paz-Ares, J.; Wienand, U.; Peterson, P.A.; Saedler, H. Molecular cloning of the *c* locus of *Zea mays*: A locus regulating the anthocyanin pathway. *EMBO J.* **1986**, *5*, 829–834. [CrossRef] [PubMed]
53. Nesi, N.; Jond, C.; Debeaujon, I.; Caboche, M.; Lepiniec, L. The Arabidopsis *TT2* gene encodes an R2R3 MYB domain protein that acts as a key determinant for proanthocyanidin accumulation in developing seed. *Plant Cell* **2001**, *13*, 2099–2114. [PubMed]
54. Terrier, N.; Torregrosa, L.; Ageorges, A.; Vialet, S.; Verriès, C.; Cheynier, V.; Romieu, C. Ectopic expression of $VvMybPA_2$ promotes proanthocyanidin biosynthesis in grapevine and suggests additional targets in the pathway. *Plant Physiol.* **2009**, *149*, 1028–1041. [CrossRef] [PubMed]
55. Gu, Z.Y.; Zhu, J.; Hao, Q.; Yuan, Y.W.; Duan, Y.W.; Men, S.Q.; Wang, Q.Y.; Hou, Q.Z.; Liu, Z.A.; Shu, Q.Y.; et al. A novel R2R3-MYB transcription factor contributes to petal blotch formation by regulating organ-specific expression of *PsCHS* in tree peony (*Paeonia suffruticosa*). *Plant Cell Physiol.* **2018**, pcy232. [CrossRef] [PubMed]
56. Goff, S.A.; Cone, K.C.; Chandler, V.L. Functional analysis of the transcriptional activator encoded by the maize B-gene: Evidence for a direct functional interaction between two classes of regulatory proteins. *Genes Dev.* **1992**, *6*, 864–875. [CrossRef] [PubMed]
57. Zimmermann, I.M.; Heim, M.A.; Weisshaar, B.; Uhrig, J.F. Comprehensive identifcation of Arabidopsis thaliana MYB transcription factors interacting with R/B-like BHLH proteins. *Plant J.* **2004**, *40*, 22–34. [CrossRef] [PubMed]
58. Wang, N.; Xu, H.F.; Jiang, S.H.; Zhang, Z.Y.; Lu, N.L.; Qiu, H.R.; Qu, C.Z.; Wang, Y.C.; Wu, S.J.; Chen, X.S. MYB12 and MYB22 play essential roles in proanthocyanidin and flavonol synthesis in red-fleshed apple (*Malus sieversii* f. *niedzwetzkyana*). *Plant J.* **2017**, *90*, 276–292. [CrossRef] [PubMed]
59. Liu, C.Y.; Long, J.M.; Zhu, K.J.; Liu, L.L.; Yang, W.; Zhang, H.Y.; Li, L.; Xu, Q.; Deng, X.X. Characterization of a Citrus R2R3-MYB transcription factor that regulates the flavonol and hydroxycinnamic acid biosynthesis. *Sci. Rep.* **2016**, *6*, 25352. [CrossRef] [PubMed]
60. Davies, K.M.; Schwinn, K.E.; Deroles, S.C.; Manson, D.G.; Lewis, D.H.; Bloor, S.J.; Bradley, J.M. Enhancing anthocyanin production by altering competition for substrate between flavonol synthase and dihydroflavonol 4-reductase. *Euphytica* **2003**, *131*, 259–268. [CrossRef]
61. Wang, F.B.; Kong, W.L.; Wong, G.; Fu, L.F.; Peng, R.H.; Li, Z.J.; Yao, Q.H. AtMYB12 regulates flavonoids accumulation and abiotic stress tolerance in transgenic *Arabidopsis thaliana*. *Mol. Genet. Genom.* **2016**, *291*, 1545–1559. [CrossRef] [PubMed]
62. Mehrtens, F.; Kranz, H.; Bednarek, P.; Weisshaar, B. The Arabidopsis transcription factor MYB12 is a flavonol-specific regulator of phenylpropanoid biosynthesis. *Plant Physiol.* **2005**, *138*, 1083–1096. [CrossRef] [PubMed]
63. Huang, W.J.; Khaldun, A.B.M.; Chen, J.J.; Zhang, C.J.; Lv, H.Y.; Yuan, L.; Wang, Y. A R2R3-MYB transcription factor regulates the flavonol biosynthetic pathway in a traditional Chinese medicinal plant, *Epimedium sagittatum*. *Front. Plant Sci.* **2016**, *7*, 1089. [CrossRef] [PubMed]
64. Luo, J.; Butelli, E.; Hill, L.; Parr, A.; Niggeweg, R.; Bailey, P.; Weisshaar, B.; Martin, C. AtMYB12 regulates caffeoyl quinic acid and flavonol synthesis in tomato: Expression in fruit results in very high levels of both types of polyphenol. *Plant J.* **2008**, *56*, 316–326. [CrossRef] [PubMed]
65. Ren, X.X.; Wang, S.L.; Xue, J.Q.; Zhu, F.Y.; Liu, C.J.; Zhang, X.X. Molecular cloning and expression analysis of cryptochrome gene *PsCRY2* in tree peony. *Acta Hortic. Sin.* **2015**, *42*, 2229–2236. [CrossRef]

66. An, J.P.; An, X.H.; Yao, J.F.; Wang, X.N.; You, C.X.; Wang, X.F.; Hao, Y.J. BTB protein MdBT2 inhibits anthocyanin and proanthocyanidin biosynthesis by triggering MdMYB9 degradation in apple. *Tree Physiol.* **2018**, *38*, 1578–1587. [CrossRef] [PubMed]
67. Verma, A.; Lee, C.; Morriss, S.; Odu, F.; Kenning, C.; Rizzo, N.; Spollen, W.G.; Lin, M.; McRae, A.G.; Givan, S.A.; et al. The novel cyst nematode effector protein 30D08 targets host nuclear functions to alter gene expression in feeding sites. *New Phytol.* **2018**, *219*, 697–713. [CrossRef] [PubMed]
68. Clough, S.J.; Bent, A.F. Floral dip: A simplified method for *Agrobacterium*-mediated transformation of *Arabidopsis thaliana*. *Plant J.* **1998**, *16*, 735–743. [CrossRef] [PubMed]
69. An, J.P.; Liu, X.; Li, H.H.; You, C.X.; Wang, X.F.; Hao, Y.J. Apple RING E3 ligase MdMIEL1 inhibits anthocyanin accumulation by ubiquitinating and degrading MdMYB1 protein. *Plant Cell Physiol.* **2017**, *58*, 1953–1962. [CrossRef] [PubMed]
70. An, J.P.; Qu, F.J.; Yao, J.F.; Wang, X.N.; You, C.X.; Wang, X.F.; Hao, Y.J. The bZIP transcription factor MdHY5 regulates anthocyanin accumulation and nitrate assimilation in apple. *Hortic. Res.* **2017**, *4*, 17023. [CrossRef] [PubMed]
71. Zhao, D.Q.; Jiang, Y.; Ning, C.L.; Meng, J.S.; Lin, S.S.; Ding, W.; Tao, J. Transcriptome sequencing of a chimaera reveals coordinated expression of anthocyanin biosynthetic genes mediating yellow formation in herbaceous peony (*Paeonia lactilora* Pall.). *BMC Genom.* **2014**, *15*, 689. [CrossRef] [PubMed]
72. Schmittgen, T.D.; Livak, K.J. Analyzing real-time PCR data by the comparative CT method. *Nat. Protoc.* **2008**, *3*, 1101–1108. [CrossRef] [PubMed]

13

Isolation and Characterization of a Green-Tissue Promoter from Common Wild Rice (*Oryza rufipogon* Griff.)

Mande Xue [1], Yan Long [1], Zhiqiang Zhao [2], Gege Huang [2], Ke Huang [2], Tianbao Zhang [1], Ying Jiang [3,*], Qianhua Yuan [2,*] and Xinwu Pei [1,*]

1 MOA Key Laboratory on Safety Assessment (Molecular) of Agri-GMO, Institute of Biotechnology, Chinese Academy of Agricultural Sciences, Beijing 100081, China; mandexue@gmail.com (M.X.); longyan@caas.cn (Y.L.); Zhangtianbao274200@126.com (T.Z.)

2 College of Tropical Agriculture and Forestry, Hainan University, Haikou 570228, China; zzq19900317@126.com (Z.Z.); Luolengshuang@126.com (G.H.); Huangke199309@163.com (K.H.)

3 Experimental Center Basic Medical Teaching, Capital Medical University, Beijing 100069, China

* Correspondence: Jiangy@ccmu.edu.cn (Y.J.); qhyuan@163.com (Q.Y.); Peixinwu@caas.cn (X.P.)

Abstract: Promoters play a very important role in the initiation and regulation of gene transcription. Green-tissue promoter is of great significance to the development of genetically modified crops. Based on RNA-seq data and RT-PCR expression analysis, this study screened a gene, *OrGSE* (GREEN SPECIAL EXPRESS), which is expressed specifically in green tissues. The study also isolated the promoter of the *OrGSE* gene (OrGSEp), and predicted many cis-acting elements, such as the CAAT-Box and TATA-Box, and light-responding elements, including circadian, G-BOX and GT1 CONSENSUS. Histochemical analysis and quantification of GUS activity in transgenic *Arabidopsis thaliana* plants expressing GUS under the control of OrGSEp revealed that this promoter is not only green tissue-specific, but also light-inducible. The ability of a series of 5′-deletion fragments of OrGSEp to drive GUS expression in *Arabidopsis* was also evaluated. We found that the promoter region from −54 to −114 is critical for the promoter function, and the region from −374 to −114 may contain core cis-elements involved in light response. In transgenic rice expressing GUS under the control of OrGSEp, visualization and quantification of GUS activity showed that GUS was preferentially expressed in green tissues and not in endosperm. OrGSEp is a useful regulatory element for breeding pest-resistant crops.

Keywords: common wild rice; Promoter; Green tissue-specific expression; light-induced

---

1. Introduction

Promoters are key regulators of transcription and also play critical roles in genetic engineering [1]. Promoters can be divided into three types: constitutive, inducible and tissue-specific. Constitutive promoters are widely used in plant genetic engineering [2]. The cauliflower mosaic virus (CaMV) 35S promoter, which drives the expression of genes in almost all tissues, is an important constitutive promoter in dicotyledonous plants [3]. The *Actin*1 promoter from rice is a classical constitutive promoter [4]. However, inducible promoters greatly increase the transcription level of genes under specific physical or chemical signals [5], this is different from constitutive promoters. At present, a great number of inducible promoters have been isolated, including light-inducible, heat-inducible and trauma-inducible [6–10]. Abiotic stress is a major obstacle for crop production, and the promoters of many genes related to abiotic stress have been cloned and applied in crop biotechnology. Five cold-inducible promoters have been isolated from rice, and these promoters can be applied to engineer plants that are resistant to cold stress [11]. The stress-inducible promoter of *TaSnRK2.8*, which is an important gene for wheat response to abiotic stress, has been isolated and characterized, and this promoter can be used to engineer plants with resistance to various abiotic stresses [12].

Because gene expression is driven in specific tissues and development stages, tissue-specific promoters, also called organ- or cell-specific promoters [13], are different from other promoters [14]. These promoters are significant because they avoid potential negative effects of using constitutive promoters, such as metabolic burden [15–19]. Five non-endosperm tissue-expressed promoters have been isolated from rice, and an exogenous *Cry1Ab* gene (*mCry1Ab*) driven by green tissue-specific promoter was expressed in all tissues except for endosperm [20]. Progress has also been made in identifying promoters that drive expression in roots, which is of interest because crops are faced with root-related pests, pathogens and abiotic stresses. The promoters of the soybean *GmPRP1* and *GmPRP2* genes, which are expressed preferentially in roots [21], were isolated and shown to exhibit root-preferential expression [22].The promoter of the serine/threonine kinase gene *ZmSTK2_USP* was isolated and with the use of a GUS reporter system was found to drive pollen-tissue-specific expression [23]. The promoter of *AtGILT* was shown to drive seed coat-specific expression in *Arabidopsis thaliana*, and further studies demonstrated that in canola this promoter drives expression, specifically in the outer integument of the seed coat, and may be useful for improving canola meal [24].

Green tissue-specific promoters have vast potential in transgenic crop breeding [18], particularly in insect-resistant or herbicide-resistant crops [25]. A growing number of transgenic crops with insect resistance gene expression driven by green tissue-specific promoters have been developed, such as cotton, rice, soybean and maize. Transgenic Bt cotton was developed by driving expression of the *B. thuringiensis* endotoxin (*Cry9C*) genes under the control of the *PNZIP* (*Pharbitis nil* leucine zipper) gene promoter, resulting in preferential expression in plant green tissues as well as lower Bt protein accumulation in transgenic cotton seeds [26]. The *rbcS* promoter is another classical green tissue-specific promoter that has been used to drive expression of the *cry2AX1* gene in rice to confer resistance to leaffolders [27]. The 731-bp 5′ flanking sequence of a potato (*Solanum tuberosum*) gene encoding ribulose-1, 5-bisphosphate carboxylase/oxygenase (rubisco) activase (RCA) was characterized, and GUS reporter gene under the control of StRCAp was expressed throughout the green tissue of light-grown transgenic tobacco seedlings. Further analysis revealed that a 220 bp fragment of StRCAp was sufficient for green-tissue-specific and light-inducible expression [7]. Common wild rice (*Oryza rufipogon* Griff.), which is an ancestor of Asian cultivated rice [28], has abundant genetic diversity and is an important germplasm resource for the improvement of cultivated rice [29]. However, the genes specifically expressed in green tissues in common wild rice have not been reported. Previously, we sequenced the transcriptome of common wild rice and identified root-specific and drought-related genes [30]. In this study, we used this dataset to screen for genes with green tissue-specific expression and cloned the promoter sequence of one of these genes, *OrGSE* (*O. rufipogon* GREEN SPECIAL EXPRESS). The full-length *OrGSE* promoter (OrGSEp) and a series of truncated promoters were fused to the GUS reporter gene to identify putative cis-regulatory elements that confer green tissue-specific and light-inducible expression in *Arabidopsis*. We found that in transgenic rice, GUS driven by the OrGSEp promoter was preferentially expressed in green tissues and not expressed in endosperm and root.

2. Results

2.1. Expression Pattern of OrGSE in Common Wild Rice

RNA-seq data from a previous study [30] were used to screen candidate common wild rice green tissue-expressed genes with higher expression in leaves than in roots (FPKM value in CL >10, FPKM value in CR is <10 and FPKM value in CL 10 times higher than in CR). A total of 1140 unigenes were identified as candidate green tissue-specific expressed genes (Table S3). A series of RT-PCR experiments were performed for confirming the RNA-seq data and combining RNA-seq and RT-PCR results, we selected a novel green tissue-specific expressed gene, comp45689_c0. Blast searches against the rice genome annotation project database (available online: http://rice.plantbiology.msu.edu/) showed that this gene corresponds to *LOC_Os08g02210* and encodes an expressed protein, named OrGSE (*O. rufipogon* GREEN SPECIAL EXPRESS), The BLASTP (Basic Local Alignment Search Tool Protein) analysis revealed that OrGSE shares homology with D27 protein in rice, and prediction of subcellular localization shows that OrGSE is located in chloroplast. Based on RNA-seq FPKM expression values from MSU, *OsGSE* is highly expressed in green tissue; the FPKM value in 20-day-old leaves is >200, but almost 0 in anthers and seeds (Figure S1).

RT-PCR and qPCR experiments confirmed that the *OrGSE* gene is expressed in leaves, stems and spikes, but not in roots and seeds (Figure 1). In addition, preferential expression in green tissue was observed at both the seedling and heading stages (Figure 1).

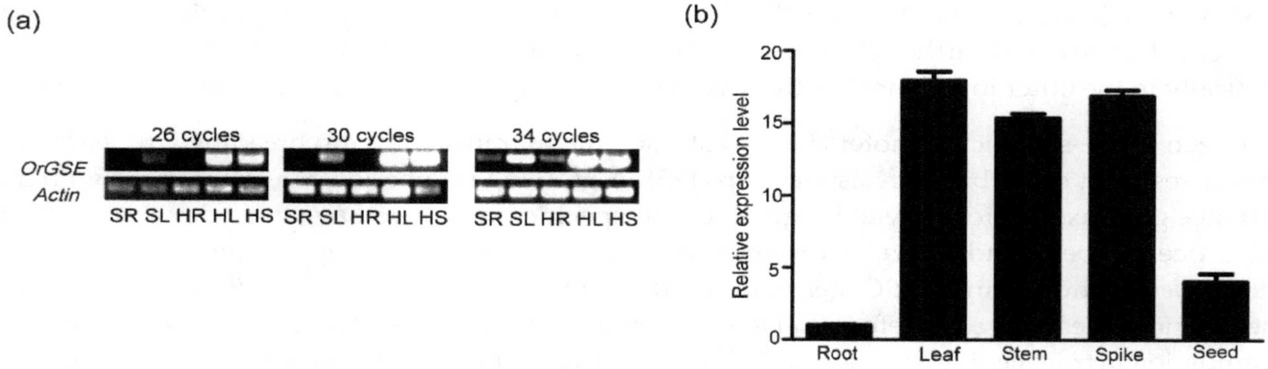

Figure 1. *OrGSE* gene expression in roots and shoots of common wild rice. The mRNA levels of *OrGSE* gene were determined in different tissues and developmental stages by RT-PCR with 26, 30 and 34 cycles of amplification. (**a**) The gel figure of RT-PCR. The rice Actin gene was used as an internal control. SR: Shooting stage Root, SL: Shooting stage Leaf, HR: Heading stage Root, HL: Heading stage Leaf, and HS: Heading stage Stem; (**b**) qPCR analysis of transcript levels of *OrGSE* in different tissues. Data are the means of three replicates, and error bars show the standard error.

2.2. Cloning and Sequence Analysis of the OrGSE Promoter (OrGSEp-374)

A 561 bp fragment including the −374 to +187 region (where the TIS is +1) upstream of *OrGSE* was isolated from common wild rice as the candidate promoter sequence. This promoter fragment, named OrGSEp-374, was submitted to PlantCARE to predict putative cis-elements involved in the regulation of gene expression (Figure 2). Potential regulatory elements were identified within OrGSEp-374 (Table S1), including core elements, such as the TATA-Box, CAAT-Box and GC-Box and cis-elements known to be involved in stress response, such as ABRE (abscisic acid responsive element), CGTCA-motif (response to methyl jasmonate), and LTR (low-temperature response).

In addition, a few elements in OrGSEp-374 have been shown to participate in tissue-specific expression, such as the CCGTCC-box (related to meristem-specific activation), the GCN4_motif (endosperm expression) and the Skn-1_motif (required for endosperm expression). Other regulatory elements are involved green tissue-specific expression; most of these elements are light-responsive, such as ACE, G-Box, box II and GT1CONSENSUS. A core element (circadian) involved in circadian control was also identified [31,32].

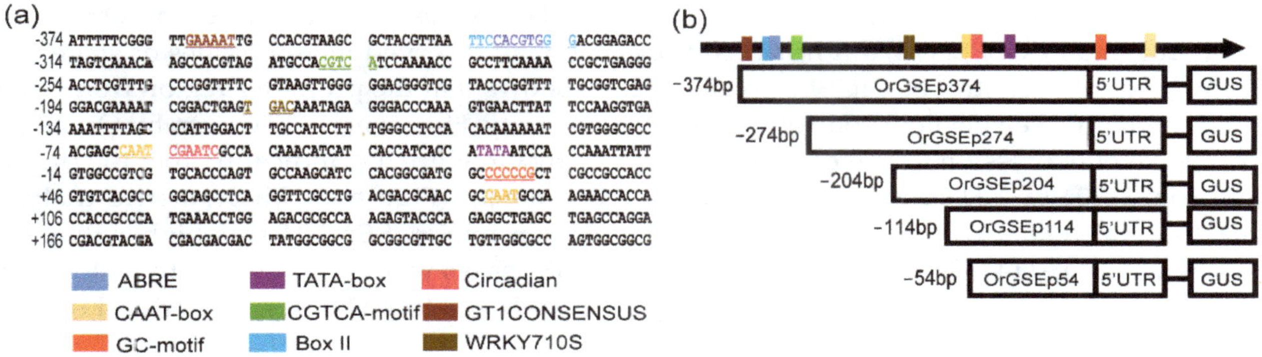

Figure 2. The location of putative cis-acting elements in OrGSEp-374 predicted by the PlantCARE database and schematic diagrams of promoter deletion constructs. (**a**) Putative cis-acting elements in OrGSEp-374, The 5′-region of the *OrGSE* gene containing the 374 bp promoter sequence and 166 bp sequence downstream of the translational start site. The transcription initiation site is defined as +1. The TATA box, CAAT box and other key cis-acting elements are underlined with and indicated by different colors as shown in the legend. The position of each element is also indicated by schematic diagrams; (**b**) The schematic diagrams of the truncated OsGSE-374 constructs. The numbers to the left of these diagrams indicate the position of the 5′-deletion.

2.3. Spatiotemporal Expression Patterns of OrGSEp-374 and 5′-Deletion Fragments in Arabidopsis

OrGSEp-374 and 5′-deletion reporter constructs were transformed into *Arabidopsis* for promoter functional analysis. OrGSEp-374-driven GUS expression was monitored during different developmental stages and in various organs by histochemical staining. GUS expression was detected in the cotyledons and hypocotyls of 3-day-old seedlings (Figure 3). GUS expression was also detected in the leaves of 5-day- and 14-day-old seedlings, and GUS expression level was higher in leaves than in cotyledons (Figure 3). No GUS expression was observed in roots at any stage (Figure 3). During the reproductive stage, GUS expression was observed in leaves, but not in roots or siliques (Figure 4), indicating that OrGSEp-374 drives expression specifically in green tissues.

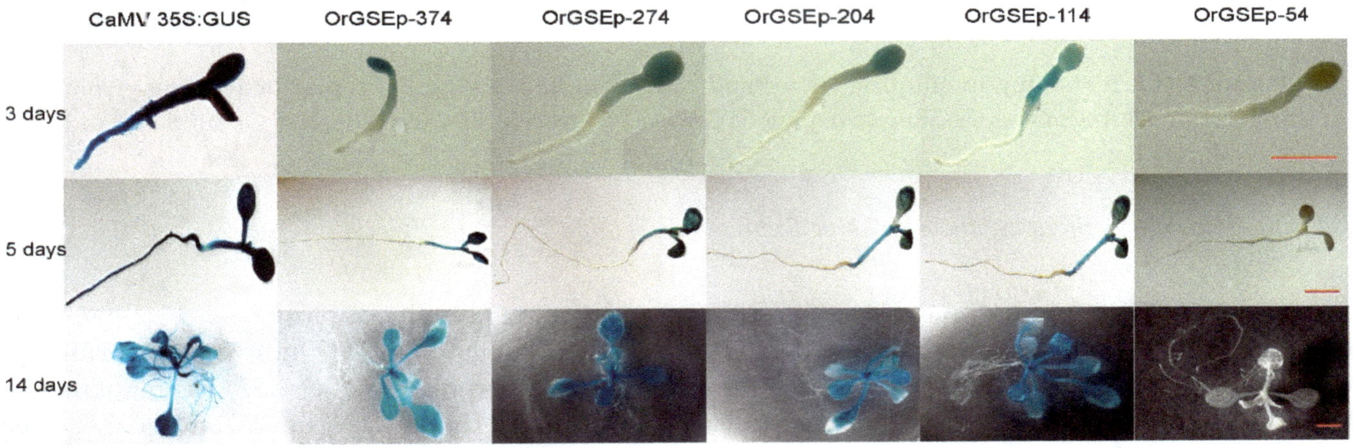

Figure 3. GUS histochemical assays in transgenic *Arabidopsis* T3 seedlings. GUS histochemical assays in transgenic *Arabidopsis* T3 seedlings harboring constructs with GUS expression driven by the CaMV 35S promoter (35S: GUS), OrGSEp-374 (OrGSEp-374) and different 5′-deletion fragments (OrGSEp-274, OrGSEp-204, OrGSEp-114 and OrGSEp-54), during vegetative growth. Photographs were taken 3 days, 5 days and 14 days after seed germination.Bar = 1 cm.

To identify the core elements responsible for green tissue-specific expression, we cloned four different 5′-deletion fragments (Figure 2) into the pBinGlyRed-GUS vector, and transformed these constructs into *Arabidopsis*. Histochemical staining was performed on 3-, 5- and 14-day-old T3 transgenic seedlings. As shown in Figure 3, strong GUS expression was driven by OrGSEp-274, OrGSEp-204 and OrGSEp-114, but no GUS expression was observed in any tissues or stages when GUS was under the control of OrGSEp-54. This result shows that the promoter region from −114 to −54 may contain a key element controlling promoter activity. However, GUS driven by OrGSEp-114 showed weaker expression in leaves than GUS driven by OrGSEp-204, indicating that the promoter region from −204 to −114 may contain an enhancer element (Figure 4). Fluorometric analysis of GUS activity also confirmed that GUS expression driven by OrGSEp-204 was higher than when driven by OrGSEp-114. The results of fluorometric GUS assays were consistent with histochemical staining (Figure 5), supporting the promoter fragment from −54 to −114 playing a critical role in the promoter activity.

Figure 4. GUS staining in siliques, leaves and roots sampled during the reproductive stage from transgenic T3 *Arabidopsis* seedlings carrying OrGSEp-374, OrGSEp-274, OrGSEp-204, OrGSEp-114 and OrGSEp-54.Bar = 1 cm.

2.4. *OrGSEp-374 Confers Light-Responsive Expression*

OrGSEp-374 sequence analysis showed that the promoter contained many cis-acting elements involved in light responsiveness, such as ACE, Box II, and GT1CONSENSUS (Figure 2). We performed GUS staining and fluorometric assays to determine whether the expression of *OrGSE* was regulated by light. GUS staining showed that *OrGSE* expression was induced by light (Figure 6), and quantitative fluorometric analysis confirmed this result (Figure 6). To find the core elements related to light responsiveness, we analyzed the light-inducible activities of 5′-deletion fragments of OrGSEp-374. GUS expression driven by OrGSEp-374, OrGSEp-274 and OrGSEp-204 was similar under light conditions; however, GUS activity driven by these promoters was significantly reduced in the dark. Interestingly, the pattern of GUS expression in OrGSEp-114-GUS lines was the same under both light and dark conditions (Figure 6). These results demonstrate that the promoter fragment from −374 to −114 contains vital elements involved in light response, but the promoter fragment from −114 to +1 does not. This is consistent with the presence of elements involved in light response in the OrGSEp-374 predicted by PlantCARE (Figure 2; Table S1).

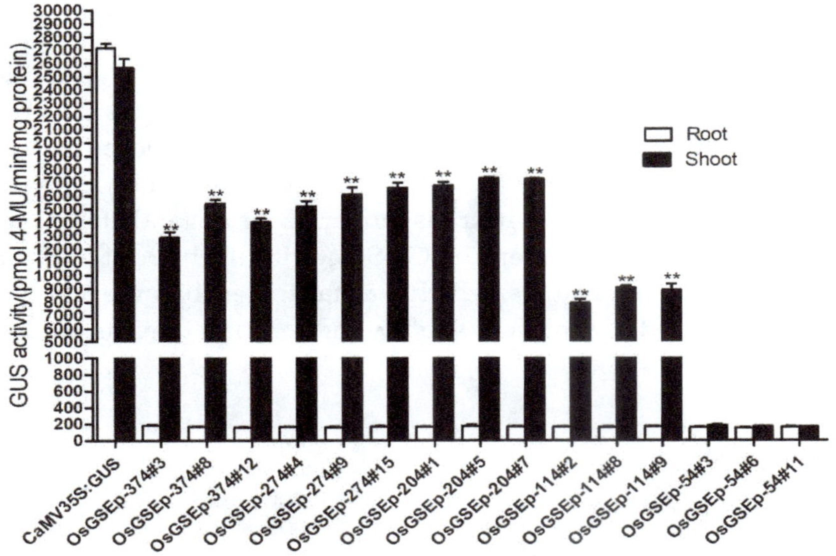

Figure 5. Quantification of GUS activity in transgenic T3 *Arabidopsis* roots and shoots carrying the CaMV 35S promoter, OrGSEp-374 and different 5′-deletion fragments constructs. Gus activity measurements are shown for three positive transgenic lines for each OrGSEp promoter construct. CaMV35S: GUS is an independent transgenic plant carrying the CaMV35S promoter construct. OrGSEp-374#3, OrGSEp-374#8 and OrGSEp-374#12 are three positive transgenic lines carrying the OrGSEp-374 construct. OrGSEp-274#4, OrGSEp-274#9 and OrGSEp-274#15 are three independent transgenic lines carrying the OrGSEp-274 construct. OrGSEp-204#1, OrGSEp-204#5 and OrGSEp-204#7 are three positive lines carrying the OrGSEp-204 construct. OrGSEp-114#2, OrGSEp-114#8 and OrGSEp-114#9 are three lines carrying the OrGSEp-114 construct. OrGSEp-54#3, OrGSEp-54#6 and OrGSEp-54#11 are three transgenic lines carrying the OrGSEp-54 fragment. Data are the means of three replicates, and error bars show the standard error. The "**" indicates that a significant difference ($p < 0.001$) in GUS activity was detected between root and shoots of plants carrying the same promoter construct.

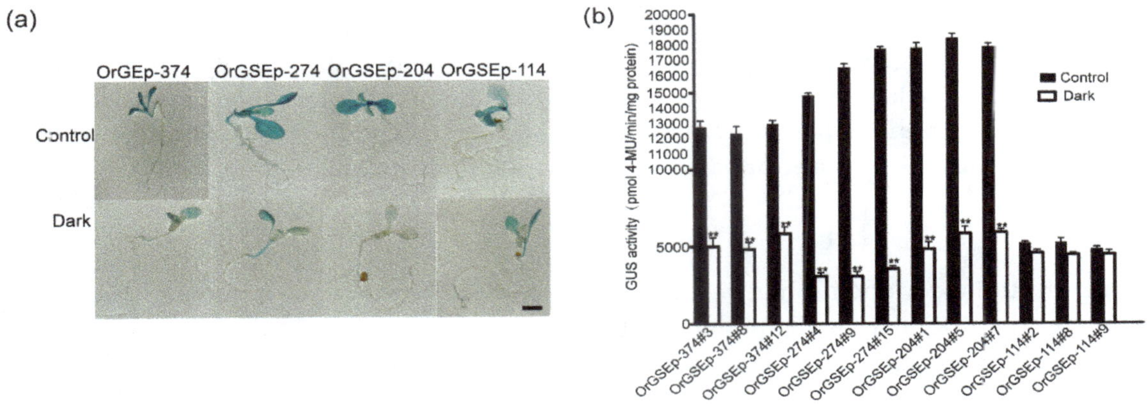

Figure 6. GUS staining and quantification of GUS activity of transgenic *Arabidopsis* seedlings containing different 5′-promoter deletion fragments grown under light and dark conditions. (**a**) Ten-day-old seedlings were stained to observe GUS expression. Bar = 1 cm; (**b**) Quantification of GUS activity in 20-day-old T3 transgenic *Arabidopsis* grown under light and dark conditions. OrGSEp-374#3, OrGSEp-374#8 and OrGSEp-374#12 are three positive transgenic lines carrying the OrGSEp-374 construct. OrGSEp-274#4, OrGSEp-274#9 and OrGSEp-274#15 are three independent transgenic lines carrying the OrGSEp-274 construct. OrGSEp-204#1, OrGSEp-204#5 and OrGSEp-204#7 are three positive lines carrying the OrGSEp-204 construct. OrGSEp-114#2, OrGSEp-114#8 and OrGSEp-114#9 are three lines carrying the OrGSEp-114 construct. Data are the means of three replicates, and standard errors are shown by error bars. The "**" indicates that a significant difference ($p < 0.001$) in GUS activity was detected between roots and shoots of seedlings carrying the same promoter construct.

2.5. The Expression Pattern of OrGSEp-374 in Rice

GUS staining and fluorometric quantification was performed on four independent lines of rice containing the OrGSEp-374 construct. Strong GUS expression was observed in the stem and leaf, weak expression was observed in the anther, ligule, spikelet and embryo, and no expression was observed in the root and endosperm (Figure 7). Further fluorometric analysis of GUS activity showed that the average specific activities in transgenic rice leaves and stems exceeded 15,000 pmol 4-MU min^{-1} mg^{-1} total proteins. However, the GUS activity in the root and panicle was less than 1000 pmol 4-MU min^{-1} mg^{-1} total proteins, which was far lower than the activity level in leaves and stems (Figure 7). GUS activity in the seed was slightly higher than in roots because GUS was weakly expressed in the embryo (Figure 7).

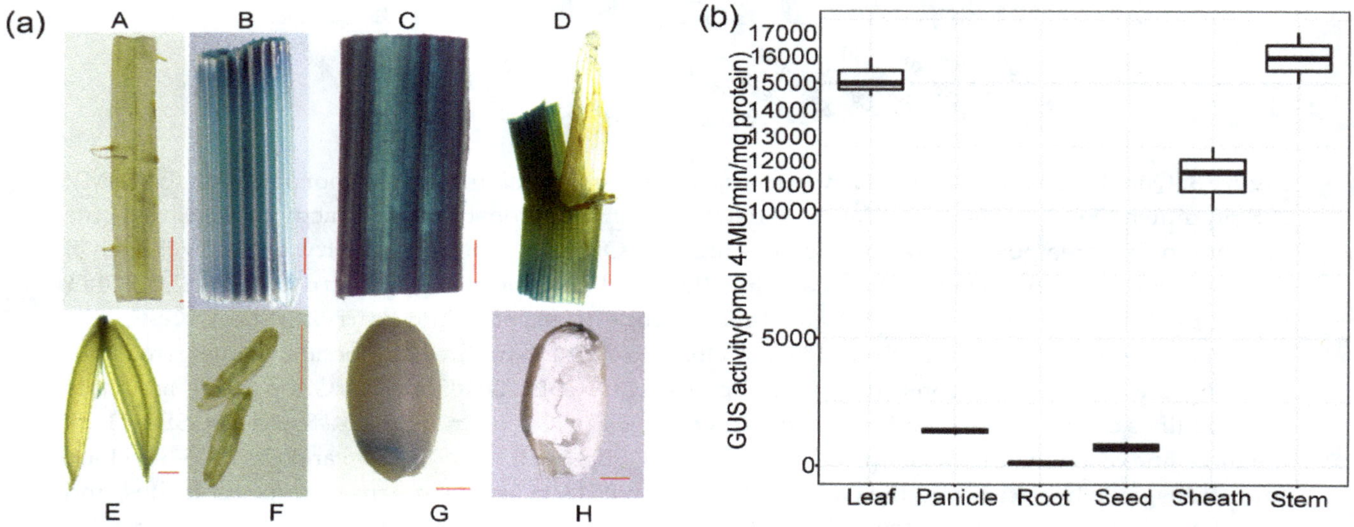

Figure 7. GUS histochemical assays and GUS activity in different tissues of OrGSEp-374 transgenic rice. (**a**) GUS histochemical assays of OrGSEp-374 transgenic rice. A. Root, B. stem, C. leaf, D. ligule, E. spikelet, F. anther, G. seed. H. endosperm. Bar = 1000 μm; (**b**) GUS activity in different tissues of transgenic rice carrying the OrGSEp-374 construct. Boxplots show GUS activity in different tissues of transgenic rice. The lower boundary of each box denotes the 25th percentile, the upper boundary of each box denotes the 75th percentile, and the solid line in the middle of each box denotes the 50th percentile. The two ends of the error bars denote the maximum and minimum values.

3. Discussion

Next-generation sequencing has become a new method for exploring transcriptional regulation and has been used to identify cis-elements involved in gene regulation. To date, many important genes and regulatory elements have been isolated by next-generation sequencing. Using RNA-seq analysis, more than 87 transcription factor genes were identified as being expressed during seed development and fatty acid accumulation. One of these genes, GmDREBL, which belongs to the DREB subfamily of the AP2 family, was shown to participate in fatty acid accumulation based on analysis of *Arabidopsis* plants transformed with the *GmDREBL* gene driven by 35S promoter [33]. These studies illustrate that high-throughput sequencing can be used to identify genes and regulatory elements. In this study, we identified 1140 candidate genes with preferential expression in green tissues by screening an RNA-seq library [30] and performing traditional RT-PCR. We screened a green tissue-specific gene, *OrGSE*, and the promoter of this gene was isolated. GUS reporter gene driven by the promoter was strongly expressed in leaves and stems.

OrGSE shares homology with D27 protein, which is located in chloroplasts. D27 is involved in MAX/RMS/D pathway, in which D27 as a member participates in the biosynthesis of strigolactones, regulating rice tiller bud outgrowth [34]. Although the function of *OrGSE* is unknown, the sequence of *OrGSE* is highly similar with D27. *OrGSE* may share similar function with D27. In *Arabidopsis*, AtD27 is required for the inhibition of secondary bud outgrowth by Strigolactones [35]. According to the sequence similarity between OrGSE and D27, we can infer OrGSE may locate in chloroplasts and participate in rice tiller outgrowth, which can account for OrGSEp drive gene expressed in green tissues.

The tissue-specific expression of transgenes plays a critical role in biotechnology crops because it avoids fitness costs caused by constitutive expression of target genes [9]. However, the information about the core cis-acting elements controlling tissue-specific expression is limited. Two novel cis-elements controlling tissue-specific expression, namely PSE1 (panicle/stem-specific element 1) and LPSRE2 (leaf, panicle/stem and root element 2), were identified in the *DX1* promoter by deletion analysis and gel mobility shift assays [35]. Five tissue-specific expression-related cis-elements in the green tissue-specific promoter PD540 were also characterized. These elements include GEAT, WRKY71OS and TGAC, and are used as references for discovering novel tissue-specific promoters. In this study, we showed that OrGSEp-114 is a short green tissue-specific promoter. However, the above described cis-elements were not found in this novel promoter. The OrGSEp-114 promoter may contain new cis-elements involved in green tissue-specific expression. Our future research will focus on screening cis-acting elements controlling green tissue-specific expression in this promoter by electrophoretic mobility shift assay (EMSA), including at least one novel green tissue-specific element. The OrGSEp-114 has promoter activity and green-specific distinguishing feature, but OrGSEp-54 cannot drive reporter gene expression, which means that 114 bp region of this promoter was sufficient to drive green tissue-specific expression. Cis-elements show that the promoter contains one CAAT-box and GC-motif from −114 to −54. This CAAT-box and GC-motif may be vital to promoter activity (Table S1). Moreover, the promoter from −114 to −54 contains one circadian cis-element, and many studies show that circadian elements participate in green tissue-specific expression. That information will support OrGSEp-114 being a short green tissue-specific promoter.

Light is essential for photosynthesis, and many green tissue-specific promoters contain light-inducible elements [36]. The *IbRbcS* gene from sweet potato shows green tissue-preferential expression and the *IbRbcS1* promoter confers light-responsive expression in transgenic *Arabidopsis* [37]. In this study, results showed that OrGSEp-374, as with the *IbRbcS1* promoter, has both green tissue-specific and light-inducible expression. Based on analysis of the activity of the full-length and truncated promoters in transgenic *Arabidopsis*, we showed that the OrGSEp-374 promoter contains cis-elements involved in light responsiveness, such as ACE, G-box and GT1CONSENSUS, which may regulate green tissue-specific and light-induced expression. Furthermore, analysis of 5′-deletions of this promoter showed that OrGSEp-274 and OrGSEp-204 also have light-induced activity, so the promoter may contain more than one core light-inducible element.

Many insect and fungal diseases influence the normal growth of rice and cause severe yield loss. Sheath blight-resistant rice has been developed by co-expressing the *chitinase* and *oxalate oxidase 4* genes in green tissues under the control of the green-specific *rbcS* promoter in rice [38]. A transgenic potato resistant to potato tuber moths with 100% tuber moth larval mortality was developed by expressing cry1Ab under the control of the *PEPC* promoter, a green tissue-specific and light-inducible promoter from maize [39]. Cry9C driven by *PNZIP* promoter, a green tissue promoter from *Pharbitis nil*, was used to develop transgenic pest-resistant cotton, and the accumulation of cry9C protein in seeds was 100 times lower than that observed for the seeds of the CaMV 35S:Cry9C line [40]. In our study, we have isolated a novel rice green tissue-specific promoter that does not drive expression in endosperm, and the GUS activity in seeds was much lower than in green tissues. Although the gene driven by this promoter is slightly expressed in embryos, the embryo is linked to bran and will be discarded with the bran during rice bran desquamation, thus alleviating concerns about food safety. OrGSEp is not only not expressed in endosperm but is also not expressed in roots. Thus, the use of this promoter in transgenic crops will ease food safety concerns and reduce the waste of resources that occurs with constitutive expression. This promoter provides a new element for developing insect-resistant and disease-resistant rice and other crops.

4. Materials and Methods

4.1. Plant Materials and Growth Conditions

Common wild rice seeds were collected from Guangdong Province in China. The collection was approved by the supervision department of Guangdong wild rice protection. Plants were grown in a greenhouse under a 12 h light/12 h dark cycle at 28 °C. *Arabidopsis thaliana* (Col-0) seeds were surface-sterilized with 75% (*v*/*v*) ethanol for 10 min and washed for 1 min with 95% (*v*/*v*) ethanol. The sterilized *Arabidopsis* seeds were spread on plates containing 1/2 Murashige and Skoog medium. After stratification at 4 °C for 2 days, the plates were transferred to a plant growth incubator, and seeds were germinated under a 16 h light/8 h dark cycle at 22 °C–24 °C.

4.2. Screening of Green Tissue-Specific Genes and Expression Analysis

Based on RNA-seq data from our previous study [30], genes with higher FPKM (Fragments Per Kilobase per Million) in CL (Control Leaf) than in CR (Control Root) were chosen as candidate green tissue-specific promoter-regulated genes. RT-PCR and Real-time quantitative PCR (qPCR) were used to confirm gene expression patterns. According to expression data, a candidate gene (*OrGSE*) was selected for subsequent analysis. The upstream sequence of *OrGSE* was regarded as a candidate green tissue-specific promoter.

Single-stranded cDNA was synthesized from the total RNA isolated from common wild rice leaves using the 5X All-In-One RT Master Mix (Applied Biological Materials, Vancouver, VAN, Canada). The sequence of *OrGSE* was obtained from a cDNA library. We analyzed sequence conservation by performing a BLAST search against the Michigan State University Rice Genome Annotation Project Database (MSU). We designed a pair of primers (OrGSE-F and OrGSE-R) to amplify the *OrGSE* sequence from leaf cDNA. The PCR amplification with 2× Phanta™ Master Mix was performed according to the manufacturer's instructions (Vazyme, Nanjing, China). The PCR products were purified, cloned into the pEASY-Blunt Cloning Vector (Transgene, Beijing, China) and sequenced. The expression level of *OrGSE* in common wild rice roots, stems and leaves was analyzed using RT-PCR and qPCR. *Actin1(LOC_Os03g61970)* was used as an internal control. Different numbers of cycles were used in RT-PCR amplification for *OrGSE* expression analysis. qRT-PCR was performed using the gene-specific primers listed in Supplemental Table S2 (Actin-F/R and OrGSE-F/R) and a real-time PCR7500 system (Applied Biosystems). Data were collected using the ABI PRISM 7500 sequence detection system. Three biological replicates with independent mRNA isolations were performed, each with three technical repeats, the rice Actin gene was used as an internal control, and the mRNA relative expression level was calculated using the $2^{-\Delta\Delta CT}$ method.

4.3. Cloning of the OrGSE Promoter and Sequence Analysis

Genomic DNA was isolated from common wild rice leaves using the EasyPure Plant Genomic DNA Kit (Transgene Biotech, Beijing, China) and used as the template for amplification of the *OrGSE* promoter. The PCR products were purified and cloned into the plant expression vector pCAMBIA1305.1 using the In-fusion method. This construct was used for sequencing and promoter activity analysis.

Core regulatory elements in the promoter sequences were predicted using the online tool plantCARE (available online: http://bioformatics.psb.ugent.be/webtools/plantcare/html/) [41].

4.4. PCR Amplification of 5′-Deletion Fragments of the OrGSE Promoter

We designed five forward primers with *Bam*H I restriction sites (F-374, F-274, F-204, F-114, F-54) and one reverse primer with an *Eco*R I restriction site (R) to obtain 5′-deletion fragments of the OrGSE promoter. These primers were designed to amplify regions −374 (F-374/R), −274 (F-274/R),

−204 (F-204/R), −114 (F-114/R), and −54 (F-54/R) upstream of the transcription initiation site (TIS), which was designed as +1. The PCR cycling parameters were as follows: 95 °C for 3 min, 35 cycles of 95 °C for 15 s, 60 °C for 15 s, and 72 °C for 1 min followed by 72 °C for 5 min. The full-length promoter and 5′-deletion fragments were cloned into a modified pBinGlyRed vector, which includes GUS plus-his6 that was inserted into the multiple cloning site using the *Eco*R I and *Xma* I restriction sites (Figure S2). The five promoter fragments were named OrGSEp-374, OrGSEp-274, OrGSEp-204, OrGSEp-114 and OrGSEp-54. The constructs were then introduced into *Arabidopsis thaliana* (Col-0), *Agrobacterium*-mediated transformation of *Arabidopsis* was performed through floral dipping [42]. We also amplificatedOrGSEp-374 fragment by 1305GSEp-F/R primers, and it was inserted into pCAMBIA1305.1 vector by *Hind* III and *Nco* I restriction sites, the construct was introduced into Nipponbare (*Oryza sativa* L. ssp. *japonica*) by *Agrobacterium*-mediated transformation.

4.5. Detection of the Expression Pattern of the OrGSEp Promoter and 5′-Deletion Fragments in Different Organs

For GUS histochemical assays, homozygous T3 transgenic *Arabidopsis* seedlings (3-, 5- and 14-day-old) were incubated in GUS staining solution overnight. During the reproductive stage, GUS histochemical assays were performed for transgenic *Arabidopsis* leaves, siliques and root. GUS activity was quantified in the leaves and roots of 3-week-old seedlings. Three independent lines were selected for each promoter deletion construct.

4.6. Inducible Activity Analysis of the OrGSEp Promoter and 5′-Deletion Fragments

To analyze the light inducible activity of the promoter and different 5′-deletion fragments, 3-week-old transgenic *Arabidopsis* plants were grown in the dark for 24 h. Then, the leaves were frozen in liquid nitrogen and stored at −80 °C for GUS fluorometric assay. The control plants were placed under natural growth conditions. For GUS staining, 10-day-old seedlings were placed in the dark for 24 h and transferred to natural condition for 24 h, and control plants were kept under natural conditions.

4.7. GUS Histochemical and Fluorometric Analysis

GUS histochemical and fluorometric analysis was performed as previously described (Wu et al., 2003). Transgenic *Arabidopsis* seedlings at different growth stages and various tissues at the reproductive stage (as described above) were incubated in GUS staining solution (Coolaber, Beijing, China) at 37 °C overnight, and then the samples were cleared with 75% (*v*/*v*) ethanol. GUS staining was observed under a ZEISS Stemi 508 microscope and photographed with a SONY camera.

Protein was extracted from 100 mg root and leaf tissue from transgenic *Arabidopsis* plants. The protein concentrations were determined using the Bradford method with bovine serum albumin (BSA) as the standard. GUS activity was determined fluorometrically by measuring the amount of 4-methylumbelliferone (4-MU) produced by GUS per milligram of total protein per minute [43]. GUS activity was measured for three lines for each promoter construct, and three replicates were performed for each line. The error bars are reported as "mean ± standard error". The boxplot was drawn by R language gglpot2 package.

Author Contributions: X.P. and Q.Y. conceived and designed the experiments. M.X., Z.Z., K.H., T.Z. and G.H. performed the experiments. M.X. performed the data analysis and wrote the manuscript. Y.L., Y.J. and X.P. revised the manuscript.

Acknowledgments: This research was supported by The National Special Program for Transgenic Research (2016ZX08011-001).

References

1. Li, Y.; Sun, Y.; Yang, Q.C.; Kang, J.M.; Zhang, T.J.; Gruber, M.Y.; Fang, F. Cloning and function analysis of an alfalfa (*Medicago sativa* L.) zinc finger protein promoter MsZPP. *Mol. Biol. Rep.* **2012**, *39*, 8559–8569. [CrossRef] [PubMed]
2. Niu, G.L.; Gou, W.; Han, X.L.; Qin, C.; Zhang, L.X.; Abomohra, A.E.; Ashraf, M. Cloning and functional analysis of phosphoethanolamine methyltransferase promoter from maize (*Zea mays* L.). *Int. J. Mol. Sci.* **2018**, *19*, 191. [CrossRef] [PubMed]
3. Pih, K.T.; Yoo, J.; Fosket, D.E.; Han, I.S. A comparison of the activity of three cauliflower mosaic virus 35s promoters in rice seedlings and tobacco (by-2) protoplasts by analysis of *gus* reporter gene transient expression. *Plant. Sci.* **1996**, *114*, 141–148. [CrossRef]
4. McElroy, D.; Blowers, A.D.; Jenes, B.; Wu, R. Construction of expression vectors based on the rice *actin 1* (*act1*) 5' region for use in monocot transformation. *Mol. Gen. Genet.* **1991**, *231*, 150–160. [CrossRef] [PubMed]
5. Christensen, A.H.; Quail, P.H. Ubiquitin promoter-based vectors for high-level expression of selectable and/or screenable marker genes in monocotyledonous plants. *Transgenic Res.* **1996**, *5*, 213–218. [CrossRef] [PubMed]
6. Eva, C.; Teglas, F.; Zelenyanszki, H.; Tamas, C.; Juhasz, A.; Meszaros, K.; Tamas, L. Cold inducible promoter driven cre-lox system proved to be highly efficient for marker gene excision in transgenic barley. *J. Biotechnol.* **2018**, *265*, 15–24. [CrossRef] [PubMed]
7. Qu, D.; Song, Y.; Li, W.M.; Pei, X.W.; Wang, Z.X.; Jia, S.R.; Zhang, Y.Q. Isolation and characterization of the organ-specific and light-inducible promoter of the gene encoding rubisco activase in potato (*Solanum tuberosum*). *Genet. Mol. Res.* **2011**, *10*, 621–631. [CrossRef] [PubMed]
8. Thomson, B.; Graciet, E.; Wellmer, F. Inducible promoter systems for gene perturbation experiments in *Arabidopsis*. *Methods Mol. Biol.* **2017**, *1629*, 15–25. [CrossRef] [PubMed]
9. Yang, F.; Ding, X.; Chen, J.; Shen, Y.; Kong, L.; Li, N.; Chu, Z. Functional analysis of the *GRMZM2g174449* promoter to identify rhizoctonia solani-inducible cis-elements in maize. *BMC Plant. Biol.* **2017**, *17*, 233. [CrossRef] [PubMed]
10. Yu, S.I.; Lee, B.H. Generation of a stress-inducible luminescent *Arabidopsis* and its use in genetic screening for stress-responsive gene deregulation mutants. *Methods Mol. Biol.* **2017**, *1631*, 109–119. [CrossRef] [PubMed]
11. Li, J.; Qin, R.; Xu, R.; Li, H.; Yang, Y.; Li, L.; Wei, P.; Yang, J. Isolation and identification of five cold-inducible promoters from *Oryza sativa*. *Planta* **2017**, *247*, 99–111. [CrossRef] [PubMed]
12. Zhang, H.; Jing, R.; Mao, X. Functional characterization of *Tasnrk2.8* promoter in response to abiotic stresses by deletion analysis in transgenic *Arabidopsis*. *Front. Plant. Sci.* **2017**, *8*, 1198. [CrossRef] [PubMed]
13. Kakrana, A.; Kumar, A.; Satheesh, V.; Abdin, M.Z.; Subramaniam, K.; Bhattacharya, R.C.; Srinivasan, R.; Sirohi, A.; Jain, P.K. Identification, validation and utilization of novel nematode-responsive root-specific promoters in *Arabidopsis* for inducing host-delivered RNAi mediated root-knot nematode resistance. *Front. Plant. Sci.* **2017**, *8*, 2049. [CrossRef] [PubMed]
14. Nanjareddy, K.; Arthikala, M.K.; Aguirre, A.L.; Gomez, B.M.; Lara, M. Plant promoter analysis: Identification and characterization of root nodule specific promoter in the common bean. *J. Vis. Exp.* **2017**, *23*, 130. [CrossRef] [PubMed]
15. Chavez-Barcenas, A.T.; Valdez-Alarcon, J.J.; Martinez-Trujillo, M.; Chen, L.; Xoconostle-Cazares, B.; Lucas, W.J.; Herrera-Estrella, L. Tissue-specific and developmental pattern of expression of the rice *SPS1* gene. *Plant. Physiol.* **2000**, *124*, 641–654. [CrossRef] [PubMed]
16. Kim, J.; Lee, H.J.; Jung, Y.J.; Kang, K.K.; Tyagi, W.; Kovach, M.; Sweeney, M.; McCouch, S.; Cho, Y.G. Functional properties of an alternative, tissue-specific promoter for rice nadph-dependent dihydroflavonol reductase. *PLoS ONE* **2017**, *12*, e0183722. [CrossRef] [PubMed]
17. Panguluri, S.K.; Sridhar, J.; Jagadish, B.; Sharma, P.C.; Kumar, P.A. Isolation and characterization of a green tissue-specific promoter from pigeonpea [*Cajanus cajan* (L.) millsp.]. *Indian J. Exp. Biol.* **2005**, *43*, 369–372. [PubMed]
18. Xu, W.; Liu, W.; Ye, R.; Mazarei, M.; Huang, D.; Zhang, X.; Stewart, C.N., Jr. A profilin gene promoter from switchgrass (*Panicum virgatum* L.) directs strong and specific transgene expression to vascular bundles in rice. *Plant. Cell. Rep.* **2018**, *37*, 587–597. [CrossRef] [PubMed]

19. Zaidi, M.A.; O'Leary, S.J.; Wu, S.; Chabot, D.; Gleddie, S.; Laroche, A.; Eudes, F.; Robert, L.S. Investigating triticeae anther gene promoter activity in transgenic *brachypodium distachyon*. *Planta* **2017**, *245*, 385–396. [CrossRef] [PubMed]
20. Li, H.; Li, J.; Xu, R.; Qin, R.; Song, F.; Li, L.; Wei, P.; Yang, J. Isolation of five rice non-endosperm tissue-expressed promoters and evaluation of their activities in transgenic rice. *Plant. Biotechnol. J.* **2017**, *16*, 1138–1147. [CrossRef] [PubMed]
21. Suzuki, H.; Fowler, T.J.; Tierney, M.L. Deletion analysis and localization of *SbPRP1*, a soybean cell wall protein gene, in roots of transgenic tobacco and cowpea. *Plant. Mol. Biol.* **1993**, *21*, 109–119. [CrossRef] [PubMed]
22. Chen, L.; Jiang, B.; Wu, C.; Sun, S.; Hou, W.; Han, T. *GmPRP2* promoter drives root-preferential expression in transgenic *Arabidopsis* and soybean hairy roots. *BMC Plant. Biol.* **2014**, *14*, 245. [CrossRef] [PubMed]
23. Wang, H.; Fan, M.; Wang, G.; Zhang, C.; Shi, L.; Wei, Z.; Ma, W.; Chang, J.; Huang, S.; Lin, F. Isolation and characterization of a novel pollen-specific promoter in maize (*Zea mays* L.). *Genome* **2017**, *60*, 485–495. [CrossRef] [PubMed]
24. Wu, L.; El-Mezawy, A.; Shah, S. A seed coat outer integument-specific promoter for *Brassica napus*. *Plant. Cell. Rep.* **2011**, *30*, 75–80. [CrossRef] [PubMed]
25. Shrestha, A.; Khan, A.; Mishra, D.R.; Bhuyan, K.; Sahoo, B.; Maiti, I.B.; Dey, N. WRKY71 and TGA1A physically interact and synergistically regulate the activity of a novel promoter isolated from petunia vein-clearing virus. *Biochim. Biophys. Acta* **2018**, *1861*, 133–146. [CrossRef] [PubMed]
26. Wang, Q.; Zhu, Y.; Sun, L.; Li, L.; Jin, S.; Zhang, X. Transgenic Bt cotton driven by the green tissue-specific promoter shows strong toxicity to lepidopteran pests and lower Bt toxin accumulation in seeds. *Sci. China Life Sci.* **2016**, *59*, 172–182. [CrossRef] [PubMed]
27. Manikandan, R.; Balakrishnan, N.; Sudhakar, D.; Udayasuriyan, V. Development of leaffolder resistant transgenic rice expressing *cry2ax*1 gene driven by green tissue-specific *rbcS* promoter. *World J. Microb. Biot.* **2016**, *32*, 37. [CrossRef] [PubMed]
28. Menguer, P.K.; Sperotto, R.A.; Ricachenevsky, F.K. A walk on the wild side: *Oryza* species as source for rice abiotic stress tolerance. *Genet. Mol. Biol.* **2017**, *40*, 238–252. [CrossRef] [PubMed]
29. Mao, D.; Yu, L.; Chen, D.; Li, L.; Zhu, Y.; Xiao, Y.; Zhang, D.; Chen, C. Multiple cold resistance loci confer the high cold tolerance adaptation of dongxiang wild rice (*Oryza rufipogon*) to its high-latitude habitat. *Theor. Appl. Genet.* **2015**, *128*, 1359–1371. [CrossRef] [PubMed]
30. Tian, X.; Long, Y.; Wang, J.; Zhang, J.; Wang, Y.; Li, W.; Peng, Y.; Yuan, Q.; Pei, X. De novo transcriptome assembly of common wild rice (*Oryza rufipogon* griff.) and discovery of drought-response genes in root tissue based on transcriptomic data. *PLoS ONE* **2015**, *10*, e0131455. [CrossRef] [PubMed]
31. Civan, P.; Svec, M. Genome-wide analysis of rice (*Oryza sativa* L. Subsp. Japonica) tata box and y patch promoter elements. *Genome* **2009**, *52*, 294–297. [CrossRef] [PubMed]
32. Conley, T.R.; Park, S.C.; Kwon, H.B.; Peng, H.P.; Shih, M.C. Characterization of cis-acting elements in light regulation of the nuclear gene encoding a subunit of chloroplast isozymes of glyceraldehyde-3-phosphate dehydrogenase from *Arabidopsis thaliana*. *Mol. Cell. Biol.* **1994**, *14*, 2525–2533. [CrossRef] [PubMed]
33. Zhang, Y.Q.; Lu, X.; Zhao, F.Y.; Li, Q.T.; Niu, S.L.; Wei, W.; Zhang, W.K.; Ma, B.; Chen, S.Y.; Zhang, J.S. Soybean *GmDREBl* increases lipid content in seeds of transgenic *Arabidopsis*. *Sci. Rep.* **2016**, *6*, 34307. [CrossRef] [PubMed]
34. Lin, H.; Wang, R.X.; Qian, Q.; Yan, M.X.; Meng, X.B.; Fu, Z.M.; Yan, C.Y.; Jiang, B.; Su, Z.; Li, J.Y.; et al. DWARF27, an Iron-Containing Protein Required for the Biosynthesis of Strigolactones, Regulates Rice Tiller Bud Outgrowth. *Plant Cell* **2009**, *21*, 1512–1525. [CrossRef] [PubMed]
35. Waters, M.T.; Brewer, P.B.; Bussell, J.D.; SmitSh, S.M.; Beveridge, C.A. The *Arabidopsis* ortholog of rice dwarf27 acts upstream of max1 in the control of plant development by strigolactones. *Plant. Physiol.* **2012**, *159*, 1073–1085. [CrossRef] [PubMed]
36. Ye, R.; Zhou, F.; Lin, Y. Two novel positive cis-regulatory elements involved in green tissue-specific promoter activity in rice (*Oryza sativa* L. ssp.). *Plant. Cell. Rep.* **2012**, *31*, 1159–1172. [CrossRef] [PubMed]
37. Eberhard, S.; Finazzi, G.; Wollman, F.A. The dynamics of photosynthesis. *Annu. Rev. Genet.* **2008**, *42*, 463–515. [CrossRef] [PubMed]

38. Tanabe, N.; Tamoi, M.; Shigeoka, S. The sweet potato *RbcS* gene (*IbRbcS1*) promoter confers high-level and green tissue-specific expression of the gus reporter gene in transgenic *Arabidopsis*. *Gene* **2015**, *567*, 244–250. [CrossRef] [PubMed]
39. Karmakar, S.; Molla, K.A.; Chanda, P.K.; Sarkar, S.N.; Datta, S.K.; Datta, K. Green tissue-specific co-expression of chitinase and oxalate oxidase 4 genes in rice for enhanced resistance against sheath blight. *Planta* **2016**, *243*, 115–130. [CrossRef] [PubMed]
40. Ghasimi Hagh, Z.; Rahnama, H.; Panahandeh, J.; Baghban Kohneh Rouz, B.; Arab Jafari, K.M.; Mahna, N. Green-tissue-specific, c(4)-pepc-promoter-driven expression of *Cry1ab* makes transgenic potato plants resistant to tuber moth (*phthorimaea operculella*, zeller). *Plant. Cell. Rep.* **2009**, *28*, 1869–1879. [CrossRef] [PubMed]
41. Lescot, M.; Dehais, P.; Thijs, G.; Marchal, K.; Moreau, Y.; Van de Peer, Y.; Rouze, P.; Rombauts, S. Plantcare, a database of plant cis-acting regulatory elements and a portal to tools for in silico analysis of promoter sequences. *Nucleic Acids Res.* **2002**, *30*, 325–327. [CrossRef] [PubMed]
42. Clough, S.J.; Bent, A.F. Floral dip: A simplified method for *agrobacterium*-mediated transformation of *Arabidopsis thaliana*. *Plant. J.* **1998**, *16*, 735–743. [CrossRef] [PubMed]
43. Wang, R.; Zhu, M.; Ye, R.; Liu, Z.; Zhou, F.; Chen, H.; Lin, Y. Novel green tissue-specific synthetic promoters and cis-regulatory elements in rice. *Sci. Rep.* **2015**, *5*, 18256. [CrossRef] [PubMed]

Permissions

All chapters in this book were first published in MDPI; hereby published with permission under the Creative Commons Attribution License or equivalent. Every chapter published in this book has been scrutinized by our experts. Their significance has been extensively debated. The topics covered herein carry significant findings which will fuel the growth of the discipline. They may even be implemented as practical applications or may be referred to as a beginning point for another development.

The contributors of this book come from diverse backgrounds, making this book a truly international effort. This book will bring forth new frontiers with its revolutionizing research information and detailed analysis of the nascent developments around the world.

We would like to thank all the contributing authors for lending their expertise to make the book truly unique. They have played a crucial role in the development of this book. Without their invaluable contributions this book wouldn't have been possible. They have made vital efforts to compile up to date information on the varied aspects of this subject to make this book a valuable addition to the collection of many professionals and students.

This book was conceptualized with the vision of imparting up-to-date information and advanced data in this field. To ensure the same, a matchless editorial board was set up. Every individual on the board went through rigorous rounds of assessment to prove their worth. After which they invested a large part of their time researching and compiling the most relevant data for our readers.

The editorial board has been involved in producing this book since its inception. They have spent rigorous hours researching and exploring the diverse topics which have resulted in the successful publishing of this book. They have passed on their knowledge of decades through this book. To expedite this challenging task, the publisher supported the team at every step. A small team of assistant editors was also appointed to further simplify the editing procedure and attain best results for the readers.

Apart from the editorial board, the designing team has also invested a significant amount of their time in understanding the subject and creating the most relevant covers. They scrutinized every image to scout for the most suitable representation of the subject and create an appropriate cover for the book.

The publishing team has been an ardent support to the editorial, designing and production team. Their endless efforts to recruit the best for this project, has resulted in the accomplishment of this book. They are a veteran in the field of academics and their pool of knowledge is as vast as their experience in printing. Their expertise and guidance has proved useful at every step. Their uncompromising quality standards have made this book an exceptional effort. Their encouragement from time to time has been an inspiration for everyone.

The publisher and the editorial board hope that this book will prove to be a valuable piece of knowledge for researchers, students, practitioners and scholars across the globe.

List of Contributors

Sandeep Kaur Dhaliwal and Pankaj Sharma
Department of Plant Breeding and Genetics, Punjab Agricultural University, Ludhiana 141004, India

Akshay Talukdar
Division of Genetics, Indian Agricultural Research Institute, New Delhi 110012, India

Ashish Gautam
Department of Genetics and Plant Breeding, G.B. Pant University of Agriculture and Technology, Pantnagar, Uttarakhand 263145, India

Vinay Sharma
International Crops Research Institute for the Semi-Arid Tropics (ICRISAT), Hyderabad 502324, India

Prashant Kaushik
Instituto de Conservación y Mejora de la Agrodiversidad Valenciana, Universitat Politècnica de València, 46022 Valencia, Spain
Nagano University, Ueda 386-0031, Japan

Jian-Zhi Jiang, Chun-Hsiang Kuo, Bo-Hong Chen, Mao-Kei Chen and Shin-Lon Ho
Department of Agronomy, National Chiayi University, Chiayi 60004, Taiwan

Choun-Sea Lin
Agricultural Biotechnology Research Center, Academia Sinica, Taipei 11529, Taiwan

Zhineng Li, Yingjie Jiang, Daofeng Liu, Jing Ma, Jing Li, Mingyang Li and Shunzhao Sui
College of Horticulture and Landscape Achitecture, Southwest University, Chongqing 400715, China
Key Laboratory of Horticulture Science for Southern Mountains Regions, Ministry of Education, Chongqing 400715, China
Chongqing Engineering Research Center for Floriculture, Chongqing 400715, China

Ling Wang, Feng Liu, Xu Zhang, Wenju Wang, Tingting Sun, Yufeng Chen, Mingjian Dai and Shengxiao Yu
Key Laboratory of Sugarcane Biology and Genetic Breeding, Ministry of Agriculture, Fujian Agriculture and Forestry University, Fuzhou 350002, China

Liping Xu, Yachun Su and Youxiong Que
Key Laboratory of Sugarcane Biology and Genetic Breeding, Ministry of Agriculture, Fujian Agriculture and Forestry University, Fuzhou 350002, China
Key Laboratory of Ministry of Education for Genetics, Breeding and Multiple Utilization of Crops, College of Crop Science, Fujian Agriculture and Forestry University, Fuzhou 350002, China

Min Wang and Biao Jiang
Vegetable Research Institute, Guangdong Academy of Agricultural Sciences, Guangzhou 510640, China
Guangdong Key Laboratory for New Technology Research of Vegetables, Guangzhou 510640, China

Qingwu Peng, Wenrui Liu, Xiaoming He, Zhaojun Liang and Yu'e Lin
Vegetable Research Institute, Guangdong Academy of Agricultural Sciences, Guangzhou 510640, China

Xuanbo Zhong, Qizheng Zhou, Nan Cui and Guixiang Tang
Zhejiang Provincial Key Laboratory of Crop Genetic Resources, Institute of Crop Science, Zhejiang University, Hangzhou 310058, Zhejiang, China

Daguang Cai
Department of Molecular Phytopathology, Christian-Albrechts-University of Kiel, Hermann Rodewald Str. 9, D-24118 Kiel, Germany

Bin Wang
National Engineering Laboratory for Resource Development of Endangered Crude Drugs in Northwest China, Key Laboratory of the Ministry of Education for Medicinal Resources and Natural Pharmaceutical Chemistry, College of Life Sciences, Shaanxi Normal University, Xi'an 710119, China
College of Chemistry, Biology and Materials Science, East China University of Technology, NanChang 330013, China

Junfeng Niu, Bin Li, Yaya Huang, Limin Han, Yuanchu Liu, Wen Zhou, Suying Hu, Lin Li, Donghao Wang, Shiqiang Wang, Xiaoyan Cao and Zhezhi Wang
National Engineering Laboratory for Resource Development of Endangered Crude Drugs in Northwest China, Key Laboratory of the Ministry of Education for Medicinal Resources and Natural Pharmaceutical Chemistry, College of Life Sciences, Shaanxi Normal University, Xi'an 710119, China

Xiaorong Sun, Jinshuai Shu, Ali Mohamed Ali Mohamed, Xuebin Deng, Xiaona Zhi, Jinrui Bai, Yanan Cui, Xiaoxiao Lu, Yongchen Du, Xiaoxuan Wang, Zejun Huang, Yanmei Guo, Lei Liu and Junming Li
Institute of Vegetables and Flowers, Chinese Academy of Agricultural Sciences, Key Laboratory of Biology and Genetic Improvement of Horticultural Crops, Ministry of Agriculture, 12 Zhongguancun Nandajie Street, Beijing 100081, China

Xue Fu, Jing Xu, Mengyu Zhou, Minmin Chen, Lan Shen, Ting Li, Yuchen Zhu, Jiajia Wang, Jiang Hu, Li Zhu, Zhenyu Gao, Guojun Dong, Longbiao Guo, Deyong Ren, Guang Chen, Jianrong Lin, Qian Qian and Guangheng Zhang
State Key Laboratory of Rice Biology, China National Rice Research Institute, Hangzhou 310006, China

De-Lin Pan, Gang Wang, Tao Wang, Zhan-Hui Jia, Zhong-Ren Guo and Ji-Yu Zhang
Institute of Botany, Jiangsu Province and Chinese Academy of Sciences, Nanjing 210014, China

Pan Wu and Wenying Lu
Key Laboratory of Plant Germplasm Enhancement and Specialty Agriculture, Wuhan Botanical Garden, Chinese Academy of Sciences, Wuhan 430074, China
University of Chinese Academy of Sciences, Beijing 100049, China

Lingling Zhang and Tao Feng
Key Laboratory of Plant Germplasm Enhancement and Specialty Agriculture, Wuhan Botanical Garden, Chinese Academy of Sciences, Wuhan 430074, China

Huayan Zhao
Applied Biotechnology Center, Wuhan Institute of Bioengineering, Wuhan 430415, China

Jianzhong Li
Tianjin Garrison hangu farm, Tianjin 300480, China

Shiyou Lü
Key Laboratory of Plant Germplasm Enhancement and Specialty Agriculture, Wuhan Botanical Garden, Chinese Academy of Sciences, Wuhan 430074, China
Sino-Africa Joint Research Center, Chinese Academy of Sciences, Wuhan 430074, China

Xinpeng Zhang
State Key Laboratory of Crop Biology, College of Horticulture Science and Engineering, Shandong Agricultural University, Taian 271018, China

Zongda Xu, Xiaoyan Yu, Lanyong Zhao, Mingyuan Zhao, Xu Han and Shuai Qi
College of Forestry, Shandong Agricultural University, Taian 271018, China

Mande Xue, Yan Long, Tianbao Zhang and Xinwu Pei
MOA Key Laboratory on Safety Assessment (Molecular) of Agri-GMO, Institute of Biotechnology, Chinese Academy of Agricultural Sciences, Beijing 100081, China

Zhiqiang Zhao, Gege Huang, Ke Huang and Qianhua Yuan
College of Tropical Agriculture and Forestry, Hainan University, Haikou 570228, China

Ying Jiang
Experimental Center Basic Medical Teaching, Capital Medical University, Beijing 100069, China

Index

Printed in the USA
CPSIA information can be obtained
at www.ICGtesting.com
LVHW080738221023
761596LV00011B/2181